Pioneers in Arts, Humanities, Science, Engineering, Practice

Volume 24

Series Editor

Hans Günter Brauch, Peace Research and European Security Studies
(AFES-PRESS), Mosbach, Germany

More information about this series at http://www.springer.com/series/15230
http://www.afes-press-books.de/html/PAHSEP.htm
http://www.afes-press-books.de/html/PAHSEP_Reychler.htm

Luc Reychler · Arnim Langer
Editors

Luc Reychler: A Pioneer in Sustainable Peacebuilding Architecture

 Springer

United Nations · UNESCO Chair in
Educational, Scientific and · Building Sustainable Peace,
Cultural Organization · KU Leuven

CENTRE FOR RESEARCH ON
PEACE AND DEVELOPMENT

Editors
Luc Reychler
Faculty of Social Sciences
Leuven International and European Studies
Centre for Research on Peace and
Development (CRPD)
University of Leuven/KU Leuven
Leuven, Belgium

Arnim Langer
Faculty of Social Sciences
Leuven International and European Studies
Centre for Research on Peace
and Development (CRPD)
University of Leuven/KU Leuven
Leuven, Belgium

Acknowledgement: The cover photograph, as well as all other photos in this volume, was taken from the personal photographic collection of the author, who also granted permission for them to be published in this volume. A book website with additional information on Luc Reychler and his major book covers is at: http://www.afes-press-books.de/html/PAHSEP_ Reychler.htm.

ISSN 2509-5579 ISSN 2509-5587 (electronic)
Pioneers in Arts, Humanities, Science, Engineering, Practice
ISBN 978-3-030-40207-5 ISBN 978-3-030-40208-2 (eBook)
https://doi.org/10.1007/978-3-030-40208-2

Copyediting: PD Dr. Hans Günter Brauch, AFES-PRESS e.V., Mosbach, Germany
Language Editing: Vanessa Greatorex, Chester, England

This Springer imprint is published by the registered company Springer Nature Switzerland AG
The registered company address is: Gewerbestrasse 11, 6330 Cham, Switzerland

*Dedicated to my dear family, friends
and to the courageous people,
who research peace
and recognize
that world peace
is not a utopia,
but humanity's
greatest challenge.*

Foreword

Luc Reychler: An Innovative Peace Thinker and Provocateur

It is a great honour for me to have been asked to write the foreword to this book, which collates Luc Reychler's main contributions to the field of peace research. I have had a privileged seat when it comes to observing Luc Reychler's academic career, given that our academic lives have been intertwined ever since I attended his lectures at KU Leuven more than twenty years ago. Further, following the completion of my political science studies at KU Leuven in 2000, I joined Reychler's research team at the Centre of Peace Research and Strategic Studies (CPRS) – a research centre at the Department of Political Science which he had established in 1983. CPRS was the first dedicated peace research centre in Belgium.

Under Reychler's direction, the Centre conducted innovative research on a range of important topics and issues, including, among others, effective strategies to prevent violent conflicts, preconditions of successful peacebuilding architecture, characteristics of transformative peacebuilding leadership, early warning systems for genocide and violent conflicts, and the role of religion in conflict and peacebuilding. While I left CPRS after a few years to conduct my DPhil in Development Studies at the University of Oxford, I took Reychler's inspiration and dedication to the field of peace research with me to Oxford, where I was to conduct research on the causes of violent conflict in multi-ethnic societies. In 2010, things came full circle, when I succeeded Luc Reychler as Professor of International Politics at KU Leuven.

Reflecting on Reychler's academic career is a challenging exercise, not only because of the large number of academic papers and publications he has produced since obtaining his Ph.D. in Psychology from Harvard University in 1976, but also because of the diversity of topics and themes he has focused on throughout the five-decade span of his academic career. Luc Reychler has made a range of important contributions to the field of peace research. And while I will discuss a number of contributions and insights below which I personally find most inspiring and insightful, I hope and encourage anyone interested in the field of peace research to read the different papers in this book in full, and to see for themselves the innovative contributions and insights he has made at different stages of his academic career.

Before highlighting Reychler's main academic contributions and achievements, it is useful to briefly expand on how his academic trajectory got started and how he became a peace researcher. Indeed, it is notable that he became a peace researcher at a time when this field of study and research was only just emerging as an independent academic discipline. Yet, it is equally true that scholars all over the world have been thinking and writing about issues of conflict, war and peace for at least 2500 years, and probably much longer. Many inspiring and insightful works and treatises have been written throughout history, including, for instance, Sun Tzu's *The Art of War* (5th century BC), Desiderius Erasmus' *The Complaint of Peace* (1517), Immanuel Kant's *Zum Ewigen Frieden* (1795) and Woodrow Wilson's *Fourteen Points Speech* (8 January 1918). These works demonstrate that people have been interested in understanding and studying these existential issues for a very long time, and it appears that people's interest has not diminished over time. Yet, it was ultimately the devastation of the Second World War that gave the essential impetus to the development of peace research – also known as 'peace and conflict studies' or 'peace science' – as an independent academic discipline or field of investigation (see Bright and Gledhill, 2018 for a detailed history of the emergence of peace research as an independent field of study and investigation). As Bright and Gledhill (2018: 130) further note in this respect: "Driven by a conviction that such destruction must never be repeated, a diverse set of scholars came together with the goal of developing academic research that could contribute to shaping a world in which conflict is managed without resort to violence".

In the late 1950s and early 1960s, a number of important events and activities started putting peace research on the academic map, including most notably the establishment of the *Peace Research Institute Oslo* (PRIO) in 1959, the founding of the *Peace Research Society* in 1963 (which was to become the Peace Science Society in 1973) and the *International Peace Research Association* (IPRA) in 1964, and the publication of the *Journal of Conflict Resolution* (JCR) in 1957 and the *Journal of Peace Research* (JPR) in 1964. While nowadays there is a range of international, peer-reviewed journals dedicated to the publication of high-level scholarship on conflict and peace research, including, for example, *Conflict Management and Peace Science* (CMPS), *Conflict Resolution Quarterly* (CRQ), *Asian Journal of Peacebuilding* (AJP), *Cooperation and Conflict*, and *International Journal of Conflict Management* (IJCM), JPR and JCR have been and remain to this day the two leading journals in the field of peace research (see in particular the annual Journal Citation Reports).

The trailblazers of the peace research discipline, including most notably Johan Galtung, John Burton, Walter Isard, Kenneth Boulding, Elise Boulding and Anatol Rapoport, were predominantly based in the United States and Western Europe. While they had very different disciplinary backgrounds, they shared a profound interest in advancing the *scientific* study and analysis of conflict and peace. Importantly, to this day, peace research remains a decidedly multidisciplinary field of study and research, which brings together researchers and scholars from a range of disciplines and backgrounds, including, among others, political science, sociology, gender studies, psychology, educational sciences, economics, philosophy and theology. Mirroring the expansion of academic research on issues of conflict and peace, especially from the 1990s onwards, a rapidly growing number of universities started offering courses and programmes on peacebuilding and conflict resolution. In 2019, there is a plethora of courses and programmes available at undergraduate, postgraduate and doctoral levels, which offer students the opportunity to learn about and study issues of conflict management, conflict resolution and peacebuilding. While initially these courses and programmes were mainly offered at American and European universities, since the turn of the century conflict and peace studies have also gained a strong foothold in the curricula of many non-Western universities.

While Luc Reychler had a profound and early interest in wanting to understand and analyse human behaviour and intergroup relations, when he began his university education in the mid-1960s no university programme dedicated to peace and conflict studies existed at Ghent University (i.e. the university where he was to study psychology) or at any other Belgian university for that matter. Had such a programme existed in Belgium, Reychler would probably have been intuitively drawn to it. Instead Luc Reychler started studying business psychology at Ghent University in 1964 and subsequently specialized in forensic psychology, a field of study which forced him to analyse murders and suicides, or what he termed 'micro violence'. Then, after reading Herbert Kelman's book *International Behavior: A Social-Psychological Analysis*, he decided to shift his focus and started studying the psychology of international politics. Eventually, he went to Harvard University to study for a Ph.D. in Psychology under the supervision of the same Herbert Kelman. Herbert Kelman was not only an internationally renowned social psychologist who was on track to become one of the foremost peace researchers of the twentieth century, but he was also, in the words of Reychler, "a wonderful Ph.D. supervisor who provided both essential academic guidance and inspiration".

At Harvard, Luc Reychler was introduced to the social psychology of international behaviour, international diplomacy and conflicts, interactive problem-solving workshops and other conflict prevention and resolution techniques and strategies.

In his doctoral thesis, titled "*Patterns of diplomatic thinking: A cross-national study of structural and social-psychological determinants*", he explained how and to what extent diplomatic thinking was influenced by power and the position of a country in the international system. He analysed four facets of diplomatic thinking: i.e. the perception of the international climate, the value systems of diplomats, their analytic styles, and their strategic approaches. While Luc Reychler obtained a Ph.D. in Psychology at Harvard, his Ph.D. co-supervisor was Professor Karl Deutsch, Harvard's famous political science professor and former President (1976–1979) of the International Political Science Association (IPSA). According to Reychler, Karl Deutsch was a thought-provoking and inspirational co-supervisor who strongly encouraged him to expand his disciplinary horizon beyond psychology. And consequently, while at Harvard, Luc Reychler also intensively studied and mastered a range of *political science* subjects, including diplomatic history, theory and methodology of international relations, comparative politics, and foreign policy analysis. Interestingly, in Reychler's subsequent academic career, political science research and teaching took centre stage.

After successfully completing his doctoral research at Harvard University in 1976, he decided to return to Belgium, where he obtained a professorial position at the Department of Political Science at KU Leuven. As a newly appointed professor at KU Leuven, a significant chunk of Reychler's time was dedicated to teaching and supervising political science students. Besides teaching political science research methodology, Reychler's initial teaching portfolio included courses such as theories of international relations, strategic studies, global security analysis and comparative foreign policy analysis. From the early 1990s, his teaching and research increasingly shifted towards conflict and peace research. Moreover, from 1993, Reychler started teaching several postgraduate courses which explicitly focused on these issues, including "Peace Research and Conflict Management" and "Mediation and Negotiation Skills". These courses became very popular with students from the Faculty of Social Science, but also attracted a large number of students from other faculties. The widespread interest in better understanding the causes of violent conflict and strategies to resolve conflicts peacefully may well have been directly related to the fact that many 'new' violent conflicts and civil wars erupted or recurred in the 1990s, for instance in Bosnia, Rwanda, Somalia, Georgia, Sierra Leone, Tajikistan, Nepal and the Democratic Republic of the Congo. In order to respond to the growing demand for a postgraduate degree in conflict and peace studies, Luc Reychler became the driving force behind the introduction of a Master of Conflict and Sustainable Peace (MaCSP) at KU Leuven, an MA programme which successfully ran from 2002 to 2007. While this programme was very timely and attracted a substantial number of highly motivated students from all over the world, it was nonetheless discontinued in 2007, three years before Reychler's retirement, because there was insufficient teaching and supervision capacity at the Department of Political Science to maintain such an intensive and interactive MA-programme. With a heavy heart, Reychler accepted the unfortunate decision to terminate MaCSP in 2007.

While studying in Leuven for a political science degree, I followed a number of Reychler's courses and also wrote my MA thesis under his supervision. Thinking back about this time, I remember Luc Reychler as a tremendously dedicated and driven lecturer who was extremely passionate about his field of study and investigation. Indeed, he never tired of pointing out to his students that not only was it essential to proactively prevent or resolve conflicts in order to reduce suffering around the world, but also that it was crucial to improve our understanding of how best to do this. In this respect, he used to emphasise that both helping without knowing and knowing without helping were equally unacceptable courses of action. I also remember Luc Reychler as a provocative lecturer, who deliberately attempted to get his students out of their comfort zone by asking difficult and provoking questions with the aim of making students reflect critically on uncomfortable truths and situations. I distinctly recall in this context that during a lecture as part of his "Peace Research and Conflict Management" course, Reychler started a debate with his students around the question of whether they could imagine killing somebody. While it was surprising to see how many students openly acknowledged that in certain 'extreme' circumstances they could indeed imagine killing somebody, the main lesson that Reychler aimed to convey to his students was the importance of preventing situations in which 'normal' people would be prepared to use lethal force to survive or achieve certain other objectives.

Reychler also did not shy away from asking 'awkward' questions or making provocative statements in order to engender debate and discussion among fellow academics and/or policy-makers. In this respect, I remember him telling me early on when I started working at CPRS about an incident that happened at the International Peace Research Association (IPRA) meeting in Groningen in 1990. At this conference, which biannually brings together a large number of a peace researchers, Reychler offered a highly critical reflection on the field of peace research. In particular, he argued that while the peace research community consisted of many serious researchers, it was in danger of being overrun by what he called "peace quacks" or, in Dutch, *"paxzalvers"*. Reychler coined this term to refer to unqualified persons who *pretend* or erroneously think that they possess the necessary knowledge and skills to advance peace. Similarly, the term may also refer to politicians and lobbyists who, in the name of peace and security, contribute to war and insecurity. In his conference presentation, Reychler further argued for a complete overhaul of the field of peace research and boldly claimed that it was time for *Peace Research II*. According to Reychler, his provocative claim engendered a constructive – albeit fierce – debate in which the attendees critically reflected on their own field of research. Mission accomplished! To this day, Luc Reychler remains convinced that in order to make progress, either politically, academically or otherwise, it is sometimes necessary to throw the cat among the pigeons.

Turning to his research, Luc Reychler is a prominent peace researcher, who has contributed enormously to the field of peace research ever since he started studying and analysing diplomatic behaviour and thinking as part of his Ph.D. research at Harvard University. He has received widespread praise and recognition from his fellow researchers for his research and contributions to the field of peace research.

For instance, he was elected Secretary General of the International Peace Research Association (IPRA) 2004–2008 by his scholarly peers and named Chair Holder of the UNESCO chair on Intellectual Solidarity and Sustainable Peacebuilding in 2007. Further, he has produced a large number of high-level publications, which include 18 books as (co-)author or as (co-)editor and more than 60 scientific journal articles and book chapters. Arguably, even more impressive than the quantity of his work, is the scope of topics and issues he has focused on during different stages of his academic career. In this respect, the current book offers a nice compilation of the wide scope of issues and topics Luc Reychler has addressed throughout his academic career. Illustratively, the current book contains a diverse set of previously published papers, which focus among other things on the analysis of diplomatic behaviour and thinking (see Chap. 4, "Values in Diplomatic Thinking: Peace and Preferred World Order"), the limitations, shortcomings and challenges of the field of peace research (Chap. 11, "Challenges of Peace Research"), and the role of time in conflict and peacebuilding (see Chap. 15, "Time, Temperament and Sustainable Peace: The Essential Role of Time in Conflict and Peace"). In my opinion, each of the chapters included in this book makes an interesting contribution and is worth a detailed read. Yet, when reflecting on Luc Reychler's main academic achievements and contributions, there are three areas of research which stand out for me.

First, Reychler's research on a Conflict Impact Assessment System (CIAS), dating back to 1996, is fascinating and remains highly relevant to the field of peace research. As Reychler argues below, the aim of CIAS is to assess in a systematic and comprehensive manner the positive and/or negative impact of different kinds of interventions (or the lack thereof) on the dynamics of conflict. More specifically, it aims to serve as a sensitizing tool for policy-makers and other stakeholders to improve their conflict prevention and peacebuilding interventions and policies. Since its introduction in 1996, the CIAS methodology has been widely used to assess peacebuilding policies in different countries. Moreover, together with Thania Paffenholz, Luc Reychler published the CIAS-handbook, titled "*Aid for Peace: A Guide to Planning and Evaluation for Conflict Zones,*" in 2006, which is still considered a milestone in the field of peace research.

Second, Luc Reychler's work on peacebuilding architecture is also groundbreaking and has not lost any of its relevance. He coined the term "peace architecture" in order to draw attention to the architectural principles and considerations that have to be addressed in sustainable peacebuilding processes (see Chap. 2 for more details). As part of his research on sustainable peacebuilding architecture, he identified and analysed the following six peacebuilding blocks: effective communication, consultation and negotiation; peace-enhancing political, economic and security structures; peace-enhancing software; peace support systems; a supporting international environment; and effective peacebuilding leadership for the installation of all the preceding peacebuilding blocks. Significantly, he also demonstrated that there is good and bad peace architecture and, subsequently, good and bad architects. According to Reychler, an example of bad peace architecture was the Treaty of Versailles at the end of the First World War, while an example of good peace architecture was the creation of the European Union. As he noted with respect to the European peacebuilding process following the carnage of WWII, "this peace architecture turned Europe, which scored all the Guinness Records of violence before 1945, into one of the most free, secure and affluent regions of the world".

Third, Luc Reychler's recent work on the role and impact of time and temporal behaviour in conflict and peace processes is also truly fascinating. His recent book, entitled *Time for Peace: The Essential Role of Time in Conflict and Peace Processes*, is in many ways the result of a 30-year interest in the essential – yet seriously understudied – role of time in conflict and peacebuilding processes. In this book, Reychler observes that despite the fact that people tend to take time seriously, a great deal of their temporal behaviour, especially in conflicts, shows gross deficiencies. He further argues that "in our fast-changing globalized world, time and the way we deal with it, will more than ever determine the success or failure of handling of global crises and opportunities for building of sustainable peace and development" (see below). An essential lesson that emerges from this work is that we need urgently and radically to change the way we deal with time in conflict and peace processes and develop "a more adaptive temporament". This lesson is also an important avenue for future research, which the next generation of peace researchers will have to systematically and innovatively investigate.

In his latest research, Luc Reychler focuses on the arguably controversial topic of the role of humour in conflict and peacebuilding. While this topic is bound to make some heads turn, I am absolutely convinced that Reychler will once more produce a fascinating, insightful and perceptive piece of research, which will surely receive widespread attention from academics, practitioners and policy-makers interested in the field of peace research. And while this latest research may well be Luc Reychler's final project in his long and distinguished academic career, it is comforting to know that his work will not only continue to be read by academics, practitioners and students alike, but also that his work will definitely continue to inspire future researchers for a long time to come.

Leuven, Belgium
July 2019

Prof. Arnim Langer
Director of the Centre for Research on
Peace and Development (CRPD)

Chair Holder of the UNESCO Chair
in Building Sustainable Peace

Professor of International Politics
KU Leuven

Acknowledgements

Numerous experiences and persons, personal and professional, over the course of time helped me to find my way. I enjoyed the peace research teams. The list of names is long. My deepest thanks.

Leuven, Belgium   Luc Reychler
August 2019   Professor Emeritus

Contents

Part I   On the Author Luc Reychler

1   My Personal and Academic Journey   3

2   Luc Reychler's Comprehensive Bibliography   47

Part II   Luc Reychler's Selected Texts

3   Introduction to the Selected Texts   63

4   Values in Diplomatic Thinking: Peace and Preferred World
    Order   85

5   The Effectiveness of a Pacifist Strategy in Conflict Resolution   129

6   The Art of Conflict Prevention: Theory and Practice   159

7   Religion and Conflict   175

8   Lessons Learned from Recent Democratization Efforts   197

9   Peacebuilding: Conceptual Framework   209

10   Peacebuilding Software   223

11   Challenges of Peace Research   241

12   Intellectual Solidarity, Peace and Psychological Walls   257

13   Peacemaking, Peacekeeping, and Peacebuilding   271

14   Raison ou Déraison d'état: Coercive Diplomacy in the Middle
    East and North Africa   301

**15 Time, Temporament, and Sustainable Peace: The Essential Role
      of Time in Conflict and Peace**   325

On the University of Leuven/KU Leuven   349

Centre for Research on Peace and Development (CRPD)   351

About the Editors   353

Index   355

Part I
On the Author Luc Reychler

Chapter 1
My Personal and Academic Journey

It was a dark blue night. I was six and tried to get out of a tall hedge maze. The full moon made me see better, but not escape. I longed to be on the moon to spot the exit. This was the first dream I remember. My academic life has been a search for points of view from which I could see the big picture and find a way out of violence and peace problems.

The small stuff or non-academic experiences have played an important role in my academic journey. The reader can skip these stories. I'll do my best to invalidate the second part of 'Old men remember things that did happen, and things that did not happen'.

1.1 Early Experiences

I was born on 30 December 1944 near the end of the Second World War. One more for the 2.5 billion world population. In my mother's belly, the war period was a good time. She told me that I was her only child with bluish-black hair. I was the second of seven (three brothers died at birth) (Photo 1.1).

Until the age of 18, I lived in the provincial town of Eeklo, the capital of the Meetjesland, a rural place between Ghent and Bruges, in the North bordering on Zeeuws Vlaanderen of the Netherlands. A highway to the North Sea painfully split the town. A boy from my class was run over. There were schools, shops, cafés and factories. On Thursdays, the center of Eeklo was transformed into a marketplace, with grunting, snorting and squalling pigs and piglets. The church bells and the sirens of a weaving factory and the brewery 'Krüger' set the time. My parents ran a tobacco shop on the Grand Market and a small tobacco factory in the Sterrestraat (Stars street), where dry tobacco leaves were misted, washed, cut, packaged and taxed, cigars rolled, snuff stamped, and chewing tobacco twisted. I enjoy the aroma of an unlit cigar and cigar brands, such as 'Help Yourself and Mephisto', but I never got into the habit of smoking; neither did my brothers and sisters. In the shop,

L. Reychler and A. Langer (eds.), *Luc Reychler: A Pioneer in Sustainable
Peacebuilding Architecture*, Pioneers in Arts, Humanities, Science, Engineering,
Practice 24, https://doi.org/10.1007/978-3-030-40208-2_1

Photo 1.1 German tanks in my home town Eeklo in September 1944 during the occupation of Belgium in World War II. *Source* Personal photo collection

two mission charity boxes stood on the counter: one for the *zwartjes* (little blacks) and one for the *chineeskes* (little Chinese). When a coin was inserted, a small head nodded thanks. I liked to move the heads with my finger. These terms were not considered politically incorrect. For us colonialism was charity. We never heard the story of greed or knew Leopold's ghost.[1] If Ai WeiWei had seen them, he would probably have sculpted two mega charity boxes as tokens of obsolete mindsets. Most of the factories have gone (Photo 1.2).

My birthplace remains a source of imagination. In the Middle Ages, it had a widespread reputation for remaking the heads of people. After detaching them from the body, they were treated with ointments and oils, rebaked and reattached. When successfully operated on, the patients could start a fresh and new life. However, when the head stayed too long in the oven, the patient became a hothead; when not baked enough, she or he became *een halve gare* (a nutcase), and when the rebaking failed, he or she ended up as *een misbaksel* (an ogre).

---

[1]*King Leopold's Ghost: A Story of Greed, Terror and Heroism in Colonial Africa* (1998) is a best-selling popular history book by Adam Hochschild that explores the exploitation of the Congo Free State by King Leopold II of Belgium between 1885 and 1908, as well as the large-scale atrocities committed during that period.

Photo 1.2 Luc carried by Mam. *Source* Personal photo collection

I had a happy family life. My parents worked hard. Sometimes, my mother got up at five o'clock to get an hour of silence before fulfilling her roles as mother, cook, house- and shop-keeper. She was serene, always uplifting, a diplomat, never complained, and wanted us to be tactful and ourselves. When my youngest sister was born, my mother looked paper-white and nearly died. After her convalescence in 1957, a green Chevrolet drove my family of nine to Lourdes in France for thanksgiving. My father was open-hearted, a person of integrity, authoritative and stood for order at the table. We had a fixed place and were not allowed to gossip about the private affairs of others. This was considered a waste of time. Admonitions, like "You are the oldest; you should be the wisest", "Use your wits", and "What you have in your head cannot be taken" reminded us to be independent and responsible. Although baptised, we were prompted to take moralising people with a grain of salt, and to check words with deeds. I often accompanied my father and our hunting dog, Frieda, to the Polders or the Creek region. To chase hares, pheasants and partridges, I dashed between walloping leaves of fodder beets which one could barely see over. I never hunted, but remained captivated by nature. The scent of stagnant creek water and reed calls up childhood memories. When teaching in Leuven, I went to the cabin at the *Bomloze put* (Bottomless pit), a sixty-metre wide circular creek in East Flanders, to write and savour natural beauty. This was my Walden Pond. It reminded me of Henry David Thoreau, whose books *Walden* and especially *Civil Disobedience* inspired me (Photos 1.3 and 1.4).

Little events mould world views and aspirations. My relation to religion drastically changed on a train, steaming to Austria for a vacation with 10- and 11-year olds. I was one of them. Some kids were pestering a boy. The monitor from the Catholic college did nothing. I urged them to stop the bullying. When not succeeding, I left the cabin with the victim to sit in the door space. The peace was short. When the train entered a tunnel and the lights dimmed, the thugs came in and continued their harassment. I tried to stop it. Then came a helping hand; the monitor from the secular athenaeum, sitting in the other cabin, ordered them to leave. This was the tipping point in my appreciation of religion. I discerned that there were good people with and without religion. From then on, I chose eclecticism. From Greek and Roman philosophy I borrowed Epicurean and Stoic lifestyles (Photo 1.5).

My personal motto became *Imo summoque concors*, or "In highs and lows listen to your heart". Much later, when visiting South Korea as part of a long-term research and teaching collaboration with Kyung Hee University, I learned that most of my Korean friends were eclectic. Some simultaneously embraced Christianity,

Photo 1.3 Family picture of the 1950s. *Source* Personal photo collection

Photo 1.4 Primary school 1953–1954. *Source* Personal photo collection

Buddhism, Confucianism and Shamanism. Each belief system fulfilled different aspirations. Another experience, as a youngster, related to language. Disconcerting was the fact that many elite persons in Belgium spoke French to peacock over Flemish-speaking people. French for them meant standing with a superior cultural fragrance. Most Flemings perceived this as disrespectful. At the beginning of the Sixties, the high schools in Flanders joined a campaign to upgrade Flemish culture by means of learning civilized Dutch (ABN) and etiquette. Although useful, the teaching was done in a stilted way. In addition, the label "civilized Dutch" insulted the Flemish dialects and the mother tongues spoken by most students. The new etiquette felt artificial and unnatural. I favoured 'tact' and likened the new etiquette to a pair of glass crutches.

Foreign countries and cultures became captivating. Every two years, the primary school visited an exhibition of missionary work in Africa and China. I craved for more. But since I had no call for missionary work, I later opted for diplomacy. A great eye-opener was the Brussels World Fair in 1958. In the miniature world, around the Atomium, you could run from country to country. Most alluring was the pavilion of the Soviet Union with the atomic icebreaker and, above all, a model of

Photo 1.5 My personal motto became 'Imo summoque concors'. *Source* Personal photo collection

Sputnik 1, the bleeps and the dog Laika. There were also pavilions for the Belgian colonies, which unexpectedly became independent in 1960. I was not impressed by the American site. Later, America's appeal was elevated by John F. Kennedy's inaugural address and the movie *West Side Story*, which was screened in the

De Leeuw cinema in Eeklo. The excitement and uplifting mood were enthralling. America was the place to go.

I treasured independence and pursued what I regarded as important. When necessary, I assumed leadership. As the oldest boy in the family, there were many occasions to practise leadership and diplomacy. In the Baden Powell scouts group in my hometown you learned to practise leadership. It also made you a member of a worldwide youth organization. I liked my totem, *zachte panter* (gentle panther). In 1963, I attended a leadership Gilwell course at De Kluis (near Leuven). The training focused on knowing and expressing yourself, listening to others, taking the initiative, teamwork, and having the courage to innovate. Later, I participated in a higher leadership course, taught by inspiring and less inspiring leaders, such as J. L. Dehaene, who became the Belgian Prime Minister from 1992 to 1999. He told journalists that you only have to solve a problem when it presents itself. I disagreed and argued that it is important to pre-empt a problem or to proactively prevent them given that the consequences might otherwise be dire. When, in 1994, the genocide in Rwanda erupted, Belgium intervened too late.

John F. Kennedy's book *Profiles in Courage* was a source of inspiration, but one event taught me that problem-solving requires more than courage and goodwill. In our living-room stood a cage with a cheerful yellow canary. One day, I noticed that the creature had difficulty balancing on its perch. Droppings had dried between its toes. I carefully tried to remove the stuff with a pair of small scissors. Suddenly a drop of blood appeared. When my father saw it, he was angry and told me that I should have soaked the little legs in lukewarm water. He was right. I learned that both helping without knowing and knowing without helping were not the best remedies, and that the confrontation was a learning experience (Photo 1.6).

Although wars and natural disasters were far from our own beds, we were often reminded of them. Older men bade us listen to their experiences as soldiers in the trenches of the First World War. Kids found the stories boring. The hobby of a townie was collecting pictures of the countless military graveyards in Belgium. In 1953, a platoon of Belgian volunteers came back from the Korean War and paraded in the Grand Market. In the attic we kept dried foods, sardines and Sunlight Soap in case the Korean war spilled over. At the age of ten, I visited *Het Gravensteen* (the castle of the Counts) in Ghent and its torture room. We associated torture with the dark Middle Ages. I was flabbergasted when I later discovered that torture is still used by democratic and undemocratic states today (Photo 1.7).

Occasionally, my father talked about the War. His stories dealt with asymmetric power situations in which courageous men and women stopped or deterred the use of physical violence during and after the Second World War. Not only stories, but also theatre confronted me with violence. In high school, the choreographer Heiko Kolt asked me to act in *Max Havelaar*. The play was a protest against the colonial deprivation of the native population of Dutch India (Indonesia) in the nineteenth century. In 1960, on the day the Congo gained independence, Patrice Lumumba, the first democratically elected prime minister, gave a political speech in which he denounced colonial rule. It was a speech from the heart. I agreed with him, but I could not say so. Criticizing the colonial civilization of Congo was unheard of. The

Photo 1.6 Luc learning leadership with the Boy Scouts. *Source* Personal photo collection

thinking at the time was that the Congolese were expected to be thankful for the Belgian civilization efforts. This was my first encounter with the violence of denial and taboo.

On the whole, Belgium was a secure place to grow up. The closest natural disaster, but at a safe distance, was the 1953 North Sea flood which struck the Netherlands and parts of Belgium. Many dykes collapsed under pressure.

Contrasting with the distant violence was the non-violent resolution of conflicts within Belgium and the European integration. The European Economic Community at the end of the 1950s made it easier to cross the Dutch border by bicycle on the way to the swimming pool at Aardenburg in the Netherlands.

I enjoyed primary school. High school was my Middle Ages. There were many good experiences, but, on the whole, the climate was conservative and suffocating. Near the end of the last year the class teacher of the Retorika advised me not to go to university. Consequently, he inquired if I had a vocation for the priesthood. I disliked his monologue. Traditionally, all the final-year Greek and Latin students went to Rome. I decided to break the tradition and travelled to Cambridge to learn English. There, I also got acquainted with rich young students from Iran and with English humour. After apples vanished from a small tree in the garden of the language school, a sentence on the blackboard said that 'Stolen apples taste like kisses'. The use of poetry to stop the forbidden fruit being stolen was sublime (Photo 1.8).

Photo 1.7 Luc at the age of 11. *Source* Personal photo collection

1.2 Becoming a Social Psychologist of International Relations and Peace

1.2.1 The University of Ghent

1.2.1.1 Industrial Psychology

To enjoy pluralism, I registered in 1963 as a student at the University of Ghent. Since international politics and diplomacy did not exist as a field of study, I chose psychology. Industrial psychology provided a wide range of analytical tools to understand individual and collective human behaviour. I also took a range of elective courses on international law, diplomatic history, international economics, and history of the sciences. The introductory psychology course was called *Zielkunde*, or the study of the soul. The teaching was *ex cathedra*. The university had plenty of facilities to take care of the second part of *Mens sana in corpore sano*. Besides jogging, I acquired a taste for kayaking, mountain climbing, speleology, fencing, yoga, horse-riding, skiing, sailing, and pantomime. I cherished them all. I graduated with *magna cum laude*, received a travel grant from the Ministry of National Education for my MA thesis, and became an assistant of Prof. Paul

Photo 1.8 Luc mountain climbing in 1964. *Source* Personal photo collection

Ghijsbrecht. He studied in Ghent, Leiden and Tübingen, and taught psychology and forensic psychiatry. My MA thesis researched some social-psychological determinants of productive participation and democratic corporate management. The research took place in the Krüger brewery, where I tasted unfiltered ice-cold beer for the first time. Tasting beer eased collecting data.

1.2.1.2 Forensic Psychology

Ghijsbrecht, a forensic psychiatrist, opened for me the world of micro violence, murder and double suicide. I attended autopsies, encountered murderers, and analysed the social and physical context in which the crime took place. The high level of disorder at the houses or apartments where violent crimes had taken place was remarkable. The first time I assisted an autopsy of a cadaver by Prof. Thomas, I was stressed, but the procedure and the scientific language attenuated these feelings and sharpened my attention. I believe that it is sometimes useful to dissect reality as a cadaver; to undress and analyse problems after the removal of labels, theories and prejudices. Writing evaluation reports of tens of student theses was a great learning experience. Ghijsbrecht was an extravagant humanist. He introduced us to Nietzsche's *Also sprach Zarathustra* and to Karl Jaspers. The latter argues that life and potentialities are shaped by four types of communication: Dasein's communication, conceptual communication, spiritual communication about the unity or the wholeness of parts, and existential communication about the meaning of life and the experience of authentic existence. The existential approach is associated with dignity in solitude, open-mindedness, existential solidarity, intellectual integrity and accepting the communication partner as equal in rank. Human beings always have two options: an option of resignation, pessimism and nihilistic despair, or the option of optimistic confidence in the meaning of life. Another mentor who gave me wings was Dr. Paul Braem, the provincial prior of the Augustinians in Belgium. He was open-minded. We discussed psychoanalysis, Freemasonry (*Les fils de la lumière*), diplomacy, Teilhard de Chardin's the Future of Man, and the brewing of Augustijn beer (Photo 1.9).

1.2.1.3 Ghent's Influences

Near the end of the 1960s, I decided to move on to the study of international politics, not only to deepen my understanding of diplomacy and war, but also to deal with *The American Challenge* of Jean Servan Schreiber (1968), and *The Americans and We* of Théo Lefèvre (1966). Schreiber saw the United States and Europe as engaged in a silent economic war, in which Europe appeared to be completely outclassed in management techniques, technological tools, and research capacity. Our Minister for Scientific Policy, Lefèvre, urged the Europeans to close the scientific gaps between the United States of America and Europe. One day, in a bookshop, I found *International Behavior: A Social-Psychological Analysis*, edited by Herbert Kelman of Harvard University (1965). This book led me to study the psychology of international politics. I became more and more committed to the prevention of armed violence. World War I on Flander's fields looked like suicidal folly in which the slaughtered soldiers were venerated as martyrs. After doing research at the University of Munich in 1967, I visited the concentration camp at Dachau and the Berlin wall. These were distressing and surreal experiences. Reading Frantz Fanon's *The Wretched of the Earth* (1961), with a foreword by Jean-Paul Sartre, widened my definition of violence and made me a fervent anti-racist. The protests against the Vietnam War and a *We shall overcome* optimism provided courage to look for the best places to study war and peace.

I became apprehensive about the uncritical face of the humanities and social sciences. My critical attitude was whetted not only by Fanon, who pleaded for a striptease of 'humanism', but also by Milton Rokeach's *Open and Closed Mind*, Herbert Marcuse's *One-Dimensional Man*, and Michel Foucault's thoughts on knowledge and its relation to power structures. There were the reductionist causal attributions of psychoanalysis. Freud attributed human and inhuman behaviour to libido and Thanatos; Adler to the will for power; Jung to the journey of individuation, etc. History is dictated by power. History should be complemented by her-story, our-story and other-stories. Not many students of human behaviour make explicit the normative, theoretical and epistemological assumptions which shape their research and conclusions. Consequently, I began to appreciate the added value of

Photo 1.9 Luc with Prof. Ghijsbrecht, a forensic psychiatrist at the University of Ghent. *Source* Personal photo collection

methodological behaviourism, or the objective observation of public and individual behaviour, without paying attention to justifications, labels, thoughts and feelings. The psychiatry course in the old Bijloke Hospital was remarkable. Every week, in a semicircular auditorium, we listened to a theoretical presentation of schizophrenia, manic-depression, psychopathy, phobias, dementia, etc. Consequently, the professor presented one or two patients who embodied the disorder. The course was fascinating, but I felt uncomfortable about the labelling and the dehumanising impact of stereotyping patients with psychiatric labels. My concern with political labelling and stereotyping was strengthened in 1973 by David Rosenhan's experiment *On being sane in insane places*. The subjects faked hallucinations in order to enter psychiatric hospitals, and behaved normally afterwards. Most of the pseudo-patients were diagnosed with psychiatric disorders. They did not succeed in being released, because the staff reframed everything they observed to fit their earlier diagnosis.

1.2.2   The London School of Economics and University of Oslo

In 1970 I left for London. The city was a cosmopolitan delight with plenty of academic and cultural opportunities. We lived in Hampstead, the Montmartre of London, close to the Hampstead Heath park. Nearby I learned horsemanship. With a research grant from the Royal Society, at the London School of Economics (LSE) I studied peace research with Hilde Himmelweit, international relations with Michael Banks and strategic studies with Philip Windsor, and at University College London (UCL) I studied international behaviour with John Burton and conflict analysis with Chris Mitchell. John Burton, a scholar-diplomat and gentleman, was director of the Centre for the analysis of conflict. He advocated, in theory and practice, conflict resolution and conflict prevention for dealing with deep-rooted international conflicts. Traditional diplomatic history and the study of international relations did not pay enough attention to the fears, hopes, perceptions and feelings of the players. To get to the heart of a conflict, he urged bringing the leaders together in a process of controlled communication. In 1969 he shared these ideas in his book, *Conflict and Communication: the use of controlled communication in international relations*.

In the summer of 1971, with a Norwegian grant, I attended the International Summer School at the University of Oslo and visited Johan Galtung's *Peace Research Institute* (PRIO). This was the year his article 'A structural theory of imperialism' was published. Galtung pioneered peace research. Towards the end of the summer, the summer students were invited to a reception at the palace by a relaxed and informal cigar-smoking King Olav V. For a Dutch student, who told me that his compatriots do not laugh in the company of their royalty, the reception was a majestic moment. For me, this was a wonderful time with plenty of academic inspiration.

1.2.3   Harvard University

1.2.3.1   Harvard Town

In 1971, the plane landed at Logan International Airport in Boston. Vic and Sue Ricardo, our host family, welcomed us. They drove to Harvard, a small town twenty-five miles east of Boston, with wooden houses in the midst of nature and loud chirping crickets. Historically, the town once housed the first Shaker settlement in Massachusetts. The place contrasted starkly with my skyscraper image of America. At the apple-blossom festival in May, people picnicked and had fun on the commons and a neighbouring green cemetery lawn. Dr. Ricardo, a specialist in medical genetics at the Harvard medical school, helped me to settle in at Cambridge, Massachusetts, and register at the university. They were cordial hosts. Towards the end of summer, we were invited to spend the weekend with them. On Sunday morning during breakfast we talked about many things, including how long can you stay under water without air. In the afternoon, we went swimming in Bare Hill Pond, a gorgeous 330-acre lake. Vic ran vigorously to the lake, took a dive and stayed under water. His hairs floated around his head. Close by, I checked the time. His timing was very good, but suddenly he slapped the water with his hand. I gently lifted him up. He had broken his neck. Instead of competing, he was drowning. I had stepped into the 'expectation trap' and learned 'the framing effect' the hard way. He remained partially paralysed.

1.2.3.2   Harvard University

The University of Harvard was my renaissance-baroque period. Prof. Herbert Kelman, my mentor, and his wife Rose, made me feel at home and guided me during the five MA and Ph.D. years. Herb was born in Vienna in 1927 and after the *Anschluss*, when he was eleven years old, his family were given asylum in Belgium and stayed in Antwerp for a year. He lived in Van Luppenstraat, where Lucas, my son, now resides. I became Kelman's head teaching fellow. Herb introduced me to social psychology, the psychology of international behaviour, and to interactive problem-solving workshops for the Israeli-Palestinian conflict. I assisted in the first 1971 workshop. I decided to adopt the conflict and stick to it until it was resolved. This decision was made in order to monitor the impact of workshops and other peace mediation efforts. I was convinced that the pursuit of genuine peace and security is possible and would be significantly better for both the Israelis and Palestinians. I have friends on both sides. As the most internationalised conflict, genuine peace would be a giant step forwards in the building of world peace. After fifty years the conflict has deteriorated. Palestine could be a place of peace, but instead it has been transformed into a place with open wounds. It has become a very asymmetric and humiliating conflict that hurts (Photo 1.10).

More than half of the courses dealt with different aspects of research methodology and design. We even learned how to write a block article, a scientific paper and a book. Instead of imbibing general principles, we listened to distinguished authors, like B.F. Skinner and Roger Brown, who told us how they did it. Skinner started by mapping the outlines and the different chapters very systematically on the back of big computer sheets. Roger Brown began typing and followed his train of thought. I embraced Skinner's mind-mapping approach. To become versed in the political sciences and the study of international relations, I attended courses,

Photo 1.10 In Harvard's William James building in 1974. *Source* Personal photo collection

seminars and guest lectures given by Karl Deutsch, Stanley Hoffmann, Thomas Schelling, John Rawls, Roger Fischer, Joseph Nye, Kenneth Boulding, Lawrence Kohlberg, Noam Chomsky, Ivan Illich, Donella Meadows, et al. Thomas Schelling urged us to resolve real problems and not confuse them with our problems, to provide operational definitions for all the key concepts, to look at the whole and not just at a fraction of the problem, and to respect deadlines. Karl Deutsch saw international relations as a complex system with interlocking sectors and levels. Stanley Hoffmann distilled the intellectual essence of peace studies and described peace as one of those fruits with an inedible core. Ionesco's *Rhinoceros* was on his reading list. It describes brilliantly how normal people slowly transform into rhinos, bullies or fascists. The play is as relevant as ever. It explores conformity, mimetics, group pressure, self-delusion, culture, fascism, responsibility, logic, mass transformations, populism, mob mentality, and morality. New concepts, such as win-win negotiations, the original position in which everyone decides principles of justice from behind a veil of ignorance, the economics of fear and hope, the international security community, stages of moral development, iatrogenic side-effects, critical pedagogy, limits to growth, civil disobedience etc., helped us discern complex reality. As a doctoral associate of the *Center for International Affairs* (CFIA), I attended lectures and discussions with researchers and practitioners on international developments. The Harvard education was exacting and very stimulating. Approximately two-thirds of the time dealt with research methodology: how to analyse a problem, design and implement a research project, and write up and evaluate results. The underlying normative, theoretical- and epistemological assumptions needed to be made explicit. The other third of time was used to overview the state of relevant and competing theories. The professors were accessible and treasured informed discussion. I learned a lot from practitioners, case studies and simulations.

1.2.3.3 Extracurricular Experiences

Every evening, on a small colour television, I saw pictures of the Vietnam War and body bags shipped to America. This war and its continuation in the name of "peace with honour" was deeply disconcerting. Power may be an aphrodisiac, but it became an orgy of unnecessary violence. To vent anger and frustration, I sketched "peace with honour" as a pot with a plant that had lost all but one little leaf. Anger can be a great source of creative energy. There were protests, the Kent State shootings and a strong demand for peace studies. Three doctoral students, including myself, convinced professors from different faculties to contribute a lecture to a new course, 'Peace, Justice and Social Change'. The course lasted several years.

Harvard was also an extracurricular experience. As a student working in the Widener periodical room, I had access to the archives and was surprised to find files of family members who were professors in different Belgian universities (Brussels, Ghent, Leuven, Louvain) and specialized in botany, chemistry, pulmonary medicine and orthodontics. I had never heard of them. At home we had been told that the first Reychler who migrated in the eighteenth century from Austria to Belgium was a saddle-maker.

To keep a healthy body, I sailed on the Charles River, played squash, and jogged around Fresh Pond. While jogging I reflected on my personal philosophy and future. At Fresh Pond someone was learning to play the trumpet and within a year he played like a professional. I was impressed by how much one can learn in a year. During one semester I tried ballet. Walking in the street after training, it felt as if my head and shoulders were lifted by strings.

To vent what words could not, I started painting. I registered for surrealistic painting, but since I was the only candidate, the course was cancelled. Later, I showed my canvasses to an old Viennese art teacher at the university. He looked, abstained from comments and advised me to start drawing glass bottles. Anyway, before leaving Cambridge, I made a painting of diplomatic thinking. It shows a backward-looking diplomat moving on.

In Boston I discovered the pillars of the temple of humanity: beauty, wisdom and strength. I was enraptured by a particular type of American humour: roasting. Roasting, or being tactfully mean, is meant to honour a person. When Mr Hall from the Harvard University Press died, he was roasted in Harvard Memorial Church. Friends told stories. The laughter made the commemoration sublime (Photo 1.11).

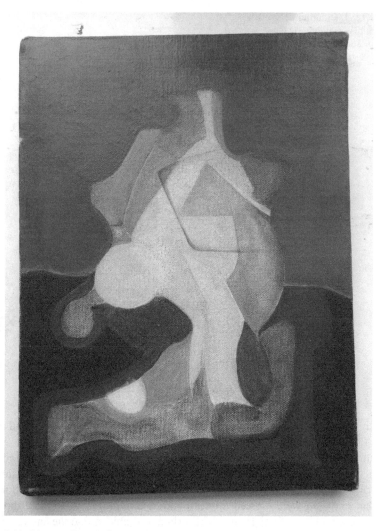

Photo 1.11 Picture of a painting on diplomatic thinking. *Source* Personal photo collection

1.2.3.4 Washington D.C.

In 1975, when my Ph.D. proposal on diplomatic thinking was approved by Herbert Kelman and Karl Deutsch, I headed for Washington D.C. As a visiting scholar at the Bureau of Social Sciences Research, I had an office in M street. I lived on Rock Creek Park Street in a room walled by books. The host was the widow of an ambassador without a country, the former state of Latvia. At the time, Washington D.C. contained the embassies of three Baltic states (Latvia, Lithuania and Estonia) which had been absorbed into the USSR and in that era did not

Photo 1.12 Luc during field research in Washington D.C. in 1975. *Source* Personal photo collection

officially exist as separate states. On average, I completed one interview and report a day (Photo 1.12).

By the end of my research, 266 diplomats from 114 countries had shared their thoughts on international politics. It was fascinating to look at the world through diplomatic eyes. China was approaching the end of the Cultural Revolution; the diplomats and staff were dressed in green and blue uniforms. A Chinese diplomat handed me a package of articles in which I should be able to find the answers to my questions. At the end of April 1975, the Vietnam War ended and the embassy closed. As usual, the summer was hot and humid, and the air was often polluted. A year later, fifty members of the Roosevelt Home for Senior Citizens performed a "Homage to Pollution Dance" inspired by Washington's low air quality index. The dancers gasped, choked and fell on the ground at Farragut Square.

1.2.3.5 Back in Cambridge, MA

When I returned to Cambridge, the data were coded, punched and analysed. It took three months to write the manuscript. At various stages of the research, I had the skilled and indefatigable assistance of D. Elaine Exum: she did the reliability coding, provided sensitive editorial advice, and helped me to deliver the study on time. Once she told me about segregation experiences. I became pale and speechless. After visiting George Washington's estate at Mount Vernon, she headed for the ticket office to ask for her money back. The history of slavery had been deleted from the site and the tour. She taught me to see black and white in colour. Unfortunately, at the end of November 1980, I had to fly back to Washington D.C. On arrival, I rushed by taxi to the Baker funeral home on 14th Street, where she lay in a coffin. She was thirty-five.

After graduation, I was not sure whether to stay in America or return to Belgium. Finally, I decided to come back because of the appeal of the European Community and the opportunity to teach in Leuven. In 1976, before leaving, I joined the bicentennial celebration in Boston on the 4th of July. Four hundred thousand people picnicked near the Charles River and relished in the Boston Pops' open-air performance of Tchaikovsky's *1812 Overture*. Heading for the exit in the midst of a powerful stream of people was a chilling experience (Photo 1.13).

Photo 1.13 Elaine Exum. *Source* Personal photo collection

1.3   Polemology and Irenology in Leuven

1.3.1   Back in One of the Hearts of Europe

Belgium was a culture shock. It was difficult to readapt to the political, cultural and academic scene. I even abstained from eating Belgian fries for a year. Luckily, I encountered inspiring colleagues and indulged in course- and research work. The rector, Pieter De Somer, was a man of the world, and my colleague in international relations, Paul Van De Meerssche, was fascinated by the paradox of European integration and disintegration. With colleagues from different faculties we created an American Studies programme. The widening and deepening European Union with nine and now twenty-seven hearts made Leuven an attractive geo-academic place to work.

1.3.2   Teaching in Leuven

I taught theories and methodology of international relations, strategy and means of power, international organization, institutions and policy of the United States, peace research, negotiation and mediation, sustainable peacebuilding architecture, and diplomatic history. Peace research, strategic studies, and negotiation-mediation formed the core of my teaching. The courses were embellished by Dutch textbooks and videos: *Notities bij vredesonderzoek en internationale conflictbeheersing* [Notes on peace research and international conflict management] (1990); *Nieuwe muren: over leven in een andere wereld* [New walls: on living in another world] (1994); *Een wereld veilig voor conflict: Handboek voor vredesonderzoek* [A World Safe for Conflict: Handbook for Peace Research] (1995), *Handboek voor terreindiplomatie* [*Handbook for Field Diplomacy*] (2000), and videos on conflict and peace with Fred Brouwers. The students were asked to read the articles from a reading list so that I could combine *ex cathedra* teaching with discussion. Discussion worked well in seminars, but was more difficult in courses. Traditionally most of the teaching was done *ex cathedra*. The students were expected to listen and take notes. I tried to engage the students by discussing current news events or opinions which related to the content of the course; by testing their background knowledge of a problem; by asking existential questions or what they personally would do in a specific context; by simulation; by imagining alternative futures; by challenging taboos and politically correct stories; and by provocative questions, answers or data. The course 'Negotiation and mediation' included negotiation exercises and invited practitioners who dealt with reconciliation. To limit abstraction of reality, students who registered for the course 'peace research' were invited to adopt a conflict during the course. This implied that they had to acquaint themselves with the parties involved, the issues, the opportunity structure, the perceptions and strategic thinking, and the dynamics.

It is a pity that practically all the courses in the humanities and the social sciences deal with the past and the present, and not the future. In contrast, in the 1960s and 1970s, there was great interest in anticipating, imagining and planning alternative world futures. Fred Polack pioneered prognostica. Saul Mendlovitz and other scholars created the *World Order Models Project* (WOMP) to design global futures which tried to reflect the diversity of interests that any world future must serve. The Club of Rome warned about the negative side-effects of current trends and policies. Alvin Toffler dealt with future shock or people and groups overwhelmed by change. Elise Boulding stressed the importance of imagining a non-violent world. In *Utopia or Oblivion* Buckminster Fuller pleaded for generalized, comprehensive and anticipatory design science.

In order to make the study of international relations and peace research more accessible, a Complementary Diploma in International Relations (Aanvullend Diploma International Betrekkingen AIB) was created in the Nineties for students with an MA from other faculties. The programme was very successful. It attracted 100–200 students. Several courses were taught in English to give Erasmus students a chance to select courses. However, the success did not translate into more structural support for the study of international relations and peace research. The extra income earned by the AIB programme was, without consultation, siphoned off to other departments of the Faculty of Social Sciences with negative balance sheets. This was a missed academic opportunity (Photo 1.14).

Photo 1.14 Prof. Luc Reychler with Prof. Haers and some Master of Conflict and Sustainable Peace students. *Source* Personal photo collection

The most challenging and rewarding project was the transdisciplinary-interfaculty Master of Conflict and Sustainable Peace (MaCSP) in 2002. This was a response to a growing demand from both Belgian and foreign students for a more in-depth and comprehensive education in peace research. Initially, the MaCSP was envisaged as a pre-doctoral programme to prepare doctoral students in a more systematic and comprehensive way. A year before, both Prof. J. Haers (a mathematician, theologian and Jesuit) and I visited the deans of all the faculties to map available expertise and raise support for the project. The MaCSP core team created the *Sustainable Peace Action Research Centre – Leuven* (SPARC-L) to develop research, teaching, and service to society. SPARC-L aimed at a more effective exchange of knowledge and know-how between researchers, practitioners, policy-makers and civil societies. Elias Lopez summarized this in his Rhombus model. Each of the actors can provide unique insights into advocacy, peace-building, adaptive management, learning and accountability. The complementarities between the different approaches tend not to be articulated and consequently create the potential for inadequately informed analyses and policies and missed opportunities. Over the years, MaCSP strengthened its relationships with various faculties and research centres. In the midst of the anti-nuclear marches in the 1980s, other faculties developed new courses and centres dealing with peace education, peace theology, peace ethics and restorative justice. These courses were also offered to MaCSP students. The Centre also developed cooperation with external partners in Belgium and abroad: the EU institutions, the *European Doctoral Enhancement Network* (EDEN), the Coimbra Group network on conflict studies, the *International Peace Research Association* (IPRA), and the United Nations Educational, Scientific and Cultural Organization (UNESCO). I was appointed to the UNESCO Chair for Intellectual Solidarity and Sustainable Peacebuilding. Intellectual solidarity helps to eradicate epistemic violence. The latter is furthered by an unlevel playing field, biased research, gaps in the research, and efforts to block critical research and restrain academic freedom. All university students should learn to identify different types of epistemic violence and its negative impact on conflict dynamics. In lectures and papers, I drew attention to fake and hidden news, misleading labelling, invalid theories, the absence of reliable evidence, self sensorship and mind guards. The pioneering experiment, MaCSP was successful.[2] Relying on the enthusiasm and goodwill of professors from different faculties, it lasted five years. Initially, we foresaw a two-year programme and expected extra structural support from the university. The programme was discontinued because of inadequate structural support. I fully agreed with the oral evaluation made by the chair of the external visitation committee, Prof. Bart Tromp of the University of Amsterdam. Alas, at the age of 62, he died a couple of months later from cardiac arrest.

---

[2]See *Master in Conflict and Sustainable Peace (MaCSP): Self Evaluation Report + Appendices* (KU Leuven, Faculteit Sociale Wetenschappen, December 2006).

1.3.3 Security and Strategy

Initially, most of my research dealt with strategy and security issues. Peace research was then not my responsibility and I had to comment in the media on security related events. In the 1970s and 1980s, Europe was upset by the terrorism of the Basque separatists ETA, the *Rote Armee* faction in Germany, the Red Brigades in Italy, the Northern Ireland troubles, the explosion of the Pan Am flight over Lockerbie, the Iran-Iraq war, and the possibility of nuclear war. The installation of cruise missiles in the Western and SS20s in Eastern Europe and Reagan's Strategic Defense Initiative or Star Wars led to anti-nuclear protest. The 1983 movie *The day after* traumatized a generation with the horrors of nuclear war. The study of strategy and security was most opportune, because security is a key component of peace; strategic analysis is essential in peacebuilding, and professional recognition in the 'hard' field of security and strategic studies raised the credibility of peace studies. In 1981, I became the chief editor of a new periodical, *Veiligheid en strategie/Sécurité et strategie* of the Defence Study Centre in Brussels, and wrote articles on: "What is security? The analysis of international conflicts", "The Iran-Iraq war: a polemological analysis", "Arms dynamics and arms control", "The power of public opinion", etc. At the time, I became disappointed with the European delegation of grand strategic thinking to the Americans. I was also amazed that, with the exception of the Revolutionary strategy of Mao-Tse-Tung, hardly attention was paid to Chinese thinking on war and peace. Think of Sun Tzu, Shang Yang, Lao-Tzu and Motse. When Prof. Werck retired in 1987, I got the opportunity to develop peace research and to create a Centre for Peace Research. The focus of my research shifted to conflict prevention, the architecture of sustainable peacebuilding, genocide, conflict impact assessment, coercive diplomacy, and the essential role of time in conflicts.

1.4 Leuven Peace Research

Both environmentalism and peace research are relatively new disciplines. They are creatures of the 1960s. The Club of Rome raised the political and ethical awareness of the relationship between the quality of the natural environment and harmful human activities. The aim of peace research was to provide education to prevent and reduce violence and ameliorate the human climate. They are the most exciting fields of study. Environmentalism became strongly anchored in politics and academia. Peace research became a distinct and unavoidable discipline. In some countries, however, including democratic ones, peace research remains fragile. Rightist regimes tend to treat human rights and peace researchers as unpatriotic appeasers and idealists. Both environmental and peace research are transdisciplinary fields of study and complementary in the prevention of existential problems:

human violence, and the breakdown of our environment and climate. Sustainable peace implies both a good natural and a good human climate. I was delighted to be able to further peace research in Leuven.

1.4.1 A History of War and Peace

The University of Leuven, founded in 1425, has a rich historical track record in war and peace. The city experienced several wars and destructions. The violence never succeeded in breaking its resilience. The *Vierge guerrière* on the façade of the university library pierces the dragon of barbarous war. In the sixteenth and seventeenth centuries, Leuven was an important node in the European university network. Three professors had an influential voice in the war and peace polemics. Justus Lipsius was an irenist (1547–1606), a tolerant thinker who longed for non-conventional primeval Christianity and wanted to end the theological wars. Nicolaus Clenardus (1495–1542) was interested in the Arab world, aspired to create an Arab college and advocated combating Islam by peaceful methods – with words, not swords. In an era of inquisition, this was a revolutionary stance. The great inquisitor Henry, the cardinal-king of Portugal, found it impossible to tolerate his fraternization with Jews in Fez and withdrew his support. Desiderius Erasmus (1466–1536) preferred to be an independent scholar and made an effort to avoid any actions or ties that might inhibit his freedom of intellect and of expression. He defined himself as a world citizen and as a critic of war. He is a pertinent role model for today's peacebuilders. The question of peace is as urgent today as it was in the sixteenth century. Academics should, from time to time, shed their academic straitjacket, and publish, as Erasmus did, forceful pamphlets and essays, like: "War is sweet to those who do not know it", "The education of a Christian prince", "The protest of peace", and "The praise of folly".

1.4.2 The Centre for Peace Research and Strategic Studies (CPRS)

Several initiatives put peace research back on the map. In 1983, the Centre for Peace Research published the first of the *Cahiers of the Centrum voor vredesonderzoek* (Centre for Peace Research): *Methodologie van het vredesonderzoek* ("Peace research methodology"). The cahiers dealt with theoretical and methodological developments and concrete security and peace issues, such as *S.D.I.: een paswoord voor of tegen de vrede* ("SDI: a password for or against peace"). The centre also developed new courses and textbooks for the supplementary Diploma in International Relations (AIB) and the Master of Conflict and Peacebuilding (MaCSP). It attracted a fine group of doctoral students, who researched different facets of conflict and peace. It also organised two global conferences for the International Peace Research Association (IPRA) in 2006 (Calgary Canada) and 2008 (Leuven).

1.4.3   International Peace Research Association (IPRA)

In 1989, for the first time I attended a global conference of the *International Peace Research Association* (IPRA) in Groningen. Hylke Tromp, Professor of Polemology, hosted the conference. He was director of the Polemological Institute, set up in 1962 by B.V.A. Röling, who held the chair for International Law. As a peace researcher, I used the occasion to critically reflect on the field and to plead for Peace Research II. Looking back, I am shocked by the bold nature of my 'constructive' critique, but above all by the misuse of my remarks by outsiders who were inimical to the peace research discipline. A major Dutch newspaper, *Het NRC Handelsblad*, wrote an article on the conference and highlighted one of my observations, namely that the peace research community had become a meeting place for serious peace researchers, but also a haven for peace quacks (in Dutch *paxzalvers*). I coined the term to refer to unqualified persons pretending to possess the necessary knowledge and skills, and most of all politicians and lobbyists who, in the name of peace and security, contribute(d) to war and insecurity. I was thinking of the Treaty of Versailles after WWI and the pursuit of the "peace with honour" in the Vietnam War. In these cases, Ambrose Bierce's definition of war, as a by-product of the arts of peace, fits. Sometimes peace researchers need to be iconoclasts. This is not an easy task. But it can be done if one shares the passionate belief that the world can be improved through dispassionate inquiry. Long after the conference, after a couple of beers, a colleague and friend, Herman De Lange, called me "*het aambeeld van het vredesonderzoek in Nederland*", or "the anvil of peace research in the Netherlands".

1.5   Field Research and Practice

Practical experience can help you to thrive under pressure and think quickly on your feet. The interviews in 1975 with professional diplomats in Washington gave me a sense of multiple realities in international politics. A stay in Brussels as advisor to the Minister of Foreign Affairs, Marc Eyskens, from 1989 to 1992 was a steep learning curve with respect to oral and written communication, time pressure and timing, the partisan competition between political parties, and the working of the Security Council in the United Nations and the Organization for Economic Cooperation and Development, etc. Most interesting was my participation in the third meeting of experts on the peaceful settlement of disputes (PSD) in Valetta, Malta in 1991. Over three weeks the experts reflected on what the Commission on

Security and Cooperation in Europe (CSCE) could add to peaceful settlement, given its long-standing experience of cooperative and regional approaches to security. It reinforced my conviction that bi-lateral and multilateral diplomacy is essential for the promotion of human rights, military security and economic cooperation. Enriching in different ways were my experiences as visiting professor at Boston University, Kent State University, the University of California in Irvine, the University of Helsinki, Kyung Hee University in South Korea, the universities of Antwerp and Brussels, and working trips to Russia, Congo, Burundi, Rwanda, Israel, North Korea, Syria, Cameroon, Nigeria, Iran, and Azerbaijan. South Korea is a country where I felt at home. The Korean experience antiquated some of my assumptions about the role of culture, democracy and legitimacy, the pace of development, eclectic thinking, nationalism, changeability, and the importance of analysing and imagining different futures. In the 1980s, the *Graduate Institute for Peace Studies* (GIP) of Kyung Hee University asked me to develop and teach a course on "Peace Research: Problems and opportunities of the 21st century" (Photo 1.15).

Personal visits and empathic conversations with the stakeholders in conflict zones are very important for students of conflict and peace. I had the opportunity to visit Northern Ireland, Yugoslavia before and after the disintegration, Central Africa (Congo, Rwanda and Burundi), Israel, Palestine, Syria and Iran. In Port Harcourt, Nigeria, I stayed three months to study the impact of EU-funded micro projects.

Photo 1.15 Luc with South Korean colleagues. *Source* Personal photo collection

The Delta Region had a paradise-like beauty, but was paralysed by insecurity. One day, while driving to a Lebanese shop, I drew the attention of my Irish companion to a naked dead body lying on the street. He begged me not to do that anymore. In the shop, I went straight to the counter to buy two meat pies, but almost uttered, "Two dead bodies, please". At that moment I became aware that I was stressed. During my stay, I was also astonished by the competition and lack of cooperation between Western NGOs. When I invited the head of a major German NGO to cooperate, he responded: "Why should we cooperate? We are competitors." With Dr. Réginald Moreels (President of Doctors Without Borders in Belgium and from 1994 to 99 Secretary of State for development cooperation) I created a non-governmental organization, Field Diplomacy Initiative (FDI), to conduct research and field diplomacy in conflict zones.

At the beginning of the 1980s, Aurelio Peccei, the founder of the Club of Rome, wanted to rejuvenate the thinking about the future and invited young scholars to create a *Forum Humanum*. Gunter Pauli took the lead. One of the first meetings for identifying research challenges took place in Oxford, at Robert Maxwell's home. I drew attention to legitimacy crises of national and international political organizations. The host quashed it. He did not consider it a relevant research topic. Forum Humanum launched a journal called *Research* (Photo 1.16).

During the period 2004–2008, as secretary general of *the International Peace Research Association* (IPRA), I organized two global conferences in Calgary, Canada and in Leuven. It was an exciting and exhausting responsibility. For the Leuven conference, Lee Cooper designed a T-shirt with "Feel the peace" written in Braille. It reminded the conference participants that people who are not in touch with peace may turn to senseless war.

1.6 Assumptions Underlying Research and Teaching

My research, teaching and praxis has been shaped by the following underlying assumptions.

1.6.1 Don't Take the World Too Seriously

I taped on my office door a poster with a picture of an armillary, or spherical framework of rings centred on Earth, with on top the message "*Pour comprendre le monde, il faut le jouer*" ("To understand the world, you have to play it"). My library contains an old book, *Le théâtre du monde*, published in Amsterdam in 1701, in which the authors, without blushing, claim that everything is reported accurately in order to give the reader a perfect knowledge of the world. Conflicts should be

Photo 1.16 My son Lucas with a kestrel in the living room. *Source* Personal photo collection

treated as a theatre in which one learns to enact and understand different actors. Analysts should not be bothered by fears, cynicism, despair, or political, moral, theoretical or methodological correctness. Different types of reality (operational, perceived, propagated, anticipated and preferred) compete for attention. A not-too-serious attitude enhances an open mind, heart and will. Stay away from grave seriousness. Lorenz (1966) believed that humour exerts an influence on the social behaviour of people, which is analogous to that of moral responsibility: it can make the world a more honest and better place. Sufficient humour may make all humanity intolerant of phony, fake and fraudulent truths and ideas (Photo 1.17).

Photo 1.17 Global IPRA conference in Leuven in July 2008. *Source* Personal photo collection

1.6.2 Try Existential Understanding

Existential understanding is essential. Dealing effectively with a conflict demands
not only experiential and theoretical understanding, but also existential understanding. The latter requires the ability and willingness to empathize with a

particular situation and to ponder what you would do to survive. In my peace research course, the students were asked if they could imagine killing another person. Most could imagine killing another person to stay alive or protect their children or loved ones when no other options were available. Consequently, they learned the importance of preventing situations in which normal and civilized people, like you and me, could imagine harming others in order to survive.

1.6.3 Be Fully Committed to Conflict Transformation

One should be fully committed to conflict transformation and sustainable peace-building. Conflict transformation is a concept that specifies how the underlying conditions which gave rise to hostility have to be transformed to ensure sustainable peace. Sustainable peace is the most cost-effective way to prevent violence. The term conflict transformation differs from conflict resolution and conflict management, which focus on reducing or mitigating outbreaks of violence. An essential part of the adoption is a sustained commitment. This is not always easy. But without perseverance one can forget sustainable peacebuilding. Of course, when the conflict becomes too overwhelming, detached commitment is better than giving up.

Commitment strengthens by (1) adopting the conflict, (2) developing and maintaining regular contact and empathic discussions with the people involved, (3) monitoring and evaluating the conflict transformation process, (4) communicating the results, (5) encouraging open communication and deep listening about positive and negative events, (6) dealing with groupthink and *ad hominem* pressures by not giving up (7) maintaining a good level of practical optimism.

In 1971 I adopted the Israeli-Palestinian conflict. The first year I participated in a week-long workshop at Harvard with six Israelis and six Palestinians. The commitment lasted and involved several visits to Israel and the occupied areas. Initially there was some hope in 1993 with the Oslo Accords, but since then peace has been silenced. I listed the negative consequences of the occupation, the territorial expansion, the segregation, the belief in *faits accomplis*, and the pursuit of military superiority and absolute security (Photo 1.18).

On television, I was asked to comment on the 9/11 suicide attacks in New York. I answered that they were a spillover from conflicts with Western involvement in the Middle East. On the way out, I was explicitly told not to say such things. The TV invitations stopped. Academic freedom is fragile. In essence, a scholar is not a diplomat. Radical honesty is important to restore conversation and to stand against the repressive political etiquette that blocks thoughts and community. Adopting conflict is exacting, but can also become frustrating. Some students called Herbert Kelman a traitor because he had invited Palestinians and had spoken to Yasser Arafat. From Machiavelli, I learned that to achieve an outcome you need to be extraordinarily self-disciplined, brave and far-sighted, and to care little for the venial rewards of ordinary life. I remain hopeful, because I believe that building

Photo 1.18 Herb and Rose Kelman in Antwerp in 1982. *Source* Personal photo collection

sustainable peace between Israel and Palestine is a vital interest for both. Peace is more precious than victory (Young Seek Choue).

1.6.4 Do not Forget the Big Picture

The fourth underlying assumption is to look at the big picture: the architecture of peace. Sustainable peace does not come alone. It is the result of the installation of several necessary conditions, outlined in the sustainable peacebuilding pentagon. The pentagon metaphor refers to the five clusters of conditions which are necessary to achieve sustainable peace. Sustainable peace requires: integrative negotiation and consultation; integrative political, economic and security structures; an integrative social political climate (the software of peace); peace-supporting institutions; and a supportive external environment. For the installation we need peacebuilding leadership in different sectors and at different system levels. Peacebuilding is the work of all of us. Sustainable peacebuilders use a broad definition of violence. The quantitative and qualitative life-expectancy of people can be reduced by different means of violence: armed, structural, psychological, environmental etc.

Peace is also about taking away emotional barriers (insecurity, hate, cynicism, despair, humiliation, distrust) and creating emotional supports (security, hope, dignity, trust, optimism, integrity). Sustainable peacebuilding demands improvements in human relations, but also in our relationship with nature. In the end, sustainable peace is about reconciling competing values, such as truth, justice, safety, social and health security, gender neutrality, care for nature, mercy, dignity, education, freedom, harmony and beauty. On the backside of his *Ginevra de' Benci* (National Gallery of Arts, Washington D.C.), Leonardo da Vinci painted *Virtutem forma decorat* or 'Beauty adorns virtue'. In his Nobel lecture, Menachem Begin called peace the beauty of life. I agree. My intuition tells me also that beauty furthers peace.

1.6.5   Discern

The fifth assumption concerns differentiating and discerning. There are different types of conflict, violence, security and peace. The perception of reality is often manipulated for political reasons. One should distinguish between at least five interlocking realities: existential, perceived, anticipated, preferred, and propagated. Our own violent behaviour tends to be denied, or framed in positive terms; the violence committed by others tends to be overblown and framed in negative terms. For instance, the coercive diplomacy of the West in the Middle East is justified in terms of defence and security, while the local resistance is called terrorism. Most concepts, such as poverty, security, violence, justice, order and democracy, are interest-biased; they are not value-neutral or defined from a fair and impartial point of view. Discernment is also recommended in the objective assessment of the costs and benefits of violence. Benefits can be security, but also prestige, credibility, material gain or the pleasure of beating someone. There are practically no data on the shadows of war: the violence profiteers and their profits. For most people in conflict zones, wars are total losses, but for a minority of conflict profiteers they are a secure form of income. Yes, there is fake news, but there is also highly selective and politically-embedded news in the mainstream media. What about the 'hidden' news which cannot be mentioned or discussed in public? More attention should be given to what is not covered or pictured in the mainstream media. The prevailing narrative that Western security must be secured through power projection, and that coercive diplomacy constitutes a "force for good" for the long-term human security of citizens of the West and the intervened states, should be challenged.

1.6.6   Take Time Seriously

The sixth assumption stresses the essential role of time in conflict and peace dynamics. Time is a precious, limited and irreversible asset. The way decision-makers and shapers handle time is paradoxical. They take time very seriously, but their temporal behaviour, or temperament, shows many deficiencies.

Photo 1.19   Temporament. *Source* Personal photo collection

Many protracted conflicts are histories of missed opportunities. The negative side-effects of the regime changes in the Middle East, such as the civil wars, the refugee streams, the expansion of the regional influence of Iran and the appearance of ISIS, were not anticipated. In the midst of the crises, groupthink prevented better alternative futures being imagined. Initially, the intervention in Libya was considered a success, susceptible to repetition in Syria. The success of conflict transformation and sustainable peacebuilding is influenced by the temporament of leaders involved. In my book *Time For Peace: The Essential Role of Time in Conflict*, I coined the term 'temporament' because it involves more dimensions of time than time management and temporal intelligence (Photo 1.19).

1.7   Conclusion

I became a peace researcher in order to contribute to the prevention of violent conflicts and to the building of sustainable peace. The study of micro violence, murder, suicide and double suicide was a shocking existential experience. Later, I was startled by the ugliness of macro violence, international and 'civil' wars, genocides and the holocaust. Equally disturbing are the accompanying dissonances. The killing of a person is fiercely condemned and punished. But when thousands or millions of casualties are caused in the name of defence, faith, security, civilization, or regime change, the perpetrators are applauded and monuments are erected to the heroes or martyrs. The First World War was suicidal folly. The Versailles peace treaty led to a twenty-year crisis and a new war. Ambrose Bierce's definition of war as a by-product of the arts of peace is cynical but often fitting. Think also of the "Peace with honour" in Vietnam. Most of the killing in wars is committed by normal people in abnormal circumstances. There is also the rationalization and fervent denial of our own violence. Democracies have been responsible for the lion's share of external violence since the second half of the twentieth century. Finally, after wars, instead of learning how to prevent new ones, most attention goes to the development of better weapons and preparations for the next war.

I was and am still convinced that a better architecture for sustainable peace-building, is indispensable to make the world a more peaceful place. This implies looking at the big picture of violence, and at the cross impacts of the variables which enhance or inhibit the peacebuilding process. An architectural mindset is essential because of the propensity to look at only part of the violence picture, and the complex interdependency of peace-inhibiting or enhancing (f)actors. Most research tends to be reductionistic and focuses on a sub-cluster of independent variables, such as diplomacy, democracy, security, education, politics, develop-ment, religion, communication and law. The architecture of sustainable peace-building examines the cross-impacts of all the variables involved. It stresses the importance of a contextual and comprehensive assessment of the available peace-building capacity, and helps to (a) develop a more coherent peace plan, (b) imple-ment the plan, and (c) monitor and evaluate the real impact of peacebuilding efforts. Sustainable peace exists where (a) violence in the broad sense of the word is nearly absent, (b) a set of peacebuilding conditions have been installed, making the chances of using war or repression to deal with conflicts low and no longer a preoccupation of the stakeholders, and (c) the commitment to establish, sustain and improve the five+1 peacebuilding blocks (the peace pentagon) is high. The building blocks are: effective communication, negotiation and consultation; integrative political, socio-economic and military structures; the software of peace; supporting institutions dealing with education, law and the environment; a supportive external environment and peacebuilding leadership. Sustainable peace is a realistic concept. Most democratic countries and the European Union have installed the necessary conditions. Sustainable peacebuilding is a transdisciplinary field of study and the responsibility of us all.

Arnim Langer invited me to conclude this chapter with 'Ten lessons learned'. I took on the challenge with a mixture of excitement and trepidation. What expe-riences can I distill from my journey that should be actively taken into account in future projects? The following are the results of a personal debriefing after more than forty years of research activity.

1.7.1 Lessons Learned

1.7.1.1 There Is More Violence Than You Want to See

In contrast to Pinker positive statistics, too many people still suffer from violence. Today, a thousand million struggle to survive. Their life expectations are significantly lower than they could or should be. For many people, life is still Hobbesian: solitary, poor, nasty, brutal and short. Violence can be committed intentionally or unintentionally. Some persons and groups profit from violence, but the victims suffer the physical, material, political, ecological, social, cultural, psychological and spiritual costs. Humiliation has been called the atom bomb of emotions (Hartling et al. 2013; Hicks 2013). It is important to monitor violence, the perpetrators and the victims. But in contrast to other fields of study related to health, the environment or the economy, comparable, comprehensive, periodical, and reliable databanks are less readily available. This stifles the raising of awareness and the development of effective violence prevention.

1.7.1.2 Denial Hurts

The denial or minimalization of the violence we commit is a serious problem. Denial hurts and can be pugnacious. It has many faces and names, such as censorship, propaganda, euphemisms, disinformation, fake news, psyops, cover up, reframing... The suffering of the other can be hidden by rank-ordering the victims. First-class victims are treated as somebodies with faces and receive a lot of attention; second-class victims get less attention; third class victims are faceless nobodies. There is also wilful blindness, or the refusal to know what could be known. Our own violence is covered up and the violence committed by others exaggerated. The term terrorism is meant for enemies. Most wars, acts of repression or state terrorism are justified as self-defence, humanitarian intervention, and regime change. The mere ugliness of consequences, however, makes a lie of the given reasons. Finally, the policy of denial impedes journalists, scholars and politicians who try to expose the less visible aspects of violence. Think of Trump's tweet, calling Ilhan Omar a sick monster. She is a young black Muslim woman and immigrant who was elected to the Congress, and opposed the prejudice against American Muslims. Whistle-blowers are called traitors. There are the mind guards of groupthink, who shield the group from dissident information. Many scholars avoid "sensitive issues and data", because that could hinder promotion and fund-raising. They transform themselves into radical methodologists, engage in the formation of theories which make abstraction of the sensitive reality, or study safe or politically correct issues. Denial is a way to reduce the stress associated with double-bind situations in which individuals or countries receive conflicting messages. Think of the regime change interventions of the West to achieve military and political supremacy in the Middle East and the "unexpected negative impacts", such as the appearance of ISIS and the refugee streams. It is important to cut through the denial and dissembling, to bare the collective self-deception for what it is.

1.7.1.3 The Ideal-Typical Peace Researcher

Peace research is an empirical and normative science. A peace researcher is expected to master descriptive and explanatory, normative, predictive, and prescriptive-strategic analysis. The approach should be transdisciplinary, multi-cultural and politically independent. The aim is not only the development of theory, but also engagement with practical concerns: how to improve peace education, how to prevent violence, how to deal with the past, how to reconcile competing values and interests, and how to increase the accountability of decision-makers and shapers. Peace research is also a normative science. Not only the underlying theoretical and epistemological assumptions, but also the normative assumptions need to be made clear and explicit. Concepts and theories have interests. The answers to the fundamental political questions are discovered by science, dialogue and ethical reasoning. Peace research challenges disinformation and checks the validity of the theories of practical persons, who believe themselves to be exempt from intellectual influences. They are, as John Maynard Keynes remarked, frequently the product of some obsolete scholarship. When justifications for violence are not credible or when one does not understand why someone committed violence, we can look at the results and be able to deduct the motivation. My research has been influenced by six strands of moral values. First, by Ken Booth's values of "global security theorizing": universalism, inclusiveness, emancipatory, progressive and critical. Peace is for all human society. Peace researchers should aim to build a human environment that is not constrained by oppressive ideas and practices. We assume that progressive change is possible in morality and politics, not just technology. Practical theory is emancipatory realism. Finally, we should stand outside "the status quo, identify the oppressions within existing structures and processes, and then develop the resources for change" (Booth 2007: 39). The second value which influenced my work is partiality-impartiality. Impartiality has always been high on my list of values. I tried to listen to all the stakeholders in a conflict and favoured impartiality. John Rawls' veil of ignorance or original position is designed to enhance fair and impartial conflict resolution (Rawls 1971). In asymmetric conflicts, involving parties with unequal power, impartiality and neutrality is problematic; it disempowers the weak. Third, there is the issue of competitive and reconciled values. Initially I favoured the principle "the more the better" to positive values, such as justice, truth, mercy and security. Later, I learned that the pursuit of too much truth, too much justice or absolute security, at the expense of the other peace-enhancing values, such as reconciliation, tends to prolong violent conflicts. Constructive conflict transformation results from the reconciliation of different and often competitive values, such as truth, justice, freedom, security, economic development, democracy, and mercy. Fourth, peace researchers need an open mind, an open heart and open will. Scharmer (2009) defines this as the capacity to suspend judgement, the capacity to let go of old identities and intentions and to tune into the future, and the capacity to use one's heart as an organ of perception. Finally, peace researchers could also profit from some faculties von Clausewitz (1832, 1968) identifies for strategists, such as physical and moral courage, presence of mind (*coup d'oeil* and resolution), searching and comprehensive minds, the power of listening in the midst of the most intense excitement, a sense of locality, and perseverance.

1.7.1.4 Sustainable Peace Is a Matter of Urgency

In 2018, the global military expenditure approximated $1.7 trillion. The 7% of the world population living in the Western World spends half of this. Institute for Economics and Peace (2019) observes that since 2008 the Global Peace Index has deteriorated by 3.78%, with 81 countries recording deterioration and 81 improving. It estimates the economic impact of violence on the global economy in 2018 at $14 trillion in purchasing power parity (PPP) terms. Trillions of subsidies are given to fossil fuels. Both the military and fossil fuel investments have enormous ancillary costs. A considerable proportion of the military expenses does not provide more security, but raises security-dilemma fears and contributes to the weaponization of the world. Such subsidies constitute a loss for sustainable peacebuilding and our natural environment. Sustainable peacebuilding is the most cost-effective way to prevent violence. Sustainable peace at state, regional and global levels is as urgent as stopping global warming. It pays attention to symptoms and to the root causes of violence and peace. It offers a view of the whole peacebuilding process and a strategy for implementing and measuring progress. It is inclusive and recommends multilevel efforts, not only at the local, national and regional levels, but also at the global level. Sustainable peace is indivisible. It requires the development of multiple loyalties, including a concern for the world community and for our planet. The human and the natural climate are strongly interlocked. Without a significant improvement in regional and international cooperation and peace, it will be close to impossible to reverse the deterioration of our natural environment and climate.

1.7.1.5 Research Within and Without the University

The university is still a unique place for conducting research and educating students. It provides access to information, research findings and communication facilities. It furthers the creation of research teams and of networks between universities and scholars with a shared sense of purpose and principles. Currently there exist more than 500 peace programmes. Many top and large-scale universities have peace research programmes. Although the number has increased in the last two decades, because of the multidisciplinary and integrative nature of the field, these programmes are more difficult to organize than traditional or 'conventional' mono-disciplinary programmes. Firstly, because universities are predominantly organized in departments and faculties, it is difficult to create and sustain trans-disciplinary research units and Master programmes. Periodically, universities stimulate the creation of transdisciplinary programmes, but without appropriate structural support and decision-making, they tend to remain fragile. The interfaculty MA in Conflict and Sustainable Peace in Leuven hinged on the voluntarism of colleagues from six faculties. It lasted five years and ended after the visitation commission correctly noticed unfulfilled promises for the structural support of the programme and a lack of commitment by one faculty to appoint a new peace research scholar after my retirement. Secondly, there is the phenomenon of reductionism. Some researchers are not interested in the big picture, and reduce peacebuilding to one of the following variables: international and restorative law, economic development, security, the environment, education, psychology, media-tion, history, and political regimes and governance. Thirdly, there is practically no transboundary coordination or relationship that furthers the identification of links between these contributions from the emerging whole. These weak relations pre-vent diverse scholars collectively forming a vehicle for seeing current possibilities and spotting emerging opportunities for teaching and researching sustainable peacebuilding. To remedy this, interested colleagues should take the time to envision a mutually beneficial joint future. A fourth problem relates to the supre-macy of the findings of scientific research at the expense of the knowledge and know-how of practitioners. The practitioners can be people in governmental and non-governmental organizations and civil society. Scharmer identifies five types of practitioners: (a) practitioners who are accountable for the results, (b) practitioners on the frontline who know the real problems, (c) people at the bottom who have no voice and no say, (d) creative outsiders, and (e) one or two activists who are wholly committed to making the project work. Practitioners can inform researchers about the relevance and the external validity of their findings and the practicality of the recommendations and proposals. Finally, universities are not always safe spaces. Political considerations can influence the allocation of research funds and promo-tion. Academic freedom should be respected by political and institutional agents and defended by scholars. This is not always easy, especially when a researcher deals with sensitive issues and political correctness prevails. A university code of ethics could not only defend academic freedom, but also highlight the visible and less visible means used to curtail the necessary academic freedoms.

1.7.1.6 The Best Way to Predict Is to Imagine and Make the Future

On the whole, the social sciences tend to give more attention to the past and the present than the future. There are practically no courses or seminars on the future, dealing with anticipating impacts, scenario planning, imagining different futures, and creating a compelling vision.

1.7.1.7 Time Is Essential and Irreversible

The t-factor is one of the most underestimated factors in conflict transformation. My study of the role of time in conflict and peace processes indicated several inadequacies in the ways we as humans deal with time, undermining the chances of constructive conflict transformation and sustainable peacebuilding. A key concept in the analysis is temporament, or the way people deal with time (Reychler 2015). An adaptive temporament enhances sustainable peacebuilding by removing temporal inadequacies, temporal misconduct and temporal violence. Temporal violence refers to the depreciation of the quantitative and qualitative life expectancies by protracted conflict, long-term sanctions, structural violence, ecological deterioration, or killing and wasting others' time. A thorough temporal analysis checks how the stakeholders in a conflict deal with the parameters of time. *Existential time*: How and to what extent is their existential or life time threatened? *Time orientations*: How do they deal with the past, the present and the future? Are they aware of the interdependence of the temporal perspectives and the problems resulting from imbalances? *Time modes*: Is an accurate analysis made of the temporal characteristics of the problem they are confronted with (nature of change, continuity, duration, viability and proximity to large and surprising changes). *Temporal management options*: Are the stakeholders acquainted with the whole spectrum of temporal policy options (proactive or reactive violence prevention, priority setting, timing and synergizing)? *Secular and religious time*: Do the stakeholders understand the impact of both temporal cultures on the behaviour of unbelievers and believers? *Temporal emotions*: Which emotions facing the past, here and now, and the future dominate the conflict? Is the situation dominated by optimism-pessimism, hope-despair, or trust-distrust? *Appreciation of time*: Do the stakeholders value the role of time and temporament in conflict transformation? *Selection of strategies*: Do the temporal strategies chosen raise or lower the building of sustainable peace? *Anticipation of impacts*: Did the conflicting parties anticipate the crisis, the opportunities and the positive and/or negative impacts? *Temporal democracy*: Is the time of the citizens and the stakeholders treated as equally valuable? One person's time is as valuable as another's. *Temporal empathy*: Do the parties have the capacity and will to discern how the others think and feel about time? *Temporal efficacy*: Do the parties have reasonable confidence in the ability to understand the significance of time in ways that further their interests? It is high time to drastically change our temporament or the way we deal with time. To build sustainable peace and security, the time is always now. Some prefer security now and peace maybe in the future. However, when no serious efforts are made to build sustainable peace, security efforts could lead to the threat of security dilemmas and intensify the level of insecurity when the security policy increases the insecurity and the fear of others and accompanies territorial expansion and the control of external resources and water.

1.7.1.8 Use Precise Language

The public discourse is saturated with imprecision, obsequiousness and mediocrity in all forms. Underneath the communication is a relentless struggle of words. Opponents are attacked by changing the meaning of words and being labelled terrorists, socialists, do-gooders, unpatriotic, etc. Terrorism is always committed by enemies; socialism is communism; do-gooders are unpatriotic people who support human rights or international law and ethics. To overcome the shunning of the use of socialism, an eminent economist, Stiglitz (2019), promotes progressive capitalism. We should use precise language. Courageous and truthful words prevent confusion and render our communication clear, clean and confident. Precise language demands that our key concepts have operational definitions. Instead of using politically loaded labels, such as 'terrorist' or 'dissident', we should get to the bottom of everyone's story and use what psychiatrists call 'formulation'. A formulation gathers all the political, historical, social, economic, and psychological factors that have led an actor to commit particular acts and examines how these factors interconnect. Most of the time, we receive a small part of the causally interrelated reality. Precise language makes also use of precise data, which speak for themselves. To obtain a better understanding of today's problems we can listen to debates and discussions or read enlightening opinion articles. They have been compared with ping-pong games, where people battle their ideas back and forth and try to win or get points for themselves. In addition, there is a propensity to stick to our existing opinions as a self protective block (Bohm 1996). Both Bohm (1996) and Kelman (1965) abide by the conflict transformation power of generous listening and dialogue. Kelman may not have experienced success with the workshops, but that does not mean that communication is not an essential building component of sustainable peacebuilding. Genuine conversation requires mutual respect and a willingness to listen to the concerns and interests of each other. The parties have to be committed to improving the situation. During the conversation, prejudices, positions, power or cynicism remain outside. This requires courage. They must be willing to articulate their concerns, identify the root causes of the conflict, assess the costs and benefits of the status quo, and explore how to build sustainable peace. Genuine conversation uses a language of precision, and "hope and possibility that is grounded in ideas and experiences of innovators in science, business and communities" (Senge et al. 2004: 218). In the 1980s, many students protested against the installation of cruise missiles in Europe. During one of the debates in my course on strategy, I tried to transform the heated theatrics into an open discussion by first inviting the pro and contra debaters to state and defend their positions and then to ask them to answer at least three 'why' questions. Somewhat surprising, most of the criteria the students used to defend their positions were similar (security, costs, escalation, peace). For example, "they will enhance or lower our security"; "the Soviets want (or do not want) peace". The questions invited the debaters to look at their underlying assumptions and at the evidence on which they based their positions. Why do you think the installation would increase or decrease our security?

Where did you get this information? Do you consider this source reliable? The questioning was well received, reduced the level of antagonism, and facilitated an open discussion.

1.7.1.9 Globalize Intellectual Solidarity

We are living in an era when many events indicate that we are in the midst of a transitional period, when the certainties of the unilateral post-Cold War have disappeared and something new is being born. The old power structure and ways of thinking are being challenged. Globalization is speeding up the process. The increased awareness of the state of the world and the planet is altering the global emotional climate, with feelings of fear, vulnerability, relative deprivation, humiliation, anger, despair, outrage and hope. In the West, we see a stark increase in the criticism of globalization and protectionist tendencies. This does not imply that globalization will go away. We will have to live in a hyper-globalized world. As in the European Union, integration and disintegration are part of the progress. Many problems, such as inequality, poverty, the loss of privacy, the post-privacy order and air pollution, do not respect boundaries. To deal with these cross-boundary threats we need international cooperation. Maximum diplomacy should replace maximum sanctions. The cost and benefits should be assessed of different scenarios, namely isolation, back to the past, inertia, defending the status quo, or making the world a better place with a sustainable environment and peace.

I believe that well integrated societies with wide and refined personal knowledge of the world will do better. Nearly 40% of the Nobel prizes in physics, chemistry and medicine awarded to Americans went to foreign-born scientists. We should endorse UNESCO's goal of contributing to peace on the basis of moral and intellectual solidarity. Mobilizing for education, international scientific cooperation and academic freedom is essential in the peacebuilding process.

1.7.1.10 Engage the Power of the Stakeholders

The transformation of conflicts implies influencing power relations and structures. Peace tends to reflect the power relations on which it is based. There is a saying that power, but also powerlessness, corrupts. Therefore, the mapping of power relations and their impact on the conflict transformation and peacebuilding process is essential analytic work. One of the prevailing assumptions in the peace negotiating literature is that constructive negotiations require a certain balance of power and an awareness that the parties need each other in order to reach their goals in a cost-effective manner. This is to a certain extent true, but the relation between power symmetry or asymmetry and peacebuilding is more complex. Despite the existence of power asymmetry, the European Union has been able to create sustainable peace in the region because of goal congruence: the prevention of another war and need for cooperative security in a new but Cold War. Another explanation uses Kenneth Boulding's distinction between destructive, productive and integrative power (Boulding 1989). The destructive power became a cooperative military alliance (NATO) and the integration relied mainly on productive and integrative power. Another prevailing assumption is that peace is causally related to democracy. They cause domestic peace and peace between democracies. That is true, but one should remember that democracies can be very aggressive and imperialistic in international relations. Sustainable peace research shows that peace is caused by many factors. Others conclude that the relation between peace and democracy is bi-directional (Reiter 2017). Peace negotiation research differentiates types of influence, such as interests, rights and power, including normative and knowledge-based power (Ury et al. 1993). Others identify different types of actors: the controllers who control the final decisions, the convincers, the coercers, the connectors, the coalitionists, the counsellors and the informers (Johnston 2012). Gene Sharp's theory of the inner nature of power is still practical. The underlying assumptions of his theory are: (a) that political power can most efficiently be controlled at its sources, (b) that these sources (for example authority or sanctions) depend on obedience, (c) that whoever knows the reasons for that obedience (for example, habit, fear of sanctions, indifference, etc.) knows the inner nature of power, and (d) that obedience is essentially voluntary and can be withdrawn by different kinds of non-cooperation (Sharp 1973). In April 2019, Oberg (Transnational Foundation, 2019) declared Greta Thunberg to be a truth teller hero of our time, because she represents the remarkable strength of non-violent, knowledge-based struggle that people worldwide so endorse. This politically incorrect statement I endorse. Research can expose the costs and benefits of violence and peace and make the decision-makers and shapers account for their policies.

1.7.1.11 Provocative Research Is the Best Revenge

One of the pleasures of peace research is the effort being made to contribute to peacebuilding. Sartre's aphorisms "Words are acts" and, "Commitment is an act, not a word" (Sartre 1988) encouraged me to go for it. As a war and peace scholar, one is blessed because the profession allows you to see all the sides of a conflict and to inhabit opposing perspectives, even when they are brutally antagonistic. In the midst of a world where peace remains fragile, truth is blurred and theatre is everywhere, I remain hopeful for the future. I have encountered violent opposition, but the experiences and the persons which impeded progress strengthened my desire to advance the theory and practice of sustainable peacebuilding. There are moments when misanthropic feelings surface, but these are brief, because you meet so many inspiring and lovely people. My optimism is based on hope and rational thinking. People hunger for peace. There is a growing awareness and disapproval of the high costs of war, military intervention and the pursuit of military dominance. The violence entrepreneurs should be exposed and made accountable. The reduction of violence has become an existential necessity. Without ameliorating the "human climate", it will be hard to impossible to stem the deterioration of the natural climate in time. Time is running out for us to achieve progress in both climates. But notwithstanding this, we need practical optimism and practical positive action to increase the probability of successful peacemaking, peacekeeping and peacebuilding. It is not just about expecting good things to happen, but about developing resilience in the face of challenges, and more creative and holistic thinking. The current situation does not wait for a probability assessment, but generates effort and creativity. One takes risks but is convinced that despair is more risky. The worst thing one can do, as a peace researcher, is to think for a moment that we are not making an impact. A positive development is the rebuke of gender-discrimination and violence. Gender-neutral leadership at national and international levels advances peacebuilding. In the end, the wisdom of sustaining the prevention of violent conflicts or the building of sustainable peace is relative to the outcomes, immediate or remote. The latter is most cost-effective and a prerequisite for our survival.

Virtues such as heroism and courage, which are traditionally regarded as manly and associated with war, should be linked to violence prevention and the pursuit of peace, which are generally regarded as weak or subversive. Turning enemies into allies requires courage and wisdom – much more than is required for making or fighting them.

References

Bierce, A. (1957). *The Devil's Dictionary*. New York: Hill and Wang.

Bohm, D. (1996). *On Dialogue*. London: Routledge.

Booth, K. (2007). *Theory of World Security*. Cambridge: Cambridge University Press.

Boulding, K. (1989). *Three Faces of Power*. Newbury Park: Sage.

Foucault, M. (2002). *The Archaeology of Knowledge*. London: Taylor and Francis.

Hartling, L. M., Lindner, E., Spalthoff, U., & Britton, M. (2013). Humiliation: A Nuclear Bomb of Emotions? *Psicología Política, 46*, 55–76.

Hartling, L. M., & Lindner, E. G. (2017). Toward a Globally Informed Psychology of Humiliation: Comment on McCauley (2017). *American Psychologist, 72*(7), 705–706.

Hicks, D. (2013). *Dignity: Its Essential Role in Resolving Conflict*. New Haven: Yale University Press.

Hochschild, A. (1998). *King Leopold's Ghost: The Plunder of the Congo and the Twentieth Century's First International Human Rights Movement*. Boston: Houghton Mifflin.

Institute for Economics and Peace (2019). *Global Peace Index 2019*. Sydney: Institute for Economics and Peace.

Johnston, P. D. (2012). *Negotiating with Giants: Get What You Want Against the Odds*. Cambridge, MA: Negotiation Press.

Kelman, H. C. (1965). *International Behavior: A Social-Psychological Analysis*. New York: Holt, Rinehart & Winston.

Kennedy, J. F. (1964). *Profiles in Courage*. New York: Harper & Row.

Lorenz, K. (1966). *On Aggression*. New York: Harcourt, Brace & World.

Marcuse, H. (1968). *One Dimensional Man: The Ideology of Industrial Society*. London: Sphere Books.

Oberg, J. (2019). Transnational Foundation. PressInfo # 510.

Pinker, S. (2012). *The Better Angels of Our Nature: Why Violence Has Declined*. New York: Penguin Books.

Rawls, J. (1971). *A Theory of Justice*. Cambridge, MA: Harvard University Press.

Reiter, D. (2017). Is Democracy a Cause of Peace? *Oxford Research Encyclopedia of Politics*. https://doi.org/10.1093/acrefore/9780190228637.013.287.

Reychler, L. (2015). *Time for Peace: The Essential Role of Time in Conflict and Peace Processes*. Queensland: University of Queensland Press.

Rokeach, M. (1960). *The Open and Closed Mind: Investigation into the Nature of Belief Systems and Personality Systems*. New York: Basic Books.

Rosenhan, D. L. (1973). On Being Sane in Insane Places. *Science, 179*(4070), 250–258.

Sartre, J. P. (1988). *What Is Literature? and Other Essays*. Cambridge: Harvard University Press.

Scharmer, C. O. (2009). *Theory U: Leading From the Futures as it Emerges*. San Francisco: Berrett-Koehler Publishers.

Senge, P., Scharmer, O., Jaworski, J., & Flowers, B. S. (2004). *Presence: Exploring Profound Change in People, Organizations, and Society*. New York: Doubleday.

Sharp, G. (1973). *Power and Struggle: Part One of the Politics of Nonviolent Action*. Boston: P. Sargent Publisher.

Stiglitz, J. (2019). Progressive capitalism is not an oxymoron. *New York Times* (19 April).

Ury, W., Brett, J. M., & Goldberg, S. B. (1993). *Getting Disputes Resolved: Designing Systems to Cut the Costs of Conflict*. San Francisco: Jossey-Bass Publishers.

Von Clausewitz, C., & Rapoport, A. (1968). *Carl von Clausewitz on War*. London: Penguin.

Chapter 2
Luc Reychler's Comprehensive Bibliography

2.1 All Publications Arranged Chronologically

Reychler, L. (1978). *Internationaal Politiek Terrorisme en Wereldveiligheid I*. *Politica, 29*(2).

Reychler, L. (1978). *Internationaal Politiek Terrorisme en Wereldveiligheid II*. *Politica, 29*(3), 263–290.

Reychler, L. (1979). The Effectiveness of a Pacifist Strategy in Conflict Resolution. *Journal of Conflict Resolution, 23*(2), 228–260.

Reychler, L. (1979). *Patterns of Diplomatic Thinking: A Cross-National Study of Structural and Social-Psychological Determinants*. New York: Praeger.

Reychler, L. (1980). *Kontra-terreur strategie* (AIB Papers 1980 No. 1). Leuven: KU Leuven, Afdeling Internationale Betrekkingen.

Reychler, L., & Riga, M. (1981). *De Legitimiteit van de Europese Gemeenschap: Voorstel van Onderzoek en Bibliografie* (AIB Papers 1981 No. 2). Leuven: KU Leuven, Afdeling Internationale Betrekkingen.

Reychler, L. (1981). *Vredesbewegingen en Wapenwedlopen: Enkele Bedenkingen* (AIB Papers 1981 No. 3). Leuven: KU Leuven, Afdeling Internationale Betrekkingen.

Reychler, L. (1981). *Wat is Veiligheid [What is Security?]* (Veiligheid en Strategie No. 1). Brussels: Koninklijk Hoger Instituut voor Defensie.

Reychler, L. (1982). *Analyse van Internationale Conflicten* (Veiligheid en Strategie No. 2). Brussels: Koninklijk Hoger Instituut voor Defensie.

Reychler, L., & Leenaards, J. (1982). *Belgian Defense Policy: A Preliminary Analysis* (AIB Paper 1982 No. 1). Leuven: KU Leuven, Afdeling Internationale Politiek.

Reychler, L. (1983). *Methodologie van het Vredesonderzoek* (Cahiers of the Centre for Peace Research and Strategic Studies No. 1). Leuven: Centre for Peace Research and Strategic Studies.

© The Editor(s) (if applicable) and The Author(s), under exclusive license to
Springer Nature Switzerland AG 2020
L. Reychler and A. Langer (eds.), *Luc Reychler: A Pioneer in Sustainable
Peacebuilding Architecture*, Pioneers in Arts, Humanities, Science, Engineering,
Practice 24, https://doi.org/10.1007/978-3-030-40208-2_2

Reychler, L. (1983). Naar een Meer Systematisch Onderzoek van het Belgisch Veiligheids-en Defensiebeleid. In J. De Clerck, W. Dewachter, & R. Maes (Eds.), *Politieke Instrumenten ter Bestrijding van Crises* (pp. 171–201). Leuven: Universitaire pers Leuven.

Reychler, L. (1983). *Vrede, Moraal en Wetenschap* (AIB Papers 1983 No. 2). Leuven: KU Leuven, Afdeling Internationale Betrekkingen.

Reychler, L., & Rudney, R. (1983). Towards a More Efficient Organization of European Security and Defense Thinking. *Studia Diplomatica, 36*(3), 289–310.

Reychler, L. (1985). *Belgische Defensie in de Peiling* (Cahiers of the Centre for Peace Research and Strategic Studies No. 11). Leuven: Centre for Peace Research and Strategic Studies.

Reychler, L. (1985). *Democratie en Vrede* (Cahiers of the Centre for Peace Research and Strategic Studies No. 7). Leuven: Centre for Peace Research and Strategic Studies.

Reychler, L. (1985). Een Belgisch Toekomstperspectief. *Studia Diplomatica, 38*(5), 623–628.

Reychler, L. (1985). *I-CBM's YOU-CBM's and WE-CBM's* (Cahiers of the Centre for Peace Research and Strategic Studies No. 9). Leuven: Centre for Peace Research and Strategic Studies.

Reychler, L. (1985). *S.D.I: Een Paswoord Voor of Tegen de Vrede* (Cahiers of the Centre for Peace Research and Strategic Studies No. 13). Leuven: Centre for Peace Research and Strategic Studies.

Reychler, L. (1985). Sterrenkrijg Zonder Vrees. *Trends, 11*(233), 240–241.

Reychler, L. (1986). *Arms Control Evaluation: A Joint Search for Objective Criteria. Search for Causes of International Conflict and Ways to Their Solutions*. Seoul: Institute of International Peace Studies, Kyung Hee University.

Reychler, L. (1986). European Peace Movement: Image and Reality. *Peace Forum, 2*(3).

Reychler, L. (1986). Hoe Denken over de Sterrenkrijg? *Kultuurleven, 4*.

Reychler, L. (1986). The European Peace Movement: Appearance and Reality. *Studia Diplomatica, 39*(3), 285–304.

Reychler, L. (1986). The Relevance of Scientific Research to Conflict Management. *Nacao e defesa, 11*(37), 77–98.

Reychler, L. (1986). *Van Polarisatie naar Consensus* (Veiligheid en Strategie No. 16). Brussels: Koninklijk Hoger Instituut voor Defensie.

Reychler, L. (1986). *Enkele Facetten van het Vredesonderzoek Sinds 1945* (Cahiers of the Centre for Peace Research and Strategic Studies No. 17). Leuven: Centre for Peace Research and Strategic Studies.

Reychler, L. (1987). Politici Verwaarlozen Defensie. *Trends*, 183–184.

Reychler, L. (1987). Querela Pacis: Vredesonderzoek op een Keerpunt. In V. Werck (Ed.), *De Soviet Unie en de Europese veilgheid* (pp. 11–38). Leuven: Universitaire Pers.

Reychler, L. (1988). *The Iran-Iraq War: A Polemological Analysis* (Veiligheid en Strategie No. 20). Brussels: Koninklijk Hoger Instituut voor Defensie.

Reychler, L. (1989). Tien Megatrends in het Europese Veiligheidsdenken. *Acco Aktueel, 2*(3), 5–10.

Reychler, L. (1990). Arms Control and Disarmament. In L. Kaplan (Ed.), *NATO After Forty Years* (p. 27). Wilmington: Scholarly Resources.

Reychler, L. (1990). *Arms Dynamics and Arms Control* (Veiligheid en Strategie No. 32). Brussels: Koninklijk Hoger Instituut voor Defensie.

Reychler, L. (1990). Het Helsinki Proces. *Persoon en Gemeenschap, 5–6*, 121–137.

Reychler, L. (1990). The Public Perception of NATO. *NATO Review, 38*, 16–23.

Reychler, L. (1991). *Het 5000-200 Probleem: Etnische en Nationalistische Conflicten* (Cahiers of the Centre for Peace Research and Strategic Studies No. 4). Leuven: Centre for Peace Research and Strategic Studies.

Reychler, L. (1991). *Het 5000-200 Probleem: Enkele Nota's over Etnische en Nationalistische Conflicten* (Cahiers of the Centre for Peace Research and Strategic Studies No. 9). Leuven: Centre for Peace Research and Strategic Studies.

Reychler, L. (1991). A Pan-European Security Community: Utopia or Realistic Perspective? *Disarmament, 14*(1), 42–52.

Reychler, L. (1991). Peace Research II. In J. Nobel (Ed.), *The Coming of Age of Peace Research* (pp. 89–96). Groningen: Styx.

Reychler, L. (1991). *The Power of Public Opinion* (Veiligheid en Strategie No. 36). Brussels: Koninklijk Hoger Instituut voor Defensie.

Reychler, L. (1991). Une Communauté de Sécurité Paneuropéenne: Utopie ou Réalité? *Désarmement, XIV*(1).

Reychler, L. (1992). De Nieuwe Wereldorde. *Noord-Zuid Cahier, 17*, 25–42.

Reychler, L. (1992). Naar een Nieuwe Veiligheidsorganisatie voor Europa. In K. Malfliet (Ed.), *De Muur Voorbij* (pp. 144–157). Antwerp: Garant.

Reychler, L. (1992). The Price is a Surprise on Preventive Diplomacy. *Studia Diplomatica, 45*(5), 71–85.

Reychler, L. (1993). De Federalisering van het Buitenlands Beleid van België. *Internationale Spectator: Tijdschrift voor Internationale Politiek, 47*(7), 394–398.

Reychler, L. (1993). Megatrends in de Internationale Politiek. In W. Dumon, G. Fauconnier, R. Maes, & E. Meulemans (Eds.), *Scenarios voor de Toekomst: Feestbundel naar Aanleiding van Honderd Jaar Sociale Wetenschappen aan de KU Leuven* (pp. 59–76). Leuven: Acco.

Reychler, L. (1993). We Don't Know Where We Are Going, But at Least We Are Heading On. *Peace and the Sciences* (March), 21–26.

Reychler, L. (1994). Buitenlands Beleid. In M. Deweerdt, C. De Ridder, & R. Dillemans (Eds.), *Wegwijs Politiek* (pp. 561–572). Leuven: Davidsfonds.

Reychler, L. (1994). Conflictdynamiek en Preventie. *Transaktie: Tijdschrift Over de Wetenschap van Oorlog en Vrede, 23*(1), 15–28.

Reychler, L. (1994). *Oorlog en Anti-Oorlog: Lessen voor de Eenentwintigste Eeuw.* Leuven: Leuven Universitaire Press.

Reychler, L. (1995). Conflict Dynamics and Prevention: A Preliminary Research of the Past Research Efforts. In J. Balazs & H. Wiberg (Eds.), *Changes, Chances and Challenges: Europe 2000* (pp. 41–52). Budapest: Akadémiai Kiadó.

Reychler, L. (1995). Hoop op Duurzame Vrede (Hope for Sustainable Peace). In R. Dillemans, B. Pattyn, et al. (Eds.), *Wegen van Hoop: Universitaire Perspectieven* (pp. 87–98). Leuven: Leuven University Press.

Reychler, L. (1996). *Beyond Traditional Diplomacy* (Diplomatic Studies Programme University of Leicester Discussion Paper No. 17).

Reychler, L. (1996). Field Diplomacy: A New Conflict Paradigm? Paper presented at *International Peace Research Association Conference* (July).

Reychler, L. (1996). In Vredesnaam voorkomen. *Knack* (July/August), *31*(26).

Reychler, L. (1997). Conflicts in Africa: The Issues of Control and Prevention. In Commission on African Regions in Crisis, *Conflicts in Africa: An Analysis of Crises and Crisis Prevention Measures* (pp. 15–37). Brussels: European Institute for Research and Information on Peace and Security.

Reychler, L. (1997). De Ontwapening van Conflicten. *Noord-Zuid Cahier, 22*(3), 99–105.

Reychler, L. (1997). Getting Rid of Human Cruelty: Law and Conflicts in Our Time. Paper presented at the *Symposium of the Red Cross Netherlands*, The Hague.

Reychler, L. (1997). Great Lakes Challenges: New Routes. *Journal of Peace Research and Action, 2,* 17–21.

Reychler, L. (1997). Religion and Conflict. *International Journal of Peace Studies, 2,* 19–38.

Boulding, E., & Reychler, L. (Eds.). (1997). Special Issue: Peace Building in Fractionated Societies: Conceptual Approaches and Cultural Specificities. *International Journal of Peace Studies, 2*(2).

Boulding, E., & Reychler, L. (Eds.). (1997). Special Issue: Rethinking Peace Building. *Peace and Conflict Studies, 4*(1).

Reychler, L. (1997). Crisissen en hun Oorzaken: Het Voorkomen van Gewelddadige Conflicten. In E. Suy (Ed.), *Conflicten in Afrika: Crisisanalyse en Preventiemogelijkheden* (pp. 39–66). Brussels: GRIP Van Halewyck.

Reychler, L. (1997). Conflicten in Afrika: Beheer en Preventie. In E. Suy (Ed.), *Conflicten in Afrika: Crisisanalyse en Preventiemogelijkheden* (pp. 15–33). Brussels: GRIP Van Halewyck.

Reychler, L. (1997). *The Great Lakes District: Forecast: More War* (From Early Warning to Early Action, Report on the European Conference on Conflict Prevention).

Reychler, L., Calmeyn, S., Craeghs, J., Dadimos, H., & De la Haye, J. (1997). *Democratic Peacebuilding and Conflict Prevention. Annex: Country Studies.* Leuven: Centre for Peace Research and Strategic Studies.

Reychler, L. (1997). Grenzen en Mogelijkheden van Conflictpreventie. In *Onze Alma Mater* (pp. 326–339). Leuven: Academische Stichting Leuven.

Reychler, L. (1997). Field Diplomacy: A New Conflict Paradigm? *Peace and Conflict Studies, 4*(1), Article 4.

Reychler, L. (1998). Lessons Learned from Recent Democracy Building Efforts. In P. Cross (Red.), *Contributing to Preventive Action*. Ebenhausen: Stiftung Wissenschaft und Politik.

Reychler, L. (1998). Beweegredenen voor Humanitaire Interventie. In J. D. Tavernier & D. Pollefeyt (Eds.), *Heeft de traditie van de mensenrechten toekomst?* (pp. 71–75). Leuven: Acco.

Reychler, L. (1998). Conflict Impact Assessment (CIAS) at the Policy and Project Level. Paper presented at *European University Centre for Peace Studies International Conference on Development and Conflict* (24–26 September), Stadtschaining, Austria.

Reychler, L. (1998). Security Issues and the Role of External Involvement. Paper presented at *Seminar on Interstate Conflict and Options for Policy* (November), The Hague.

Reychler, L. (1998). Democratic Peace Building: A Major Confidence Building Measure. In *Meeting Human Needs in a Cooperative World: Proceedings of the Commission on Conflict Resolution and Peace Building, June 1998, Durban, South Africa* (p. 50). Leuven: International Peace Research Association (IPRA).

Reychler, L. (1998). Conflict Prevention and Regional Integration. Paper presented at *8th Persian Gulf Seminar* (February).

Reychler, L. (1998). Conflict Impact Assessment (CIAS): An Essential Tool for Conflict Prevention. In *Meeting Human Needs in a Cooperative World: Proceedings of the Commission on Conflict Resolution and Peace Building, June 1998, Durban, South Africa*. Leuven: International Peace Research Association (IPRA).

Reychler, L. (1998). *Democratic Peace Building: The Devil is in the Transition.* Leuven: Leuven University Press.

Reychler, L., Huyse, H., & Verleyen, H. (1998). *Conflict Effect Rapportering. Rapport Beleidsvoorbereidend onderzoek Departement Ontwikkelingssamenwerking.* Leuven: Centre for Peace Research and Strategic Studies.

Reychler, L. (1998). Proactive Conflict Prevention: Impact Assessment? *International Journal of Peace Studies, 3*(2), 87–98.

Reychler, L. (1999). The Conflict Impact Assessment System (CIAS): A Method for Designing and Evaluating Development Policies and Projects. In P. Cross (Ed.), *Conflict Prevention Policy of the European Union. Yearbook 1998–99 of the Conflict Prevention Network, European Commission* (pp. 144–163). Baden-Baden: Nomos.

Reychler, L. (1998). Lessons Learned from Recent Democracy Building Efforts. In P. Cross (Ed.), *Contributing to Preventive Action* (pp. 165–184). Ebenhausen: Stiftung Wissenschaft und Politik.

Reychler, L. (1998). Democratic Peace Building and Conflict Prevention. In K. Wellens (Ed.), *International Law: Theory and Practice: Essays in Honour of Eric Suy* (pp. 83–106) Leiden: Martinus Nijhoff.

Reychler, L. (1998). Beweegredenen voor Humanitaire Interventie. In J. De Tavernier & D. Pollefeyt (Eds.), *Heeft de traditie van de mensenrechten toekomst?* (pp. 71–75). Leuven: Acco.

Reychler, L., Huyse, H., & Verleyen, H. (1998). *Conflict Effect Rapportering: Beleidsvoorbereidend Onderzoek* (Centre for Peace Research and Strategic Studies Working Paper 2). Leuven: Centre for Peace Research and Strategic Studies.

Reychler, L. (1999). *Vrede is geld waard*. Amsterdam: Studium Generale.

Reychler, L. (1999). A Safer World Through a Better Coordinated European Foreign Policy. Challenges and Perspectives for Europe, Ten Years After the Fall of the Wall. *Peace and Conflict Studies, 1*, 1–12.

Reychler, L. (1999). De Ethische en Juridische Rechtvaardiging van Vredeshandhaving. In L. Cumps (Ed.), *Arbeid en Ethiek: Liber Amicorum Dr. R. Claeys* (pp. 217–232). Antwerp: Handelshogeschool.

Reychler, L. (1999). Terreindiplomatie: een ander denkkader voor conflicten. *Geweldloos Aktief, 1*, 11–19.

Reychler, L. (1999). Conflict Effect Rapportering: Conflict Impact Assessment (CIAS). *Rapport voor het Departement Ontwikkelingssamenwerking*. Leuven: Centre for Peace Research and Strategic Studies.

Reychler, L. (1999). *Democratic Peace Building: The Devil is in the Transition*. Leuven: University Press.

Reychler, L., Huyse, H., & Verleyen, H. (1999). *Conflict Effect Rapportering. Conflict Impact Assessment (CIAS)* (Cahiers of the Centre for Peace Research and Strategic Studies No. 58). Leuven: Centre for Peace Research and Strategic Studies.

Kennes, E., Laakso, L., Reychler, L., Schraeder, P., & Vandeginste, S. (Eds.). (1999). *The Congo Crisis. Background and International Dimensions*. Helsinki: Department of Political Sciences, University of Helsinki.

Reychler, L. (1999). Field Diplomacy. In *World Encyclopedia of Peace*. Seoul: Kyung Hee University.

Reychler, L. (1999). De Ethische en Juridische Rechtvaardiging van Vredeshandhaving. In: R. Claeys (Ed.), *Arbeid en Ethiek* (pp. 217–232). Antwerp: Handelshogeschool.

Reychler, L. (1999). Conflict Impact Assessment. In *World Encyclopedia of Peace*. Seoul: Kyung Hee University.

Reychler, L. (1999). Vrede is geld waard. *InterAxis, 4*(11).

Reychler, L. (2000). The Promotion of Peace Building. In Y. S. Choue (Ed.), *Proceedings of the International Peace Conference on Global Governance in the 21st Century* (pp. 53–74). Seoul: Kyung Hee University.

Reychler, L. (2000). An Evaluation of the International Efforts in Burundi. In M. Lund & G. Rasamoelina (Eds.), *The Impact of Conflict Prevention Policy: Cases, Measures and Assessments* (pp. 46–62). Baden-Baden: Nomos.

Reychler, L. (2000, August). Democratization in Africa and Eastern Europe. Paper presented at *IPRA Biannual Conference on Challenges for Peace Research in the 21st Century*.

Reychler, L. (2000). Peace Pays Off. *Dialogue & Reconciliation, 1*, 9–16.

Reychler, L., Musabyimana, T., & Calmeyn, S. (2000). *Le Défi de la Paix au Burundi: Théorie et Pratique*. Paris: L'Harmattan.

Paffenholz, T., & Reychler, L. (Eds.). (2000). *Handboek voor Terreindiplomatie*. Leuven: Garant.

Reychler, L., & Paffenholz, T. (Eds.). (2000). *Construire la Paix sur le Terrain: Mode d'Emploi*. Bruxelles: GRIP.

Reychler, L. (2000). Women in Violence Prevention. *Globaal, 15*, 37–41.

Reychler, L. (2000). The Promotion of Sustainable Peace Building. In Y. S. Choue (Ed.), *Global Governance in the 21st Century* (pp. 53–73). Seoul: Kyung Hee University.

Reychler, L. (2001). Beter Voorkomen dan Genezen. *Tertio* (June), 2(6).

Reychler, L. (2001). De vijand elimineren neemt de voedingsbodem voor terrorisme niet weg. *P Magazine* (26 September), 20–24.

Reychler, L. (2001). European Security Architecture. Paper presented at *Belgo-Korean Conference on 100 Years of Diplomatic Relations* (September), Seoul.

Reychler, L. (2001). Evaluacion del Impacto del Conflicto: Una Herramienta Esencial en la Prevencion de Conflictos. In Ú. Oswald (Ed.), *Estudios para la Paz desde una Perspectiva Global*. Cuernavaca, Mexico: Centro regional de investgaciones multidisciplinarias.

Reychler, L. (2001). *Guidelines for a Sectorial Impact Assessment* (Cahiers of the Centre for Peace Research and Strategic Studies No. 19). Leuven: Centre for Peace Research and Strategic Studies.

Reychler, L., & Paffenholz, T. (Eds.). (2001). *Peacebuilding: A Field Guide*. Boulder: Lynne Rienner.

Reychler, L. (2001). Rumor Control. In L. Reychler & T. Paffenholz (Eds.), *Peacebuilding: A Field Guide* (pp. 467–471). Boulder: Lynne Rienner.

Reychler, L. (2001). Monitoring Democratic Transitions. In L. Reychler & T. Paffenholz (Eds.), *Peacebuilding: A Field Guide* (pp. 216–221). Boulder: Lynne Rienner.

Reychler, L. (2001). Vredesarchitectuur. In J. Van Der Lijn (Ed.), *Conflictonderzoek in het Nederlandse Taalgebied: Problemen bij Onderzoek naar het Ononderzoekbare?* (pp. 75–88). Nijmegen: Studiecentrum voor Vredesvraagstukken.

Reychler, L. (2001). Listening. In L. Reychler & T. Paffenholz (Eds.), *Peacebuilding: A Field Guide* (pp. 453–460). Boulder: Lynne Rienner.

Reychler, L. (2001). From Conflict to Peace Building: Conceptual Framework. In L. Reychler & T. Paffenholz (Eds.), *Peacebuilding: A Field Guide* (pp. 3–15). Boulder: Lynne Rienner.

Reychler, L. (2001). Field Diplomacy Initiatives in Cameroon and Burundi. In L. Reychler & T. Paffenholz (Eds.), *Peacebuilding* (pp. 90–96). Boulder: Lynne Rienner.

Reychler, L. (2002). Geweld tegen Geweld. In B. Pattijn & J. Wouters (Eds.), *Schokgolven: Terrorisme and Fundamentalisme* (pp. 47–52). Leuven: Davidsfonds.

Reychler, L. (2002). Peace Architecture. *Peace and Conflict Studies, 9*(1), 26–35.

Reychler, L., & Langer, A. (2002). *The Software of Peace Building* (Cahiers of the Centre for Peace Research and Strategic Studies No. 20). Leuven: Centre for Peace Research and Strategic Studies.

Reychler, L. (2002). *Initial Conflict Impact Assessment (I-CIAS) Bayelsa, Delta and Rivers* (Report: European Union Micro Project Programme (MMP3)).

Reychler, L. (2003). Ethnic Conflicts and Conflict Prevention: The State of the Art. In R. Thakur (Ed.), *Conflict Prevention: The Secretary-General's Report: The Way Forward*. Tokyo: United Nations University.

Reychler, L., & Jacobs, M. (2003). *Het Geweld Vierkant*. Leuven: Centre for Peace Research and Strategic Studies.

Reychler, L. (2003). *Het Fenomeen Genocide*. Leuven: Centre for Peach Research and Strategic Studies.

Reychler, L., & Renckens, S. (2003). *Een Existentiële Kijk op Genocide*. Leuven: Centre for Peace Research and Strategic Studies.

Reychler, L. (2003). *CIAS Conflict Impact Assessment. Field Diplomacy Initiative (FDI)*. Leuven: FDI.

Reychler, L. (2003). *Conflict Impact Assessment* (Cahiers of the Centre for Peace Research and Strategic Studies No. 67). Leuven: Centre for Peace Research and Strategic Studies.

Reychler, L., Calmeyn, S., De la Haye, J., Hertog, K., & Renckens, S. (2004). *De Volgende Genocide*. Leuven: Universitaire Pers.

Reychler, L., & Stellamans, A. (2003). *Peace Building Leaders and Spoilers* (Cahiers of the Centre for Peace Research and Strategic Studies No. 66). Leuven: Centre for Peace Research and Strategic Studies.

Reychler, L., & Langer, A. (2003). Peace-Building Software: The Creation of an Integrative Climate. *Peace Research, 35*(2), 53–73.

Reychler, L. (2003–2004). Welke Lessen uit de Ervaringen met de Genocidewet? Welke Bijdrage kan België Leveren aan de Strijd Tegen de Straffeloosheid? *Jura Falconis, 40*(1), 149–152.

Reychler, L. (2004). CFSP: Myth or Reality. Paper presented at *International Conference on Russia, its Regions and EU Enlargement* (January).

Reychler, L. (2004). Strengthening the United Nations System with Peace Research. *The Journal of Peace Studies, 11*(1), 169–206.

Reychler, L. (2004). Peace Architecture: The Prevention of Violence. In A. H. Eagly, R. M. Baron, & V. L. Hamilton (Eds.), *The Social Psychology of Group Identity and Social Conflict: Theory, Application and Practice* (pp. 133–146). Washington, DC: American Psychological Association.

Van Dijck, N., & Reychler, L. (2004). *Understanding the Role of the European Union in the Middle East: An Analysis of the Position of the European Union on the Gaza Withdrawal Plan* (IIEB Working Papers No. 13). Leuven: Institute for International and European Policy.

Reychler, L., Jacobs, M. (2004). *Limits to Violence: Towards a Comprehensive Violence Audit* (Cahiers of the Centre for Peace Research and Strategic Studies No. 22). Leuven: Centre for Peace Research and Strategic Studies.

Reychler, L. (2004). Welke Bijdrage kan België Leveren aan de Internationale Strijd Tegen de Straffeloosheid? *Ethische Perspectieven: Nieuwsbrief van het Overlegcentrum voor Christelijke Ethiek, 14*(2), 123–130.

Reychler, L. (2004). The EU Violence Prevention and Peace Building Pentagon. Paper presented at *Symposium: A New Military for a New World* (December).

Reychler, L., & Van Dijck, N. (2004). Common Foreign and Security Policy: Myth or Reality? In V. Papacosma (Ed.), *EU Enlargement and New Security Challenges in the Eastern Mediterranean*. Nicosia: Intercollege Press.

Paffenholz, T., & Reychler, L. (2005). Towards Better Policy and Programme Work in Conflict Zones: Introducing the 'Aid for Peace' Approach. *Journal of Peacebuilding and Development, 2*(2), 6–23.

Reychler, L. (2005). Peace Research After 9/11. *Peace Forum, 18*(31), 1–5.

Reychler, L., & Stellamans, A. (2005). *Researching Peace Building Leadership* (Cahiers of the Centre for Peace Research and Strategic Studies No. 71). Leuven: Centre for Peace Research and Strategic Studies.

Reychler, L. (2006). *Beyond Peace and War: On Violence Control and Sustainable Peace Building in the Middle East* (Cahiers of the Centre for Peace Research and Strategic Studies No. 73). Leuven: Centre for Peace Research and Strategic Studies.

Reychler, L. (2006). Challenges of Peace Research. *International Journal of Peace Studies, 11*(1), 1–16.

Reychler, L. (2006). De Waandacht voor Zelfmoordterrorisme. *Karakter, 14*, 12–15.

Reychler, L. (2006). Geweldpreventie en Vredesopbouw: Een Onderzoeksagenda. In J. Hovelynck (Ed.), *Relationeel Organiseren: Samen Leren en Werken In en Tussen Organisaties*. Leuven: Lannoo.

Reychler, L., & Carmans, J. (2006). *Violence Prevention and Peace Building* (Cahiers of the Centre for Peace Research and Strategic Studies No. 74). Leuven: Centre for Peace Research and Strategic Studies.

Reychler, L., & Langer, A. (2006). *Researching Peace Building Architecture* (Cahiers of the Centre for Peace Research and Strategic Studies No. 75). Leuven: Centre for Peace Research and Strategic Studies.

Reychler, L. (2006). *Beyond Peace and War: On Violence Control and Sustainable Peace Building in the Middle East* (Cahiers of the Centre for Peace Research and Strategic Studies No. 73). Leuven: Centre for Peace Research and Strategic Studies.

Reychler, L. (2006). Een handleiding voor vrede in het Midden-Oosten. *De Standaard* (20 July), 21.

Reychler, L. (2006). Humanitarian Aid for Sustainable Peace Building. In P. Gibbens & B. Piquard (Eds.), *Working in Conflict – Working on Conflict* (pp. 134–154). Bilbao: University of Deusto.

Reychler, L. (2006). Globalization and Human Security. In *Proceedings of the Inaugural Conference on Peace & Policy Dialogue in Northeast Asia* (pp. 37–74). Seogwipo City: Jeju Peace Institute.

Reychler, L. (2007). *Globalization and Human Security. Peace & Policy Dialogue in Northeast Asia.* Seogwipo City: Jeju Peace Institute.

Reychler, L. (2007). Beyond Peace and War. *Communication for Development and Social Change, 1*(3), 1–21.

Paffenholz, T., & Reychler, L. (2007). *Aid for Peace: A Guide to Planning and Evaluation for Conflict Zones.* Baden-Baden: Nomos.

Reychler, L. (2007). *Globalization and Human Security. Peace & Policy Dialogue in Northeast Asia* (pp. 37–73). Seoul: International Peace Foundation.

Reychler, L. (2007). Is Peace Research Relevant? Paper presented at *Cades conference,* Leuven.

Reychler, L. (2007). Preventing War Crimes and Genocide: A Look from the Balcony. In J. Bec-Neumann (Ed.), *Darkness at Noon: War Crimes, Genocide and Memories* (pp. 226–241). Sarajevo: Centre for Interdisciplinary Postgraduate Studies.

Reychler, L. (2008). Aid for Peace: Planning and Evaluation in Conflict Zones. In G. Scheers (Ed.), *Assessing Progress on the Road to Peace: Planning, Monitoring and Evaluating Conflict Prevention and Peacebuilding Activities* (pp. 9–17). The Hague: European Centre for Conflict Prevention.

Reychler, L. (2007). Women in Violence Prevention. *Globaal, 15,* 37–41.

Reychler, L. (2007). Researching Violence Prevention and Peace Building. In F. Ferrandiz & A. Robben (Eds.), *Multidisciplinary Perspectives on Peace and Conflict Research* (pp. 147–195). Bilbao, Spain: Humanitarian Net University of Bilbao.

Reychler, L. (2007). Violence Prevention and Peace Building: A Rough State of the Art. In H. Isak (Ed.), *A European Perspective for the Western Balkans* (pp. 67–106). Wien-Graz: Neuer Wissenschaftlicher Verlag.

Reychler, L. (2008). Research in the Midst of a Mega Crisis. Paper presented at *Fulbright 60th Anniversary. Academic Session U.S.-Belgian Relations: Past, Present and Future* (10 October), Brussels.

Reychler, L., Renckens, S., Coppens, K., & Manaras, N. (2008). *A Codebook for Evaluating Peace Agreements* (Cahiers of the Centre for Peace Research and Strategic Studies No. 83). Leuven: Centre for Peace Research and Strategic Studies.

Reychler, L. (2009). Moral Report of the International Peace Research Association (July 2006–July 2008). In L. Reychler, J. Funk, & K. Villanueva (Eds.), *Building Sustainable Futures: Enacting Peace and Development* (pp. 379–386). Bilbao: Humanitarian Net University of Bilbao.

Reychler, L. (2009). Als vrede herleid wordt tot nationale veiligheid. *De Standaard* (9 January), 18–19.

Reychler, L. (2009). An Emerging Sustainable Peace Building Paradigm. Paper presented at *International Conference on History and Peace* (19–23 August), Seoul.

Reychler, L. (2009). Peace Research: An Inconvenient Field of Study. In L. Reychler, J. Funk, & K. Villanueva (Eds.), *Building Sustainable Peace:*

Enacting Peace and Development (pp. 25–29). Bilboa, Spain: Humanitarian Net University of Bilbao.

Reychler, L., & Migabo Kalere, J. (2009). *R. D. Congo Pays de l'avenir. Questions pour construire une paix durable* (Cahiers of the Centre for Peace Research and Strategic Studies No. 84). Leuven: Centre for Peace Research and Strategic Studies.

Reychler, L. (Ed.). (2010). *DR Congo: Positive Prospects: Building Sustainable Peace Together.* Leuven: Centre for Peace Research and Strategic Studies.

Reychler, L. (Ed.). (2010). *DR Congo: Pays de l'Avenir: Construisons Ensemble une Paix Durable Pour un Meilleur Destin.* Leuven: IPRA.

Reychler, L. (2010). The Power of Intellectual Solidarity. Paper presented at *GCS Commemoration of UN International Day of Peace and NGO Forum on Territorial Issues and Conflict Resolution* (27–29 September), Seoul, South Korea.

Reychler, L. (2010). Intellectual Solidarity, Peace and Psychological Walls. Paper presented at *International Conference on Peacebuilding, Reconciliation and Globalization in an Interdependent World* (6–10 November), Berlin, Germany.

Reychler, L. (2010). *Failing Foreign Policy – Foreign Policy Failures: Assessing the Impact of the West on Peace and Development in the MENA.* Leuven: KU Leuven.

Reychler, L. (2010). Peacemaking, Peacekeeping, and Peacebuilding. In R. A. Denemark (Ed.), *The Oxford Research Encyclopedia of International Studies* (Vol. 9, pp. 5604–5626). Oxford: Oxford University Press.

Reychler, L. (2010). Sustainable Peace Building Architecture. In N. Young (Ed.), *The Oxford International Encyclopedia of Peace* (pp. 2027–2044). Oxford: Oxford University Press.

Reychler, L. (2010). Absolute Veiligheid of Duurzame Vrede? *De Standaard* (19 June), 72.

Reychler, L. (2011). Time for Peace. Europe's Responsibility to Build Sustainable Peace. Paper presented at *European Peace Research Association Conference* (20–22 July), Tampere, Finland.

Reychler, L. (2011). Strategie met toekomst: van absolute veiligheid naar duurzame vrede: Israël en Palestina in het nieuwe Midden Oosten. Actueel Denken en Leven. Brasschaat, Belgium (28 March), 27.

Reychler, L. (2011). De Keerzijde van de Oorlog. *De Standaard* (29 March), 55.

Reychler, L. (2011). Raison ou Déraison d'état: Coercive Diplomacy in the Middle East and North Africa. In *Reason of State and State of Reason in the Global Era: History and Present* (pp. 679–706). Beijing: Tsinghua University.

Reychler, L. (2012). Europese Diplomatie in een Onzekere Wereld. *Streven, 79*(6), 564–571.

Reychler, L. (2012). Raison ou Déraison d'état: Coercive Diplomacy in the Middle East and North Africa. In A. Fabian (Ed.), *A Peaceful World is Possible* (pp. 217–242). Sopron, Hungary: University of West Hungary Press.

Loeckx, A., Oosterlynck, S., Kesteloot, C., Leman, J., Pattyn, B., Reychler, L., & Vanbeselaere, N. (2012). *Naar een nieuwe gemeenschappelijkheid voor Brussel* (Metaforum Visietekst No. 8). Leuven: Werkgroep Metaforum Leuven.

Reychler, L. (2012). Realistische Diplomatie voor Syrie en Iran. *De Tijd* (27 February).

Reychler, L. (2013). Tijd voor Diplomatie en Vrede. In J. Kustermans & T. Sauer (Eds.), *Vechten voor de Vrede* (pp. 89–102). Leuven: Lannoo.

Reychler, L., & Haers, J. (2013). Versöhnung: Was lehrt uns die Konfliktforschung über Versöhnung? *Concilium: Internationalen Zeitschrift für theologie, 49*(1).

Reychler, L. (2014). Haal diplomatie van onder het stof. *De Standaard* (25 July), 38–39.

Reychler, L. (2015). *A Strategy for Extending Trust: The Northeast Asia Peace and Cooperation Initiative and Trust Building Process on the Korean Peninsula.* Seoul: KSCES Korean Society of Contemporary European Studies.

Reychler, L. (2015). Secular and Religious Time in Divided Societies. In M. Mollica (Ed.), *Bridging Religiously in Divided Societies in the Contemporary World* (pp. 17–46). Pisa: Pisa University Press.

Reychler, L. (2015). Verviers en de Europese terroristen. *De Wereld Morgen* (16 January). Retrieved from www.dewereldmorgen.be.

Reychler, L. (2015). *Time for Peace: The Essential Role of Time in Conflict and Peace Processes.* Queensland: University of Queensland Press.

Reychler, L. (2015). Time, "Temporament", and Sustainable Peace: The Essential Role of Time in Conflict and Peace. *Asian Journal of Peacebuilding* (May), *3*(1), 19–41.

Reychler, L. (2017). On the Ethics of Education in Sports. Paper presented at *International Symposium for Taekwondo*, Muju, Korea.

2.2 Supervision of Doctoral Dissertations

Lappin, R., & Reychler, L. (sup.). (2013). *Post-Conflict Democracy Assistance: An Exploration of the Capabilities-Expectations Gap in Liberia, 1996–2001 & 2003–2008,* 379 pp.

De Vuysere, W., & Reychler, L. (sup.). (2012). *Neither War nor Peace. Civil Military Cooperation in Complex Peace Operations.*

Al-Fattal, R., Keukeleire, S. (sup.), & Reychler, L. (cosup.). (2011). *Transatlantic Trends of Democracy Promotion in the Mediterranean: A Comparative Study of EU, US and Canada Electoral Assistance in the Palestinian Territories (1995–2010).*

Hertog, K., Reychler, L. (sup.), & Verstraeten, J. (cosup.). (2008). *Religious Peacebuilding: Resources and Obstacles in the Russian Orthodox Church for Sustainable Peacebuilding in Chechnya.*

Okemwa, J., & Reychler, L. (sup.). (2007). *Political leadership and democratization in the Horn of Africa (1990–2000)*, 275 pp.

Chun, K., & Reychler, L. (sup.). (2006). *Democratic Peace Building in East Asia in Post-Cold War Era. A Comparative Study*, 319 pp.

Yasutomi, A., & Reychler, L. (sup.). (2006). *Alliance Enlargement: An Analysis of the NATO Experience.*

Mollica, M., & Reychler, L. (sup.). (2005). *The Management of Death and the Dynamics of an Ethnic Conflict: The Case of the 1980–81 Irish National Liberation Army (INLA) Hunger Strikes in Northern Ireland.*

De la Haye, J., & Reychler, L. (sup.). (2001). *Missed Opportunities in Conflict Management: The Case of Bosnia-Herzegovina (1987–1996).*

Sauer, T., & Reychler, L. (sup.). (2001), *Nuclear Inertia. US Nuclear Weapons Policy After the Cold War (1990–2000).* Leuven: Departement Politieke Wetenschappen, K.U. Leuven, 2001, 358 blz.+ bijlagen.

Wets, J., & Reychler, L. (sup.). (1999). *Waarom onderweg? Een analyse van de oorzaken van grootschalige migratie en vluchtelingenstromen*, 321 pp.

Hong, K. J., & Reychler, L. (sup.). (1996). *The C.S.C.E. Security Regime Formation from Helsinki to Budapest.* Leuven, 350 pp.

Ramezanzadeh, A., & Reychler, L. (sup.). (1996). *Internal and International Dynamics of Ethnic Conflict. The Case of Iran*, 273 pp.

Huysmans, J., & Reychler, L. (sup.). (1996), *Making/Unmaking European Disorder. Meta-Theoretical, Theoretical and Empirical Questions of Military Stability After the Cold War*, 250 pp.

Bolong, L., & Reychler, L. (sup.). (1988). *Western Europe-China. A comparative Analysis of the Foreign Policies of the European Community, Great Britain and Belgium Towards China (1970–1986)*, 335 pp.

Part II
Luc Reychler's Selected Texts

Chapter 3
Introduction to the Selected Texts

The sole objective of my work was to contribute to the analysis and the development of architecture for sustainable peacebuilding. My research was influenced by the work of Karl Deutsch on system analysis and model-thinking. Deutsch used cybernetics and the application of dynamic models to study economic, social and political problems, known as wicked problems. Peacebuilding can be a wicked experience. Coherent peacebuilding demands good time management and the creation of a synergy between peacebuilding efforts in different domains, at different levels and layers of the conflict. I also preferred the term 'architecture' to indicate that peacebuilding is not only a science but also an art, since imagination and creativity are an essential part of the building process. To get there, I researched a broad range of subjects and disciplines. They are not presented in a sequential order. The first strand of research focuses on diplomacy. Diplomacy supports or undermines peacebuilding. Special attention was given to diplomatic thinking, field diplomacy and coercive and adaptive diplomacy. The second strand involves conflict, violence and non-violence. Methods were developed for analysing conflicts, differentiating phases in conflict dynamics and assessing the difficulty of conflict transformation. I use a broad definition of violence, which includes all the types of violence which can reduce the quality and quantity of the life expectancies of people. A narrow definition obscures a great part of the violence fabric and furthers denial. The third strand encompasses the security and strategic studies. It deals with arms dynamics, arms control, national and human security and different ways of organizing security, such as supremacy, balancing power, deterrence, cooperative security and the creation of a security community. The fourth strand deals with the prevention of violent conflicts and Conflict Impact Assessment. Proactive conflict prevention is more cost effective than reactive conflict prevention. An essential part is the anticipatory assessment of the impacts of intervention and non-intervention on conflict and peace dynamics. The fifth strand focuses on democratic peace. The sixth studies the role of time in conflict and peace processes. Time plays an essential role in conflict transformation. Most decision-makers consider time very important, but fail to use time effectively. Time is an instrument,

L. Reychler and A. Langer (eds.), *Luc Reychler: A Pioneer in Sustainable Peacebuilding Architecture*, Pioneers in Arts, Humanities, Science, Engineering, Practice 24, https://doi.org/10.1007/978-3-030-40208-2_3

but also a frequently overlooked part of the context. The seventh strand looks at the role of international organizations, religion, ethics and humour. My next research project explores how humour can facilitate or inhibit constructive conflict trans- formation. The eighth and last strand involves the most integrative work. It deals with peace architecture. This is the most challenging and valuable part of the research. It helps the analyst to (a) assess the elements required to achieve sus- tainable peace in a specific context, (b) develop a peace plan, (c) anticipate positive and negative impacts of peace efforts, (d) implement a peacebuilding plan, and (e) monitor and evaluate the real impact of peace efforts.

3.1 Diplomacy and Diplomatic Thinking

3.1.1 Patterns of Diplomatic Thinking

In our turbulent world, the quality of diplomacy remains a main concern. Diplomacy has many faces. There is bilateral and multilateral diplomacy, sophis- ticated diplomacy and coercive diplomacy, integrative and disintegrative diplo- macy, official and track II diplomacy. Jean Monnet was the first diplomat of interdependence. Diplomatic history includes diplomatic practice that successfully advanced interests and created peace, and diplomacy that undermined national interests and ended the peace. My first book, *Patterns of Diplomatic Thinking: A Cross-National Study of Structural and Social-Psychological Determinants* (1976), deals with four facets of diplomatic thinking: the perception of the international climate, the value systems of diplomats, their analytic styles, and strategic approaches. It also describes how and to what extent diplomatic thinking is influ- enced by power or the position of the country in the international system. My attention to structural variables was sharpened by Johan Galtung and Helge Hveem of the *Peace Research Institute* (PRIO) in Oslo. Finally, I wanted to find out, how and to what extent the academic theories of international relations reflected the practitioner's perceptions and world-views. From January 1975 to August 1975 I collected information from 266 diplomats representing 116 countries in Washington DC., by means of semi-structured interviews which lasted somewhat more than 1 h each. In addition, most diplomats responded to written question- naires. The study showed significant differences between diplomats of developing and developed countries. Diplomats stand where they sit. With respect to the assessment of peace, the predominant indicator of peace was, not unexpectedly, the absence of violence. On a comparative basis, however, it's significantly more salient for diplomats from developed countries. Their second and third indicators of peace are international cooperation and stability and order. For the respondents of developing countries, they are "the absence of structural violence" and "no domestic interference". These results tell us that diplomats use different sets of indicators to assess the degree of peacefulness of international relations. The same situation can be constructed and experienced as peaceful or violent. The defence of peace requires not only changes in the minds of people, but also of violent structures. The study affirms that the diplomat's conceptualization of the world is a key variable in international relations, and that conceptual and structural variables are strongly correlated.

3.1.2 Field Diplomacy

The development of the theory and practice of field diplomacy was intended to remediate some of the weaknesses of unofficial diplomacy. The term *terrein diplomatie* (Field Diplomacy) was coined by R. Moreels, past president of *Médicins Sans Frontières* in Belgium. Field diplomacy is characterized by a credible presence in the field, a serious commitment to conflict transformation, multi-level engagement, elicitive conflict analysis and resolution, a broad time perspective, attention to the deeper layers of the conflict, preference for an integrative conflict-prevention policy, and recognition of the interdependency between seemingly different conflicts. Field diplomats assume that one needs to be in the conflict zone to get a better insight into the dynamics of the conflict and to facilitate the transformation of the conflict more effectively. This contrasts sharply with the official and parallel diplomatic activities which mostly operate within the capitals or from abroad. The building of a trust bank or a network of people who can rely on each other is essential to elicit measures to prevent a destructive transformation of the conflict. Building trust takes a great deal of time and effort. When a conflict erupts it is too late. The elicitive approach contrasts with the prescriptive approach, which underscores the centrality of the trainer's models and knowledge. It is process-orientated and gives people a chance to participate in the way the conflict is handled. Another element of the conflict paradigm is the attention to deep conflict. War engenders a mental environment of desperation in which fear, resentment, jealousy and rage predominate. Consequently, building peace requires not only attention to the hard layers of the conflict (the political-diplomatic, military, legal, economic, ecological), but also to the softer layers of the 'deep conflict'. A publicly signed peace agreement does not guarantee a sustainable peace. Peace also requires reconciliation at the psychological and emotional levels. Peace must feel good. Also important is the spiritual level. At this level peacebuilding means transforming despair into hope, hate into love, nihilism into meaningfulness, condemnation into forgiveness, and alienation into relationship. In the social sciences, those levels of the conflict tend to be neglected. Finally, field diplomats recognize the complex interdependence between seemingly different conflicts. Problematic in the transformation of conflicts is not only the artificial legal distinction between internal and external conflicts, but also the propensity to conceptually isolate or quarantine closely interwoven conflicts. Think of the interdependence of internal and international democracy. In the West, neoconservatives promote regime change, but strongly resist international democracy. Most of the conflicts cannot be reduced to pure internal conflicts. They are – or were at one time or another – influenced by conflicts at a regional or global level. So, for example, a peace policy in Rwanda or Burundi requires not only efforts to deal with the conflicts within the boundaries, but, equally, peace efforts at the sub-regional, Euro-African, and global levels.

3.1.3   Coercive Diplomacy

In 2011, at Tsinghua University in Beijing, I voiced my dissatisfaction with the regression of Western diplomacy to coercive diplomacy. The title of the paper was "Raison d'état and déraison d'état". Coercive diplomacy seeks to weaken or change an adversary by means of diplomatic isolation, humiliation, propaganda, economic and financial sanctions, threats, cyber warfare, and military force, short of full-scale war. Most disturbing are the interventions in the *Middle East and North Africa* (MENA). The invasion of Iraq is now called a blunder or a colossal mistake. It was a crime. The pundits and consultants who sold the war still go on doing what they want and are free from account giving. Think of Paul Bremer or John Bolton. The acidic H. L. Mencken once described American democracy as the worship of jackals by jackasses. This and the other wars have benefited particular interest groups and are foreign policy failures. The policy of America and her closest allies after 9/11 turned the Middle East and North Africa into a wasteland of buried reason. It is high time we stop pursuing national interests in an unreasonable way. A more adaptive diplomacy implies: more sophisticated realism, transparency and accountability, calling out those in the media and think tanks who become the mouthpiece of politicians co-responsible for the *déraison d'état*, maintaining the health and vigour of our own society, and dealing more effectively with enemies and friends who undermine our values and interests.

3.1.4   Negotiation and Mediation

Recent developments in the world have strengthened the need for constructive negotiation and mediation. Many problems are so intertwined that isolation or denial have become counterproductive. Coercive diplomacy is hard, and the requirements for success are difficult to create. Cooperation is not always an option. During the 1980s and 1990s we saw a steep development of conflict management systems in which constructive negotiation skills are embedded. In the course "Techniques of Negotiation and Mediation", the students learned to distinguish between distributive and integrative negotiation and to assess the financial, human and temporal costs of might and flight responses to conflict. They learned how to negotiate effectively by using methods developed by Fisher/Ury (1983), Mastenbroek (1993, 2002), Ury et al. (1993), and Costantino/Merchant (1996). At the end, they received a practical guide to mediating disputes. One of the exercises involved recognizing and quickly responding to the use of tricks and manipulations. They also learned to assess the 'difficulty' of conflict and the impact of a peace agreement on the conflict transformation. At the end of the course the students participated in a day-long simulation exercise dealing with an international conflict.

3.1.5 Defining Violence and Non-violence

Defining what is violent or not has always been the subject of fervent domestic and international competition. The denial of one's own violence and the exaggeration of the violence of others is commonplace. The best way to resist such distortions is to use a broad and comprehensive definition of violence. Violence is not just the absence of war, but also refers to the qualitative and quantitative reduction of the life expectancies of people by means of different modes of violence: physical, psychological, cultural, structural, legal, political, economic, gender, temporal and ecological. There is the slow violence of poverty, segregation, and the rapid violence of war and displacement. Most violence prevention efforts are reactive rather than proactive. Measures are taken after violence erupted and the costs accumulated. In order to survive, the world has to improve both the natural and human climate. Conflicts need to be resolved in constructive and cost-effective ways. As I become older, I find it more difficult to look violence in the face. In 1971, I saw *A Clockwork Orange*. I was awed by Stanley Kubrick's dystopian violence embellished by Wendy Carlos' synthesizer versions of music by Elgar, Purcell and Beethoven. Today it is still a stunning movie, but the ugliness of violence, with advanced or less advanced weapons, remains repulsive. The "shock and awe" bombardment of Baghdad in 2003 was a rocky horror show. Bush and Blair's fireballs, followed by white flashes, lit up the blue sky. Orange plumes of flames rose up. The five million residents of the capital and the whole world watched.

3.1.6 The Effectiveness of a Pacifist Strategy

Although I am not an absolute pacifist,[1] I became fascinated by Gene Sharp's book *Politics of Non-Violence, Part One: Power and Struggle* (1973). His theory of non-violent control of political power is still valid, but needs to be contextualized. Thomas Schelling, a specialist in strategy of conflict and the international policy of threats and deterrence, wrote the foreword. He recommended detailed comparisons

---

[1]Ceadel (1987) distinguishes five attitudes towards war: militarism (war is a positive good), crusading (it is legitimate to use war to promote order or justice), defencism (only defensive war is acceptable), pacificism (war should be abolished, but some defensive wars can be justified), and absolute pacifism (the absolute view that participating in or supporting war is always impermissible.

of violence and non-violence in a multitude of contexts, to learn the strengths and weaknesses of both kinds of action in differing circumstances. This, and the fact that previous experiments suggested that pacifist strategies have little to recommend them as effective strategies in conflict resolution, led me to conduct an experimental study on the (in-)effectiveness of pacifist strategists. The experiment, held in Harvard, asked 64 subjects (32 male and 32 female undergraduates) to transport energy through a passage. When they succeeded they were rewarded. The same passage, however, was also used by an opponent who pursued the same aim. The students could make use of threats or electricity to pressure the other to let them transport everything first. The opponents (who behaved like pacifists) could, but never made use of threats or shocks. The experiment allowed me to assess the impact of four controlled and four uncontrolled variables on the use of physical (deliverance of shocks) and psychological violence by the subjects who interacted with a pacifist opponent. The experiment indicated that the variables accounted for 50% of the behaviour of the opponents of the pacifist. I found that a pacifist strategy tends to be most effective in reducing violence and exploitative behaviour when the human distance between the subject and the opponent is small, the subject is well-informed about the intentions of the pacifist, the subject is required to justify his behaviour post-facto, and when an impartial third party is present. The strongest predictors of the effectiveness of the pacifist strategy were the participant's interpretation of the experimenter's purpose (demand characteristics) and the image of the pacifist held by the opponent.

3.1.7   The Next Genocide

In 1968 during my research at the University of Munich, I visited Dachau. I was alone in the camp. The barracks were destroyed and a small museum told the story of the first Nazi concentration camp and the medical experiments (Photo 3.1).

Later, I visited Birkenau-Auschwitz, the largest of the German Nazi concentration and extermination camps. Rwanda and Bosnia Herzegovina came later. I also visited many sites of remembrance, such as the Holocaust museum in Washinghon DC, the genocide monuments in Berlin, the Murambi school genocide centre, and the Kazerne Dossin in Mechelen, Belgium. Genocidal behaviour did not stop after WWII. More than thirty serious civil conflicts led to episodes of genocidal violence between 1955 and 2004. Are the genocidal events in Rohingya and Yemen the last examples? Europeans, who believed that Europe had become immune to genocides, were perplexed by the war and ethnic cleansing in Yugoslavia. Before the violent disintegration of Yugoslavia, each year in Dubrovnik I attended an international seminar on violent conflicts. The students came from different parts of the globe and visited Mostar and Sarajevo as examples of a multicultural peace culture. In 1990, in the Grand Café, I asked my Yugoslav colleagues if a civil war could break out. They resented the question and replied that this was a typical narrative from the Western press, and asserted that they could handle conflicts in a

Photo 3.1 Entry to the concentration camp in Dachau, Germany. *Source* Personal photo collection

reasonable manner. To better anticipate and prevent genocides, the Leuven peace research team reviewed the literature on democides, politicides and genocides; looked more in depth at Bosnia and Herzegovina, Rwanda and Burundi, and sought how to anticipate and proactively prevent genocidal behaviour. The results can be found in *De Volgende Genocide* [*The Next Genocide*] (Reychler et al. 2004), and *Darkness at Noon* (Bec-Neumann 2007). The research identified different types of genocide and distilled the (f)actors which increase the probability of genocidal behaviour. Genocide is an old recipe with six ingredients: an environment or region characterized by political, economic, cultural and physical insecurity and a high level of frustration; an authoritarian regime which attributes the responsibility and insecurity to a particular group; an identity group which can be clearly distinguished from the rest of the population and therefore be targeted and dehumanized; a plan to get rid of the target group; relatively powerless victims; and an informed international community which expresses moral indignation but does not effectively intervene to prevent or stop the genocidal behaviour. The study also highlighted the role of the victims before and during the genocide; the role of justice during and after the genocide; and the importance of conflict effect monitoring and reporting. The genocides in Bosnia and Herzegovina, Rwanda and Burundi were histories of missed opportunities. *The Next Genocide* ends with a plea for an existential understanding of the behaviour of victims and perpetrators in order to build an effective prevention policy. After seeing *Rape: A Crime of War* (1996: Shelley Saywell), a documentary of a concentration camp near Omarska, in Bosnia, students in Leuven, Antwerp and Venice were asked to respond to a series of sensitive questions which put them in the role of imaginary victims or perpetrators. They experienced the existential questioning as an uncomfortable but enriching learning experience. They learned that prevention is the key, and that everything needs to be done to prevent the development of conditions in which even normal people can imagine committing or undergoing violence and humiliation. In *Darkness At Noon* I argue for the dismantlement of "sentimental walls" and for the use of both memorial mindsets and peacebuilding mindsets to prevent war crimes and genocide. In "Preventing war crimes and genocides: A look from the balcony", both mindsets are compared on eleven dimensions. Sentimental walls are attitudes, feelings, perceptions, mindsets, expectations, causal analysis and attributions, strategic analyses, values, preferences, taboos, and social psychological pressures (such as pressures to conform, groupthink and political correctness) that stand in the way of sustainable peacebuilding.

3.2 Strategic and Security Studies

Security is a universal need. Objective and subjective security is essential for sustainable peace. Therefore peace researchers should be acquainted with security and strategic studies and learn to assess the costs and benefits of different military security strategies. Initially I listened to Sun Tzu's *The Art of War*, Machiavelli's *The Prince* and Clausewitz's *On War*. They inspired me to write on the art of peacebuilding. They also enriched my thinking on peace by differentiating contexts, by identifying the natural qualities of good warriors, and by claiming that supreme excellence is to subdue the enemy without fighting (Sun Tzu). Later polemologists, or analysts of war, taught me that most wars are not a continuation of policy by other means, but a failure of policy (Rapoport 1968); that a victor's peace is seldom lasting and neither is total defeat (Stoessinger 2004); and that the folly of military interventions and wars is unrelated to the type of regime: monarchy, oligarchy or democracy (Tuchman 1984). My research into security and strategy synthesized and evaluated three aspects of security and strategy: the analysis of security, arms dynamics and control, and the international organization of security. In the midst of conflicts, the perception of security is frequently manipulated. Therefore a meaningful discussion about security requires clarification of (a) whose interests need to be secured, (b) against what type of threats, (c) how much security is wanted, and (d) who will secure the interests with what instruments. Key concepts are: objective and subjective security, the levels of security (human, national, regional, global security), and the dimensions of security (military and non-military). Later, my attention turned to arms dynamics, arms control efforts and agreements. The success or failure of arms control can be attributed to: characteristics of the international system, the political-strategic thinking, technological developments, domestic variables, aspects of the art of negotiation, and leadership. Key concepts are the Military Industrial Complex (MIC), the security dilemma, and stability. Most of my research centred on international organization of security: the UN, the OSCE/CSCE, NATO and the European Union. My publications focused on Belgian security policy, European security and defence research, European security beyond the year 2000, the Helsinki process, and NATO after forty years. Key concepts are: individual security (UNDP), collective defence, collective security, cooperative security, security dilemma and securitization. The latter refers to the framing of non-military issues as existential threats and security questions. The greatest challenge in the twenty-first century is to energize critical security thinking and take decisions about world security. The human and the natural climate are inextricably intertwined.

3.3 Violence Prevention and Conflict Impact Assessment

3.3.1 A Comparative Study of Conflict Prevention in International Organizations

A major part of my research deals with the proactive prevention of violent conflicts. The attention was jolted by the Gulf War and the Yugoslavian civil war which destroyed the initial peace-euphoria of the Post-Cold-War. The Bosnia-Sarajevo effect, or the fear of becoming entangled in ethnic and nationalist disputes, rekindled in the international community interest in the prevention of destructive conflicts. As advisor to the Belgian Minister of Foreign Affairs (Marc Eyskens), at the beginning of the 1990s I supervised a comparative study and evaluation of the state of conflict prevention in five international governmental organizations: the United Nations, the *Organization of Security and Cooperation in Europe* (CSCE), the *North Atlantic Treaty Organization* (NATO), the *Western European Union* (WEU), and the European Community. The study searched for a more cost-effective organization of international security. It was driven by the rational belief that one ounce of prevention is far better than a ton of delayed intervention. After evaluating the state of the art of conflict- or violence-prevention in international organizations, recommendations were made to remediate the inadequacies. The aim of the publication was to sensitize decision-makers and shapers to the cost-effectiveness of preventive measures. In spite of the slow pace of the requisite security reforms and the preference for coercive diplomacy, an investment in conflict prevention remains prudent in the long term. In 1994, I had the feeling that the West was losing the battle for the future. Missing were: foresight, the adaptation to the rapidly changing strategic landscape, a well-orchestrated international structure for dealing with the flood of turmoil around the world, and proactive conflict prevention. Despite the fact that the chances to prevent anarchy and violence in the MENA may already seem to have passed us, more proactive and reactive preventive measures should be taken.

3.3.2   Conflict Impact Assessment System (CIAS)

In July 1996, at the IPRA conference in Brisbane I pleaded for a *Conflict Impact Assessment System* (CIAS). There was some reticence about the use of the term, because of the possible confusion with the CIA. I decided to add an 's' not only to have a politically correct acronym, but also to stress that it deals with a complex whole. It aspires to be an early warning system. CIAS tries to anticipate the positive and negative impacts of military and other interventions. Notwithstanding progress, four problems continue to hamper the development of an effective early warning system. First, the tendency to focus on hard, tangible, quantifiable variables and to overlook the soft factors which influence the conflict dynamic, such as perceptions, expectations, values and preferred world orders. The second problem is the predominant attention given to the anticipation of threats, dangers or worst-case developments. Practically no attention is given to the development of early warning systems which identify the points in conflict processes at which particular interventions would enhance the constructive transformation of conflicts. The result is missed opportunities. The third problem is the rudimentary cost-effect analysis. Finally, impact assessments tend to be one-dimensional and to overlook the impacts on other domains, levels, time frames and layers (public layers: public behaviour/opinion; and deeper layers: private opinion, perceptions, wishes, expectations, emotions, historical memory). The aim of CIAS is to assess in time the positive and/or negative impacts of different kinds of interventions (or the lack thereof) on the dynamics of conflict; to contribute to the development of a more coherent conflict prevention and peacebuilding policy; to serve as a sensitizing tool for policy-shapers and policy-makers, helping them to identify weaknesses in their approach (such as blind spots, incoherence, bad timing, inadequate priority setting, etc.) and to further the economy of development and peacebuilding efforts. CIAS is about reminding or changing the mindsets of decision-makers. Since its inception, the theory and methodology has been further developed and CIAS has been applied in several countries, for example in 2001 to assess the European Union *Micro Project Programme* (MP3) in Bayelsa, Delta and Rivers states of the Niger Delta of Nigeria (Reychler 2002) (Photo 3.2).

In 2003, a field guide (with didactic cartoons) was made for the *Field Diplomacy Initiative* (FDI), an NGO that organized training and performed CIAS in several countries. In 2006, Thania Paffenholz and I synthesized the state of theory and practice in *Aid for Peace. A guide to planning and evaluation for conflict zones*. Thania has been an EU peacebuilding expert in Africa and is now director of the Inclusive Peace and Transition at the Graduate Institute in Geneva (Photo 3.3).

Photo 3.2 Conflict impact assessment of EU micro projects in the Niger Delta, Nigeria in 2001. *Source* Personal photo collection

3.4 Democracy and Legitimacy

Writing *Democratic Peacebuilding and Conflict Prevention: The Devil is in the Transition* (1999) was driven by the importance of democracy and legitimacy in sustainable peace, the euphoric academic-diplomatic discourse about democracy and peace after the Cold War, the neo-conservative transformation of the American-Western foreign policy, and the inconsistency of people in democratic countries who simultaneously support domestic and oppose international democracy. Many students in my classes expressed reservations about the democratization of the global community or equal rights for all global citizens. Both Reagan and Bush argued that American security depended on regime change in other countries, especially in Central America and the Middle East. The neo-conservative crusader foreign policy was justified by noble lies or alluring siren concepts, such as benevolent hegemony, American exceptionalism, responsibility to protect, etc. These policies came at enormous human political and financial cost, especially to the people at the receiving end. Since 1970, I have tried to make sense of the cross-impact between peace, democracy and legitimacy. Many scholars share the premise that a "democratic constitution" in each state decreases the likelihood of violent conflict and war. They also believe it reduces structural violence. When

Photo 3.3 Cartoon of competing values in peacebuilding. *Source* Personal photo collection

thinking about the relationship between sustainable peace and democracy, I become more and more interested in legitimacy. Legitimacy comes in when looking for 'stable democracy'. S. Lipset's definition of legitimacy = democracy × effectiveness, was a revelation (Lipset 1959). He considered the existence of a soft cleavage in society to be a hallmark of legitimate democracy, and argued that the stability of democracy was determined by the interaction between effectiveness and democracy. I joined the democratic peace discourse by criticizing the use of reductionist operational definitions of democracy and the exclusive focus on the state level. A state with colonies cannot be labelled a fully-fledged or genuine democracy. Sustainable peace at international level requires a legitimate international system. Two of the greatest challenges of the twenty-first century will be to integrate whole regions democratically and to initiate the democratic transition of the world order.

3.5 Time and Temporament in Conflict

3.5.1 The Time Paradox

For more than thirty years, I hoarded articles and books on the role of time in human behaviour and its environment. But I did not find time to write about it. Finally, as emeritus, it took me three years to compose a book on the essential role of time in conflict and peace processes. Time is the most precious resource we have. It is irreversible and unrenewable. My attention was triggered by the paradox of temporal behaviour. Despite the fact that people take time seriously, a great deal of their temporal behaviour, especially in conflicts, shows gross deficiencies. I decided not to start with theses and hypotheses, but to read everything I could find on time and try to map it in a meaningful way. In many MA and PhD theses, the role of time tends to be studied in narrow, partial and superficial ways. In our fast-changing globalized world, time and the way we deal with it will more than ever determine the success or failure of the handling of global crises and opportunities to build sustainable peace and development. It is high time to radically change the way we deal with time and to develop a more adaptive temperament. The term 'temporament' refers to the way persons or organizations deal with time in conflict management and peacebuilding situations. Time is a multi-dimensional phenomenon that can be divided into six principal and six transformative dimensions.

3.5.2 Principal Dimensions

The principal dimensions (P) distinguish the fundamental components of time in conflict and peace:

- Existential time is about life and death. The duration and quality of a lifetime can be reduced by fast (armed) or slow (structural) violence.
- Orientation to the past, present, and future. After a bloody conflict, a peace process must deal with the past, present and the future.
- Modes of time: change, succession, continuity, turning points, duration and viability.

- Temporal management options. This cluster identifies four types of temporal decisions which characterize conflict and peace behaviour: proactive or reactive decisions, priority-setting, timing and synergy.
- Secular and religious time. As long as there are religiously inspired people, religious, sacred, divine or transcendental senses of time cannot be overlooked.
- Temporal emotions. The impact of time cannot be fully comprehended without the attendant temporal emotions.

3.5.3   Transformative Dimensions

The transformative dimensions (T) draw attention to variables that can alter temporal perception and behaviour.

- The appreciation of time.
- The selection of the temporal strategy, including the framing and manipulation of perceptions of time and the use of time to pressure people in negotiation settings.
- The anticipation of crises, opportunities and positive and/or negative impacts.
- Temporal equality or inequality. In a genuine democracy, the time of each citizen is equally valuable. There are many ways of respecting or disrespecting the lifetime of others.
- Temporal discernment and empathy is the capacity and will to discern how others think and feel about time.
- Temporal efficacy is the opposite of determinism, fatalism, powerlessness, defeatism and despair. It indicates a reasonable confidence in one's ability to understand the significance of time in ways that further one's interests.

Assessing the 'temporament' or temporal behaviour of the stakeholders in conflict and peace should be an essential part of monitoring and evaluation. It could help to identify the temporal inadequacies more systematically and steepen the learning curve of peacebuilders and peacespoilers. It could advance the accounting for the costs of temporal misconduct and the accountability of policy-makers. Above all, it should make us more aware that today's prevailing temporament stands in the way of sustainable peace and security. Thus, let us take the time to know our own temporament and start to make it more adaptive.

3.6 Religion, Ethics and Humour

3.6.1 Religion and Conflict

In 1998, after teaching at the University of California in Irvine, I stopped over in Virginia to meet some scholar-practitioners in Harrisonburg, Virginia. The Eastern Mennonite University had several outstanding people dealing with peacebuilding and the elicitive approach (John Paul Lederach), reconciliation (Hizkias Assefa), restorative justice and forgiveness (Howard Zehr), mediation (Ron Kraybill) leadership (Jayne Docherty) and religious diplomacy (Cynthia Sampson) (Photo 3.4).

The discussions were very enriching and inspiring. They filled gaps in my thinking about peacebuilding architecture. Cynthia, who co-edited *Religion, the Missing Dimension of Statecraft* (Sampson 1997), conducted pioneering research on religion and conflict, and observed the American tendency to separate political from spiritual life and consequently to neglect a powerful tool in negotiations: religion. This observation is contradicted by the fact that most American presidents have been highly religious (or at least pretended to be), and in speeches have made references to God or begged God to bless America. Her research attempted to restore this missing dimension in the conduct of international diplomacy. After participating in a Bellagio Center seminar on the role of religion in conflict

Photo 3.4 John Paul Lederach and Hizkias Assefa. *Source* Personal photo collection

transformation and peacebuilding, I drafted *Religion and Conflict*. Religious organizations represent two-thirds of the world's population. They have a negative and positive reputation and history. Religious differences have been one of the causes of conflict. But they have also contributed to peacebuilding by empowering the weak, influencing the moral political climate, developing cooperation and providing humanitarian aid. Several religious organizations distinguish themselves through peacemaking efforts – both traditional diplomacy and track II or field diplomacy. The world cannot survive without a new global ethic, and religions play a major role.

In "Secular and religious time in divided societies" (Reychler 2015), I argue that paying more attention to religious time is highly recommendable. First, because religion promotes temporal values, which are foreign to secular institutions dealing with academic, political, economic or legal issues. Love, sobriety, forgiveness, faith and hope do not appear in international law or economics. Second, the study of religious time is recommended because religions highlight the importance of unsecular time, such as sacred time or Kairos. Sacred time is devoted to the heart, to the soul, to nature, to eternity. Third, globalization furthers religious eclecticism and the search for the common essence of different religions. As long as there are religions, they will play a role. The same can be said of humanists and non-believers.

3.6.2 Ethics

Ethics are a central part of the study of international relations and peace. Ethical principles relate to just war, just peace, impartiality, responsibility to protect, judging from the original position, good governance, pursuing adequate or absolute security, pre-emption and prevention, and should be pondered carefully before making a decision or drawing a conclusion. The intellectual products of scholars are not only the result of the underlying theoretical and epistemological assumptions, but also of their normative assumptions. They determine priorities in research but also how poverty, violence, radicalism, peace, justice, democracy, etc. are operationalized. For the students of MaCSP, "Ethics of Peace and International Relations", taught by Johan Verstraeten, was a mandatory course. My course on "Negotiation and Mediation" acquainted the student with the ethical standards of professional conduct in dispute resolution.

Since 2008, I have been involved in the development of ethics in an international sports organization with 70 million participants: The *World Taekwondo Federation* (WT). In "Ethics of Education in Sport", I identified several ethical challenges. First, the ethical discourse of the educational claims and ethical aspirations in sport are paradoxically easy and difficult. The majority of people know or claim to know intuitively what ethics, like time, encapsulates, yet they fail to define it successfully. Nevertheless, moral instincts should be consulted but not trusted in the face of moral controversy. Second, despite the fact that ethics should not be confused with law, most organizations legalize ethics by paying more attention to legal issues than ethical issues. Third, sport organizations, like other societies, also frequently behave like "moral tribes" when ethical issues are raised. Greene (2014), a psychology professor at Harvard, observes that our moral brains, which are reasonably good at enabling cooperation within a group, are not nearly as good at enabling cooperation between groups (Us versus Them). Fourth, the ethical discourse in sport tends to be limited to the prevention and/or remediation of the usual suspects of unethical behaviour, such as: performance-enhancing drugs, corruption, discrimination, gambling, boundary-crossing, sexual behaviour, and the temptations of the sport agent. Most attention goes to dealing with ethical failures or to negative or negatively defined ethics, and this is at the expense of investment in cooperation, good governance, personal development, conflict resolution, leadership, community building, peace, or positively defined ethics. Sport organizations can make a tangible difference if they pay attention to both the negative and positive sport ethics. Athletes should not only learn how to win and deal with loss but also how to evade lose-lose and enhance win-win outcomes.

3.6.3   The Role of Humour in Conflict Transformation

As for time, over more than thirty years I also collected books and articles on oral and visual humour. Good analysis and humour go hand in hand. Humour can help to prevent conflict. These intuitions convinced me to research and write about how humour could help to find a way out of deep-rooted conflicts. What type of humour could be used in the different phases of conflict dynamics? What about absurd humour? In certain circumstances, you need a sense of the absurdity. Is black humour OK? Some analysts answer that black humour can be empowering when it is used to overcome humiliation. Is cynicism appropriate? What about defining happiness in Ambrose Bierce's dictionary of the devil, as an agreeable sensation arising from contemplating the misery of another? The pursuit of new frontiers, the spread of civilization or regime change has almost always come at the expense of others. The path is littered with casualties and destruction. At the end 2017, Oscar Murillo responded to a *New York Times* question, "if you were to send one last message out into the cosmos that summed up the beauty of life on Earth, what would it be?", with an old joke. The astronauts of the Apollo 11 mission trained in the desert of the Western United States – formerly Native American land – in order to prepare for the final frontier. A native chief asked the astronauts if they could pass a message to the holy spirits on the moon. The man spoke some words in his own language, and when they astronauts asked what the message meant, the chief told them it was a secret between his tribe and the moon spirits. But the astronauts managed to find someone who could translate the words. The message was: "Don't believe a single word these people are telling you. They have come to steal your land." What kind of humour caters to an open mind, open heart and open will?

Photo 3.5 Luc Reychler and Fred Bervoets in Antwerp. *Source* Personal photo collection

For me, *Reynard the Fox* and *Tijl Uilenspiegel*, two mythical figures, are part of Flemish culture. Both aspired to freedom and prevailed not by armed force, but by exposing violence, poking fun at and fooling the (foreign) power (Photo 3.5).

3.7 Sustainable Peace Architecture

3.7.1 Sustainable Peacebuilding Architecture

In the 1960s, the green and the peace movement alerted the international community to the deterioration of the environment and the dangers of nuclear conflict. The first publication of the Club of Rome in 1972, *Limits to Growth*, had a catalysing effect on environmental awareness and policy in the world. The green movement has been transformed into political parties, departments, jobs, environmental impact assessments and several international regimes. The peace movement evolved differently. There were some peak moments, like the ending of the Vietnam War and the anti-nuclear protests in the 1980s, but the impacts were weaker and less decisive. Immediately after the Cold War and the accompanying euphoria and peace dividends, peace research was depicted as dispensable. One explanation is that the peace movement had to cope with the strong bureaucracies of foreign policy offices and defence departments. At the political level, peace movements and peace research were – and are – resisted by the rightist and conservative parties. Another problem is that the term 'peace' can refer to different realities; it can be repressive or emancipatory. Although the number of research centres grew, the research remained fragmented. There was competition between IPRA and the Political Science Association. The field was in need of better architecture to advance sustainable peacebuilding. I preferred to use the metaphor "peace architecture" because it draws attention to the architectural principles/considerations that have to be addressed in sustainable peacebuilding processes; it emphasises the need to identify the necessary building blocks for different types of violence and peace; it could shorten the learning curve by providing a methodology for comparative analysis and evaluation for conflict transformation, and it could enlighten the vital role of peace architects. The architecture of sustainable peacebuilding, also called "the peacebuilding pentagon", consists of five plus one necessary building blocks: effective communication; consultation and negotiation; peace-enhancing political, economic and security structures; peace-enhancing software; peace support systems; a supportive international environment; and peacebuilding leadership for the installation of all the above conditions. In 2001, *Peacebuilding: a Field Guide* assembled the best thoughts and experiences of peacebuilding work in the field.

3.7.2 Operationalization of the Peacebuilding Blocks

To facilitate the work of practitioners, the *Centre of Peace Research and Strategic Studies* (CPRS) developed several instruments to analyse and evaluate the state of five+1 building blocks of sustainable peace. In each of the code books different facets of the building block are operationalized. *Researching Peace Building Architecture* (2006) deals with four of the five buildings blocks (an effective system of communication, consultation and negotiation; peace-enhancing political, economic and security structures; the integrative climate; and the existence of international support) and peacebuilding leadership. *The Software of Peacebuilding* (2002) deals with the integrative political psychological climate. Climate refers to the total experience of the social-psychological environment in which the conflict transformation and peacebuilding take place. Constituent elements of an integrative climate are: (1) the expectation of an attractive common future, (2) reconciliation or the perception and feeling that the conflicting parties have been reconciled. Normally this involves reconciling competing needs for justice, restoration or compensation of damage, forgiveness, reassurance and commitment to future cooperation, (3) multiple loyalties, (4) human security, (5) social capital, and (6) the absence of other "senti-mental walls" which inhibit the peacebuilding process. The *Codebook for Evaluating Peace Agreements* (2008) was applied to the Dayton Peace Agreement and the Oslo Accords, and assesses the difficulty of the conflict and the chances of successful peace agreements. *Researching Peacebuilding*

Leadership (2006) assesses the existence of a critical mass of peacebuilding leadership by looking at the preferred future, the leader's analytic style, the change in behaviour, and the personality and motivation(s).

3.7.3   Coherence and Synergy

In our globalized world many people feel insecure. They become cynical, alienated, misanthropic, aggressive, or are determined to make a better future. The latter feel that that they must do something, even though what and how may be far from evident. At least they need a sense of coherence (Antonovsky 1987), consisting of (1) comprehensibility or a sense that one can understand events and reasonably anticipate what will happen in the future; (2) manageability or the belief that the skills, ability, help and resources to build peace can be made available, and (3) meaningfulness or the belief that the pursuit of peace is worthwhile and that there is a good reason or purpose to care about what happens.

The article "Peacemaking, Peacekeeping, and Peacebuilding" (Reychler 2010) tries to contribute the development of a more coherent peacebuilding architecture. Peacebuilding is about complex change: it involves concurrent activities by many people in different sectors, at several levels, in different time-scopes, and in different layers. Peacebuilding looks for synergies between multiple transformations in diplomatic, political, economic, security, ecological, social, psychological, legal, educational, and many other sectors.

The coherence of a peacebuilding intervention relates to six major components of the theory and praxis of peacebuilding architecture. First, the end state or the set of required conditions that defines the achievement of the peace one wants to build. As Mark Twain observed, "If you don't know where you are going, any road will get you there." Second, the baseline or the starting point of the peacebuilding intervention. Before planning the intervention, it is important to conduct an accurate analysis of the conflict and of the peacebuilding deficiencies and potential. Third, the contextual features: scope, time, preservation, diversity, capability, capacity, readiness for change, and power. Fourth, the planning and peacebuilding process. This is one of the most fascinating and complex areas of study. It concerns framing time, entry and exit, pacing the peacebuilding process, setting priorities, synchronicity and sequencing, and negative and positive impacts or synergies. Fifth, the peacebuilding coordination. The dimensions of coordination are: spaces of coordination, the participation, the elements of coordination, the degree of coordination, and strategy formation. Finally, there is monitoring and evaluation.

Despite the progress made, there remain big gaps and challenges. For example, many analysts leave the end state vague and implicit. With respect to the context, two issues need more attention; the qualities of the peacebuilder and the role of integrative power. In *Three Faces of Power* (1990), Kenneth Boulding distinguishes it from threat power and exchange power. Initially he called it the power of love. It relies on respect, legitimacy, friendship, community, affection and love. The widest gaps are found in the planning of the peacebuilding process. Higher quality information and a methodology for analysing complex dynamic behaviour are urgently needed. A better exchange between researchers, practitioners, and decision-makers could raise the learning curve. This involves overcoming four obstacles: diplomats and politicians often deride the academic's lack of first-hand experience when it comes to the practice of managing conflicts and building peace; the slow institutional learning; the diminution of academic freedom; and the foreign policy-making process which is low on democratic checks and balances. All of this makes critical and sustainable peace theorizing essential.

References

Antonovsky, A. (1987). *Unraveling the Mystery of Health: How People Manage Stress and Stay Well.* San Francisco: Jossey Bass Publishers.
Bec-Neumann, J. (Ed.). (2007). *Darkness at Noon: War Crimes, Genocide and Memories.* Sarajevo: Centre for Interdisciplinary Postgraduate Studies.
Bierce, A. (1957). *Dictionary of the Devil: The Cynic's Word Book.* New York: Hill and Wang.
Bok, S. (1989). *Lying: Moral Choice in Public and Private Life.* New York: Vintage Books.
Booth, K. (2007). *Theory of World Security.* Cambridge: Cambridge University Press.
Boulding, K. (1990). *Three Faces of Power.* Newbury Park: Sage Publications. Burton, J. (1969). *Conflict and Communication: The Use of Controlled Communication in International Relations.* London: Macmillan.
Ceadel, M. (1987). *Thinking About Peace and War.* Oxford: Oxford University Press.
Cohen, R., & Mihalka, M. (2001). *Cooperative Security: New Horizons for International Order* (Marshall Center Papers No. 3). Garmisch-Partenkirchen-Germany: George Marshall European Center for Security Studies.
Costantino, C. A. & Merchant, C. S. (1996). *Designing Conflict Management Systems: A Guide to Creating Productive and Healthy Organizations.* San Francisco: Jossey-Bass.
De Bono. (1979). *Future Positive: A Book for the Energetic Eighties.* London: Penguin Books.
Deutsch, K. (1963). *The Nerves of Government: Models of Political Communication and Control.* New York: The Free Press.
Deutsch, K. (1968). *The Analysis of International Relations.* New York: Prentice Hall.
Diamond, J. (2005). *Collapse: How Societies Choose to Fail or Succeed.* New York: Viking.
Duchêne, F. (1994). *Jean Monnet: First Statesman of Interdependence.* New York-London: W. W. Norton.
Fisher, R., & Ury, W. (1983). *Getting to Yes.* London: Penguin Books.
Galtung, J. (1967). *Theories of Peace: A Synthetic Approach †to Peace Thinking.* Oslo: International Peace Research Institute.
Greene, J. (2014). *Moral Tribes: Emotion, Reason and the Gap Between Us and Them.* East Rutherord, NJ: Penguin Books USA.
Kelman, H., & Hamilton, L. (1989). *Crimes of Obedience: Toward a Social Psychology of Authority and Responsibility.* New Haven: Yale University Press.
Kelman, H. (1965). *International Behaviour: A Social Psychological Analysis.* New York: Holt Rinehart & Winston.
Lipset, S. M. (1959). Some Social Requisites of Democracy: Economic Development and Political Legitimacy. *American Political Science Review, 53*(1), 69–105.
Mainelli, M., & Harris, I. (2011). *The Price of Fish: A New Approach to Wicked Economics and Better Decisions.* London-Boston: Nicholas Brealey Publishing.

Mastenbroek, Willem. (1993). *Conflict Management and Organization Development*, 2nd edn. San Francisco: Wiley.

Mastenbroek, Willem. (2002). *Negotiating as Emotion Management*. Amsterdam: Holland Business Publications.

Meadows, D. H., Meadows, D. L, Randers, J. A., & Behrense, W. W. (1972). *The Limits of Growth: A Report for the Club of Rome's Project on the Predicament of Mankind*. New York: University books.

Mecken, H. L. (2006; first published 1956). *On Politics: A Carnival of Buncombe*. Baltimore: John Hopkins University Press.

Rapoport, A. (1968). *Clausewitz: On War*. London: Penguin Books.

Reychler, Luc. (2002). *Initial Conflict Impact Assessment (I-CIAS): Bayelsa, Delta and Rivers*. European Union Micro Project Programme (23 January).

Reychler, L., Calmeyn, S., De la Haye, J., Hertog, K., & Renckens, S. (2004). *De volgende genocide*. Leuven: Leuven University Press.

Senge, P., Scharmer, C. O., Jaworski, J., & Flowers, B. S. (2005). *Presence: Exploring Profound Changes People, Organisations and Society*. London: Nicholas Brealy Publishing.

Sharp, G. (1973). *Power and Struggle: The Politics of Non-Violent Action, Part One*. Boston: Porter Sargent publishers.

Tuchman, B. (1984). *The March of Folly: From Troy to Vietnam*. London: Abacus.

Twain, M. (2010). *Autobiography of Mark Twain, Vol. 1*. Los Angeles: University of California.

Ury, W., Brett, J. M., & Goldberg, S. B. (1993). *Getting Disputes Resolved: Designing Systems to Cut the Costs of Conflict*. San Francisco: Jossey-Bass Publishers.

Wilkinson, R., Yuan, S., & Duyvendak, J. J. (Eds.). (1998). *Art of War – The Book of Shang. By Wu Sun Tzu and Wei Shang Yang*. Ware, UK: Wordsworth Editions.

Chapter 4
Values in Diplomatic Thinking: Peace and Preferred World Order

To understand diplomatic thinking, we need information not only about the work environment, but also about the value systems of diplomats.[1] To evaluate or orient himself/herself toward his/her environment, the diplomat, like any other decision maker, must relate it to his/her values. Hence, if we want to understand his/her response to the international environment, we need to investigate the criteria by which he/she judges international conduct. In this chapter we will inquire into two major values: international peace and preferred world order.

To gain some understanding of diplomatic value systems, we will group the diplomats according to the degree of economic development of their countries and study the differences in their value operationalizations. We decided to compare the diplomats in relation to only one systemic position to stay within the space limitations of this study. But before we focus on the diplomat's thinking about peace and preferred world order, we would like to say a word about each of these issues: the ambiguities and paradoxes inherent in peace thinking in general, different conceptions of peace in the relevant literature of international relations, the linkage between the concept of peace or other international values and international relations, and the advantages of systemic study of international values, such as peace and preferred world order.

4.1 Ambiguities and Paradoxes in Peace Thinking

Today, to contribute to peace is the declared goal of all countries in the world. What does this declared goal mean? An operational definition of it would significantly enhance our understanding of international behaviour. The study of this

---

[1]This text was first published as: Reychler, L. (1979). Values in Diplomatic Thinking: Peace and Preferred World Order. In L. Reychler, *Patterns of Diplomatic Thinking: A Cross-National Study of Structural and Social-Psychological Determinants* (pp. 112–162). New York: Praeger Publishers. Based on the PhD thesis of the author. Reprinted with permission.

© The Editor(s) (if applicable) and The Author(s), under exclusive license to Springer Nature Switzerland AG 2020
L. Reychler and A. Langer (eds.), *Luc Reychler: A Pioneer in Sustainable Peacebuilding Architecture*, Pioneers in Arts, Humanities, Science, Engineering, Practice 24, https://doi.org/10.1007/978-3-030-40208-2_4

phenomenon turns out to be an inquiry into paradoxes. Peace is undoubtedly one of the most frequently used words in the vocabulary of diplomats; yet it is also one of the most ambiguous and least communicative words. Furthermore, there is a pluralism of concepts of peace; rather than wanting peace, people want 'peaces'. Finally, when the record of diplomatic dialogue is weighted against the record of international events, it looks as if peace is at one and the same time the most wanted and the most elusive state of international affairs. "The central paradox in the study of peace lies in its continuing appeal for lasting peace and the actuality of continuing war" (Puchala 1971: 141).

These paradoxes and the conceptual ambiguity of peace could be accounted for by several factors. First, the vagueness of the term allows it to be used as a cover. This particular use was recognized by Dulles (1955: 6) when he attacked the kind of peace that can "be a cover whereby men perpetrate diabolical wrongs". Second, the use of unclear language makes it possible to evade critical problems. This function was well described in Janik/Toulmin's (1973: 269–273) analysis of political language in Austria before 1914. They claim that there was a consistent attempt to evade the social and political problems of Austria by the debasement of language – by the invention of "bogus language games" based on the pretence that the existing norms of life were different from what they really were. The authors perceive the present scene as not so different from the pre-World War I Austria. "Nowadays as much as in the years before 1914, political dishonesty and deviousness quickly find expression in debased language. ... Counterargument is no weapon against this tactic, since issues are always blurred by translation into officialise" (Janik/Toulmin 1973: 269). Third, an international environment characterized by competition and distrust does not help in making values or interests explicit. Not only political considerations, but also cultural factors account for the low quality of conceptualization of international peace. In the field of social science there has been, for a long time, a "flight from tenderness" (Rubin 1973: 11–12). In the research and theorizing of social scientists, much more attention tends to be devoted to the study of aggressive, hostile, and conflictual behaviour, at the expense of so-called softer behaviours, such as cooperation, peace, and justice, which are equally important ingredients of international life. Peace research is a very recent arrival in the study of international relations. A more important explanation of the relative uselessness of the term 'peace' in international dialogue is the nation-centric tendency in peace thinking, evidenced by each country's claim to having the only true definition of peace. Nations tend to claim a universally recognized patent for their operational definition and conceive any other definition as highly improper. A final explanation of misunderstandings in the dialogue about international peace is the differences in meaning associated with the universally used symbol of peace. This explanation recognizes and accepts a pluralism in definitions, and it agrees with the fact that people's values and judgements are likely to differ, especially when their interests are engaged (Rawls 1971: 196–97).

4.2 Different Conceptions of Peace

Having traced some of the sources of lack of consensus in the conceptualization of peace, let us now look at some of the differences in peace thinking. We will scan the relevant literature on peace in international relations to see which different concepts of peace can be distinguished and how these can be organized into a useful taxonomy. To compare different conceptions of peace in terms of their unique contents, we shall use a general schema for describing the cognitive structure of the concept of peace. The assumption is that the same cognitive structure is applicable to all concepts of peace, regardless of their specific contents. The basic units of conceptual structure are the dimensions or attributes used to define a given object – in this case, international peace. Conceptual differentiation, then, is a function of the specific location of the object in a multidimensional space. This approach to the comparative analysis of peace concepts has also been used in the social-psychological study of international images (Scott 1965). Several dimensions are needed to define most of the concepts of peace that have been proposed.

4.2.1 Domain of Peace

This dimension refers to the size of the geographic area to which a particular peace arrangement applies. Over the course of history, Aron (1966: 232) observed a progressive widening of the domain of peace. "If we consider wars down through history, we cannot fail to see in them the elasticity of movement, more precisely, of the progressive widening of the zones of sovereignty, hence the zones of peace." Another example of a reference to domain can be found in the concept of Pax Africana: the specifically military aspect of the principle of continental jurisdiction, which explicitly asserts that peace in Africa is to be assured by the exertions of Africans themselves (Mazrui 1967: 203).

4.2.2 Basis of Peace

This dimension classifies different peace arrangements according to the degree of agreement about the conditions for peace between the parties to the arrangement. At one end of the continuum we would place imposed peace, that is, a peace agreed upon in an asymmetrical conflict situation where the stronger power imposes its interpretation of peace on the weaker or defeated party. At the other end of the continuum, agreement on peace is characterized by consensus based on mutual satisfaction. A good example of an imposed peace is the peace legitimized by the League of Nations after World War I: "The League was not so much to keep peace, but to keep a specific peace – to legitimize and stabilize a particular world settlement based upon victory" (Claude 1959: 49). All peace arrangements based on power can be placed here. Aron distinguishes between three forms of peace based on power. The first is peace by equilibrium, which can also be called peaceful coexistence or cold peace. This type of peace refers to a prolonged absence of active hostility among states that are, however, so suspicious and dissatisfied with each other that they continuously engage in hostile activities short of actual war. A variant of peace by equilibrium is peace by terror or impotence.

> The combined progress of techniques of production and destruction introduces a principle of peace, different from power, which usage has already been baptized. Peace by terror is that peace which reigns (or would range) between political units each of which has (or would have) the capacity to deal mortal blows to the other. In this sense, peace by terror could also be described as peace by impotence (Aron 1966: 159).

The second form of peace based on power is peace by hegemony. In this case, the asymmetry between the parties is such that "the unsatisfied states despair of modifying the status quo, and yet the hegemony state does not try to absorb the units reduced to impotence. It does not abuse the hegemony, it respects external forms of state independence, it does not aspire to empire" (Aron 1966: 151). This form of peace is also very much alive and part of the present state of international relations. The third kind of peace based upon power is imperial peace. Over the course of human history we find such arrangements as *Pax Romana*, *Pax Britannica*, *Pax Germanica*, *Pax Americana*, *Pax Russica*, and so on. In each of these cases the terms of peace were dictated by the imperial power. The aims of imperial peace were very pungently expressed by Joseph Chamberlain in 1897:

> In carrying out this work of civilization we are fulfilling what I believe to be our national mission, and we are finding scope for the exercise of those faculties and qualities which have made us a great governing race. In almost every instance in which the rule of the Queen has been established and the great *Pax Britannica* has been enforced, there has come with it greater security to life, and property, and material improvement in the condition of the bulk of the population (Bennett 1953: 138).

Hitler explained his recipe for peace even more blatantly:

> Who really would desire the victory of pacifism in this world, must work with all his power for the conquest of the world by the Germans. ... Actually the pacifist humanitarian idea will perhaps be quite good, when once the master-man has conquered and subjected the world to a degree that makes him the only master of the world (Kohn 1942: 141–42).

4.2.3 Value of Peace

Here we are looking at the comparative value of peace, defined as the absence of physical violence. The relevant question is: Does the author give the highest priority in his value scale to the preservation of life? Deutsch (1972: 9), for example, in a strikingly explicit statement, affirms the absence of physical violence to be the predominant value in international relations since all other values are meaningful only among the living: "There is no pursuit of values among the dead. If mankind does not survive, discussions about how to make life freer, nobler, most just, or more aesthetically satisfying, would become worse than academic." There are, however, other thinkers about peace who think otherwise. For Jaspers (1971), freedom has priority, for neo-Marxists, justice, and for classical peace theorists, the protection of a power position.

4.2.4 Peace as a Normal or Abnormal State of International Relations

Distinctive beliefs about the nature of international peace are another source of variation in peace thinking. In his book *The Troubled Partnership*, Kissinger (1965: 244) criticizes the belief held in the United States that peace is natural and crises are caused by personal will and not by objective conditions. Instead he warns that "no idea could be more dangerous than to assume that peace is the normal pattern of relations among states. A power can survive only if it is willing to fight for its interpretation of justice and its conception of vital interests."

4.2.5 Peace as Means or End

For some, peace is considered a means; for others, it is the goal of foreign policy. The conception of peace as a means is clearly illustrated in the following statement by Ben-Gurion in 1936:

> An agreement with the Arabs is necessary for us, not in order to create peace in the country. Peace is indeed vital for us. It is not possible to build up a country in a situation of permanent war, but peace for us is a means. The aim is the complete and absolute fulfilment of Zionism. Only for this do we need an agreement (Chomsky 1974: xii).

At the other end of the dimension we find peace conceptualized as the main goal towards which international relations should be directed. The description of the purpose of the United Nations in its Charter (1945) illustrates this approach to peace:

> The purposes of the United Nations are … to maintain international peace and security, and to that end: to take effective collective measures for the prevention and removal of threats to peace, and for the suppression of acts of aggression or other breaches of peace, and to bring about by peaceful means, and in conformity with the principles of justice and international law, adjustment or settlement of international disputes or situations which might lead to a breach of the peace.

4.2.6   Scope of Peace

This dimension distinguishes different peace concepts in terms of the range of conditions that have to be present in order to define a situation as peaceful. At one end of the scale we would locate the definition of peace as the absence of physical violence, which is sometimes referred to as negative peace. Here peace is defined as the absence of war, and peace policy is aimed at the prevention of wars. Peace seen in this light is primarily concerned with questions of strategy, balance of power, peacekeeping, pacification, arms races, formation of alliances, and collective security systems. Research aiming at offering a solution to the problems of war and peace concentrates primarily on the conditions of war, giving only minimal attention to the conditions of peace: "A mentality prevails which tends to neglect the possibility that wars may result, not from factors present in the war situation, but from the absence of factors safeguarding peace" (Levi 1964: 23). A second blinder of the negative peace definition is its built-in conservative bias. Negative peace is the kind of peace that is envisaged by the law-and-order-orientated person (Schmid 1968: 223). As Galtung (1967: 2) has noted, peace research defined solely in terms of negative peace

> will therefore easily be research into the conditions of maintaining power, of freezing the status quo, of manipulating the underdog so that he does not take up arms against the topdog. This concept of peace will obviously be in the interests of the status-quo powers at the national and international level, and may equally easily become a conservative force in politics.

In summary, we could say that negative peace is quite a narrow definition, treating qualitatively different situations as peaceful as long as no symptoms of physical violence are observed.

At the other end of the scale we find concepts of peace that arose in protest to the confining character of the "negative peace" paradigm. Hveem (1973) distinguishes three schools, each of which proposes an alternative definition of peace. The first school, which he calls "early protest peace research", adds positive peace to the absence of violence as an ingredient of the concept. The term "positive peace" refers to the unification of mankind into a cooperative enterprise on a world scale (Rapoport 1971: 92). In this school of peace thinking, the absence of war still has priority over the positive component of the definition. Military force is still considered the *condition sine qua non* for the achievement of positive peace. The primary means advocated for the achievement of positive peace are impartial mediation, non-violent strategy, peace expertise, and conflict management or resolution. In contrast to the two other concepts that we will discuss below, the term "positive peace" is defined in subjective, rather than structural, terms. The definition of positive peace as cooperation on a world scale or brotherly harmony of all humankind seems to be devoid of meaning and allows people to endow it with their own subjective values.

A reaction against this subjective character of the concept of peace was crystallized in the "constructivist school of peace" researchers. They defined peace as the absence of physical as well as structural violence. Structural violence refers to a "structure which perpetuates a situation where some members of society are permitted to realize their aspirations at the same time as others are dying from non-existent health facilities, lack of protein or generally the most basic standards necessary for staying alive" (Hveem 1973: 106). To put it in other terms, structural violence is inequality in the distribution of power and resources. Therefore, for this school of thought, peacemaking implies structural transformations, mediation in symmetrical conflicts, and support for the underdogs in asymmetrical conflicts. The supporters of this structural definition of peace are split on two issues: the objects of their primary loyalty and their strategic approach. The 'constructivists' take the individual in world society and the victims of violence as their objects of loyalty and non-violence as their strategy. The 'neo-Marxians' take the proletariat of the world as their main object of loyalty and do not see violence as an a priori negative. They argue that violence can play a dual role: a reactionary one and a progressive one. That is why Marxists reject all theories or views that categorically repudiate violence: "Marxism recognizes the justification of the use of violence provided such use is relevant to historical progress. At the same time, however, Marxism rejects all theories that overestimate the role of violence in history, notably its sharpest form – war" (Kara 1968: 4–5).

One of the most recent contributions to peace research is characterized by an enlargement of the scope of peace and greater stress on the existence of different international perspectives in peace research. This new strand of research is represented most heavily in the World Order Model Project (Mendlovitz 1975). Participants in the project believe that peace cannot be realized unless improvements are made with respect to poverty, social justice, ecological instability, and alienation or identity crises. Apart from recognizing the mutual impact of these five values, they also stress that the weights attached to these values vary as a function of the social class of the proponent (Mendlovitz 1975: 296). In the growth process of the project, "It became increasingly clear that while peace, in the sense of elimination of international violence, might have a very high priority with individuals in the industrialized sector of the globe, economic well-being and social justice received a much higher rating in the third world" (Mendlovitz 1975: x). Therefore, instead of defining their object of study as 'peace' they chose to inquire into preferred world orders. This concept implies the recognition of multiple values and of the existence of different preferences.

A preferred world order is defined as "the relevant utopia, selected by a proponent because it is most likely to realize his or her goals" (Mendlovitz 1975: xiv).

4.2.7   Peace as a Perpetual Process or as an End State

Peace at one end of the dimension is equated with a relatively well-defined world order. People who subscribe to this kind of peace thinking believe that the realization of a particular world order would bring peace. For them, peace is a specific end state that should be realized. At this end of the dimension we could locate, for example, most of the world government models of world order.

At the other end of the dimension, peace is perceived as a relative concept; not as a particular end state, but as a continuous process. In contrast to the rather static quality of the first peace research group's definition of peace, the definition of the latter is much more dynamic. For them, the content of peace is always susceptible to change, and its realization is a perpetual task of all the partners involved. They also reject all abstract normative formulas for peace, which they consider to be unrealistic. Instead they recognize the existence of different levels of peace and propose that the achievement of one level generates claims for a new level of peace (t'Hart 1955).

4.2.8   Use or Non-use of Physical Violence to Achieve Peace

People concerned with the phenomenon of peace can also be distinguished according to their attitude towards the use of violent or non-violent strategies in its realization. Fanon is an example of a believer in physical violence as necessary to abolish structural violence and bring about peace of mind. He clearly prefers victory to survival and sees violence as a cleansing force: "It frees the native from his inferiority complex and from his despair and inaction; it makes him fearless and restores his self-respect" (Fanon 1968: 94). At the other end of the dimension we find proponents of non-violence who base their argument on moral grounds, together with non-violence strategists like, for example, Sharp (1973).

The foregoing set of dimensions does not do justice to all the variations in peace thinking, but it gives us an idea of the pluralistic approaches in the field. The next section will look at some effects of the existence of different conceptions of peace on the state of international relations.

4.3   Conceptions of Peace and International Relations

The impact of different conceptions of peace is clearly indicated by several students of international relations. First, these conceptions account for the unstable state of the international system. The precariousness of the world order is not a mere temporary de facto failing, but rather follows from divergent conceptions of that order. Hoffmann (1968: 16) points out clearly that the order of the world is a product of people's creation, and that "the diversity of conceptions, that is, of value preferences, concerning the ultimate directions and purposes accounts for the fragility of our order; and the existence of common directions accounts for its possibility". Further, differences in the conceptualization of peace account for the existence of opposition. The quotation from Mo-ti, a classic Chinese author, selected by F. Northrop as an epigraph for his book, *The Meeting of East and West* (1946), still retains its validity: "Where standards differ, there will be opposition."

Third, the existence of different ideas about peace and what constitutes a right world order accounts for the fact that any world order that is not based on the consensus of all the interested parties has no insurance for permanence. This is in part due to the various ideas about what constitutes the right world order (Gadamer 1968: 325). All of these authors agree that some measure of agreement is a prerequisite for a viable human community. In the absence of some degree of consensus on what is peaceful and not peaceful, just and unjust, it is clearly more difficult for countries to coordinate their plans effectively and ensure that mutually beneficial arrangements are maintained.

4.4   Advantages of a Systematic Study of International Values

The linkage between the concept of peace and other international values, such as preferred world order and international relations, suggests the importance of knowing the precise content of different conceptions of peace and different visions of a preferred world order. Let us highlight some advantages for a systematic description of the content of differing conceptions of peace and their comparative analysis. (The same considerations apply to the description of different visions of a preferred world order.) First, concepts of peace can be used as indicators in the study of international relations. Changes in concepts of peace could reflect changes in international relations. We would expect a world where most diplomats defined peace as the absence of structural and physical violence to be different from one in which peace was predominantly defined as the continuation of struggle by other means. Second, concepts of peace could be used as criteria for the evaluation of the performance of international systems. A better knowledge of peace thinking in the diplomatic world could help us understand, for example, the different appraisals of the performance of the United Nations, expressed by different countries or groups of countries. A third contribution of conceptual clarification is the elimination of the frequent dialogue between the deaf in international relations: "Peace is so emotive a term, one which lends itself so easily to political propaganda and abuse, that if it is to be used as a tool in intelligent discussion – especially international discussion – it must be precisely defined" (Howard 1971: 225). Fourth, the clarification and recognition of differences in the operationalization of the value of peace would promote a more pluralistic attitude. "Pluralism encourages both scepticism and innovation and is thus inherently subversive of the taken-for-granted reality of the traditional status quo" (Berger/ Luckmann 1966: 125). Fifth, a comparative study of peace would enhance the understanding of the phenomenon. The object of thought becomes progressively clearer with the accumulation of different perspectives on it (Mannheim 1936). Sixth, information about differences in the concepts of peace or preferred world order can be used in the assessment of the future of international relations.

To determine the feasibility of alternative international or supranational structures, it is useful to know the precise content of different aspirations for the world and the images of the future which they generate. Having highlighted some of the advantages of studying international values, let us now look at how an important group of international actors, the diplomats, operationalize two prime international values: peace and preferred world order.

4.5   Peace and Preferred World Order

Data on the conceptions of peace and visions of preferred world order among diplomats will be presented in four parts. In the first part we shall explore the indicators used by diplomats to assess the existence of peace in international relations. To that end, we have analysed the diplomats' answers to the following question posed during the interview:

> The term peace is one of the most loosely used terms in international relations; to you personally, what conditions do you think have to be present in order to call international relations peaceful – or, in other words, what do you consider to be indicators of peace?

We will review the answers given by the total sample, and then contrast the indicators of peace used by diplomats from developing and developed countries. We will focus on differences in the rank-ordering of the indicators and test the significance of these differences. In the second part we shall turn to the diplomats' views of peace in the present state of international relations. Analyses are based on the question:

> What are the three or four most important problems in the world today which cause dissatisfaction and tension, and are directly or indirectly a threat to peace?

This question was intended to elicit a more concrete definition of peace, in terms of conditions in the world today, that the diplomats perceived as inhibitors of the realization of a peaceful world order. In the third part we shall examine the diplomats' views on the possibilities for peace. We presented the diplomats with a representative set of peace proposals and asked them to evaluate each in terms of its importance for international peace and its probability of being carried out in the short or long term. In contrast to the analysis in the second part, where we looked at the inhibitors of peace, here we inquired into the factors the diplomats considered to be conducive to peace. We will focus first on the results for the total sample and then contrast the responses of the diplomats from developed and developing countries. Among the variables to be compared are the rank-ordering of the proposals in terms of importance, the most- and least-trusted proposals, the level of orientation (global, interregional, intraregional, national, subnational/individual), and the degree of incongruence between the importance assigned to the proposals and the assessment of the probability of their being realized. The final set of data is relevant to what we consider to be a necessary condition for the creation of a satisfactory international peace, namely consensus among all countries about a preferred world order. Information about the diplomats' visions of a preferred world order was evoked by the question:

> How would the world you'd like to see for your children differ from that of today, or, how would you describe your preferred world order?

4.5.1 Some Propositions

In our data we will check the validity of the following six propositions:

Proposition 1 There is no consensus among diplomats on such basic values as peace or preferred world order. In other words, these two values cannot be considered universal values, at least not at this stage in the history of diplomatic thinking.

Proposition 2 Values held by a particular diplomat are a function of the systemic position of his/her country. For example, we would expect the "absence of structural violence and the feeling of deprived status or atimia" (Lagos 1963) to be more central in the thinking about peace among diplomats from developing countries than among those from developed countries.

Proposition 3 The gap between the present state of international peace and the ideal state is considerable. Hence we predict a marked difference between the optimal and realistic peace profiles of the diplomats. The term "optimal peace profile" refers to the diplomat's ranking of a set of peace proposals in terms of their importance for the realization of peace; the term "realistic peace profile" refers to the ranking of the same proposals according to whether or not they might be carried out in the long or short term.

Proposition 4 The systemic position of a country does not affect or affects only marginally the assessment of the likelihood of a peace proposal being carried out; it does, however, have an impact on the rank-order of peace proposals in terms of their importance or preference.

Proposition 5 The discrepancy between optimal and realistic peace profiles is greater for diplomats from developing countries than from developed countries. This proposition is based on the following two expectations: that both groups of diplomats will differ less with respect to their assessment of what they can get (realistic peace profile) than in relation to their wants (optimal peace profile), and that diplomats from developing countries will tend to articulate a greater number of wants than diplomats from developed countries.

Proposition 6 Middle-range peace proposals (proposals about changes at the interregional and intraregional level of the international system) tend to be more trusted and hence perceived as operational than proposals at the global or national or subnational levels.

4.5.2 Indicators of Peace

Responses to the question about what the diplomat would consider to be indicators of peace – i.e., conditions that must be present if international relations are to be called peaceful – were grouped into meaningful categories in Table 4.1. This table presents the categories into which the indicators were grouped and rank-orders them according to the frequency of responses in each category for the total sample of diplomats.

The predominant indicator of peace was, not unexpectedly, the absence of violence. It topped the list of the total sample and of the two subsamples. A t-test

Table 4.1 Indicators of peace for total sample of diplomats (frequencies). *Source* Compiled by the author from his own research

Absence of violence	114
International cooperation	47
No domestic interference	37
Absence of structural violence	33
Economic welfare	28
Understanding	26
Respect and dignity	24
Justice	22
Symmetrical relations	22
Detente	21
Military security	21
New economic order	18
Equality	17
Stability and order	15
Dialogue	13
Conflict management	13
Friendly relations	13
Arms control and disarmament	11
Existence of common interests	10
Pluralism	9
Empathy	9
Freedom	9
Integration	9
International law	5
Ecology	2

for the significance of the difference between the means on this indicator for the two groups of diplomats yields a t-value of 2.07 (p < .02), This finding suggests that, despite the fact that the absence of violence is most salient for both groups, on a comparative basis it is significantly more salient for diplomats from developed than from developing countries.

We would also expect diplomats from those countries to place greater emphasis on "stability and order" among their indicators of peace, along with the absence of violence. As Galtung (1967: 2) has noted: "This concept of peace will obviously be in the interests of the status-quo power at the national and the international levels, and may equally easily become a conservative force in politics." This expectation was corroborated by the finding that the stability and order indicator ranks third among indicators for the developed countries and is not among the top ten indicators for developing countries. A t-test for the means of this indicator yields a t-value of 1.74 (p < .05), indicating a significantly higher salience of this indicator for diplomats from developed than from developing countries.

The second-ranking indicator for the total sample of diplomats was international cooperation. All responses that referred explicitly to cooperation or coordination in different areas were placed in this category. This indicator ranks high for both groups of diplomats – in second place for developed, and in fourth place for developing countries. Further, the difference between the two groups, although large, falls short of significance. This finding indicates that diplomats from both developing and developed countries perceive the existence of patterns of cooperation as important indicators of peace. The meaning of the international cooperation indicator is less clear than the meaning of the absence of violence or stability and order indicators, because we lack information about the terms of cooperation. However, we could say that this indicator enriches the definition of peace of the diplomats by adding a component that is positively valued by all.

The indicator on which the diplomats from developing and developed countries differ most significantly is called "absence of structural violence". Examples of responses that were so categorized are "peace is the absence of: exploitation, economic imperialism, colonialism, imperialistic and hegemonistic attitudes, or a dominated race in the world". All of these statements explicitly affirm a negative association between the existence of an exploitative hierarchical structure and peace. What distinguishes diplomats from developing countries most from their colleagues from developed countries is the high saliency of the "absence of structural violence" indicator. The t-test between the means of the two groups yields the highest t-value (see Table 4.2), with a significance level of less than .005. Further, the absence of structural violence indicator ranks second in the list for developing countries and is not among the top ten indicators for developed countries. This implies that a world with built-in structural violence, but also characterized by the absence of physical violence and by cooperation, would be perceived as less peaceful by diplomats from developing countries than their colleagues from better-off countries.

Consistent with the high ranking of the absence of structural violence (2nd) by diplomats from developing countries is the almost equally high saliency for them of

Table 4.2 Indicators of peace for diplomats from developed and developing countries. *Source* Compiled by the author from his own research

	Means*			Means*
Developed countries, N = 102 (>8)		Developing countries, N = 117 (<8)		
1 Absence of violence	.529	1	Absence of violence	.367
2 International cooperation	.254	2	Absence of structural violence	.162
3 Stability and order	.098			
4 Justice	.078	3	No domestic interference	.153
7 No domestic interference	.068	4	International cooperation	.136
7 Economic welfare	.068	5	Economic welfare	.128
7 Understanding	.068	6	Symmetrical relations	.119
7 Symmetrical relations	.068	7.5	Understanding	.111
7 Military security	.068	7.5	Respect and dignity	.111
10 Dialogue	.058	9.5	Justice	.102
		9.5	Equality	.102
Developed versus developing countries		t-value		p-value <
1 Absence of structural violence		−2.99		.005
2 Absence of violence		−2.07		.02
3 Friendly relations		−2.02		.02
4 Respect and dignity		−1.82		.05
5 No domestic interference		−1.78		.05
6 Stability and order		−1.74		.05
7 New economic order		−1.66		.05
8 Freedom		−1.64		.05
9 Equality		−1.58		.10
10 Empathy		−1.56		.10

*Obtained by dividing the total number of responses provided by the diplomats from each type of country by the number of diplomats in the subsample

Note In the interpretation of all scores arrived at by means of content analysis based on mention versus non-mention (rather than intensity of mention) as measures of saliency of the measured object in the minds of diplomats, I invite the reader to join me in exercising reasonable caution. The difference between the scores of the diplomats from developed and developing countries may have been influenced by the fact that the latter group tended to talk more so that the theme was more likely to emerge. The data are not controlled for this factor

"no domestic interference" (rank 3). Comparison of the means for developed and developing countries yields a t-value of −1.78 (p < .05). To clarify the meaning of no domestic interference, let us illustrate some of the responses that were placed in this category. The statements included "peace means non-interference in internal affairs of countries; respect for sovereignty; self-determination; abstaining from criticizing the internal policy of other countries; and live and let live". The high saliency of this indicator among diplomats from developing countries could be explained by their greater sensitivity to and dissatisfaction with the experience of being asymmetrically penetrated. "Centre countries are comparatively better off than periphery nations in avoiding strong asymmetric penetration" (Hveem 1972: 72).

Having highlighted some of the most important indicators of peace, let us now quickly look at the other indicators and their similarities or differences in saliency among the two sub samples of diplomats. In the top-ten list of both subsamples we find four indicators which show only non-significant differences: justice, economic welfare, understanding, and symmetrical relations. The meanings of the first three are self-evident. The fourth, symmetrical relations, refers to all the responses of diplomats which stressed mutuality and equilibrium as distinct qualities of inter-national peace. Although the means of these four indicators are not significantly different, we notice that they are consistently higher for developing countries, which would suggest that these indicators are relatively more important for diplomats from this group of countries. The two items on the list for developed countries that do not appear on the other list are military security and dialogue. Despite their inclusion in this top-ten list, their means do not differ significantly from the means of the same indicators for developing countries. Therefore, we would tend not to consider them as sources of variation among diplomats in their assessment of the peacefulness of the international environment. The two indicators on the list of diplomats from developing countries that do not appear on the previous list are: respect and dignity; and equality. The saliency of the respect and dignity indicator is significantly higher for diplomats from developing countries than for their colleagues from developed countries (t = −1.82, p < .05). The mean of the equality indicator, though not significantly different, is clearly higher and therefore suggests that diplomats from developing countries also assign greater importance to this indicator. Not included in either top-ten list but clearly different are the friendly relations, freedom, and empathy indicators. The first two are significantly more salient for diplomats from developing countries.

What do these results tell us? They illustrate without doubt that diplomats have different "sets of indicators" for assessing the degree of peacefulness of interna-tional relations, depending on the systemic position of their countries. This means that different appraisals of the degree of peacefulness of a particular international situation are probably based on the use of different sets of indicators; and similar appraisals could be based on the use of different or similar sets of indicators. Our understanding of the perceived peacefulness of the world should become clearer when the existence of different sets of indicators is taken into account.

4.5.3 Inhibitors of Peace

We now turn to some of the conditions in the world today that diplomats consider to be antagonistic to peace. As mentioned earlier, the data for this analysis are derived from responses to the following question:

Looking at the world today, what are the three or four most important problems which cause dissatisfaction and tension, and are directly or indirectly a threat to peace?

As in the case of the indicators of peace, we listed all the answers and grouped them into meaningful categories. An overview of these categories for the total sample of diplomats can be found in Table 4.3, and all the data that we will use for contrasting diplomats from developed and developing countries are presented in Table 6.4. For diplomats from both developing and developed countries, the

Table 4.3 Sources of dissatisfaction and direct or indirect threats to international peace for the total sample (frequencies). *Source* Compiled by the author from his own research

Middle East conflict	76
North-South polarization	60
Structural violence	40
Energy problem	35
Domestic and regional interference	34
State of the international economic system	33
Big-power relations	33
Unequal distribution of wealth	33
Racism and discrimination	30
Economic conditions	29
Food crisis	27
Ideological approach in foreign politics	26
Population growth	20
Limits of growth	20
Nationalism	20
Economic development	19
Cultural conflicts	18
Detente	18
Cyprus	18
East-west relations	17
Misperception	16
Relations between communism and capitalism	16
Expansionism	15
South Africa	14
South-East Asia	13
Power politics	11
Communication problems	10
China-Soviet Union	10
Boundary conflicts	10
Internal problems	9
Balance of power	8
Inadequate adaptation to new realities	8

Middle East conflict is perceived as the condition most antagonistic to peace. This could be explained by the fact that the Middle East is seen by most diplomats as the cockpit for multiple competitions between the superpowers, North and South, different races, Eastern and Western cultures, and different religions. The second condition that is perceived as a source of dissatisfaction and tension is North-South relations. The frequency with which this problem is mentioned is similar for both developing and developed countries. The reasons for the dissatisfaction expressed by the two groups of diplomats are, however, different. From several interviews I could infer that in addition to humanitarian considerations, the basic reason for the dissatisfaction of the Western developed countries was the impact of the North-South conflict on the energy market and the consequent economic and financial problems (Table 4.4).

This serious concern with the energy crisis is reflected in the high ranking of this problem in the list of diplomats from developed countries. On the other hand, for diplomats from developing countries, the North-South conflict is perceived as antagonistic because of its association with structural violence, domestic and regional interference, the unbalanced state of the international economic system, and their general dissatisfaction with their economic condition and state of development – problems that are most salient for diplomats from developing countries. Structural violence, the unbalanced state of the international economic system, and domestic and regional interference all appear in the top-ten list for diplomats from developing countries; their means are also significantly different from those of the diplomats from developed countries.

I will clarify the meaning of each of these categories by specifying which responses they include. All statements that explicitly disapprove of economic imperialism, penetration, intervention, colonization, or a system of domination with controllers and controlled were classified as structural violence. Under the "unbalanced state of the international economic system" category, we grouped such problems as inequitable terms of commerce and trade, artificial barriers to free exchange of goods, unfair prices for commodities, cartels formed to control exports, biased market, degradation of the terms of exchange, the impact of economic difficulties experienced by industrialized countries in their economic relations with developing countries, and so on. The category of "domestic and regional interference" includes all statements that explicitly mentioned domestic or regional interference as a threat to international peace. The dissatisfaction of diplomats from developing countries with their state of economic development, although it does not appear in their top-ten list, has a mean of .111, which is significantly higher than for diplomats from developed countries (t = −2.43, p < .01).

Having highlighted some of the most important conditions that are perceived as antagonistic to international peace, let us now quickly look at the other problems mentioned by one or both groups of diplomats. In the top-ten lists of both sub-samples of diplomats, we find three problems that are more or less equally salient for both: big-power relations, unequal distribution of wealth, and food crisis. The meaning of the last two problems is self-evident, but the problem of big-power relations requires some clarification. This problem refers essentially to relations

Table 4.4 Sources of dissatisfaction and direct or indirect threats to international peace for diplomats from developed and developing countries. *Source* Compiled by the author from his own research

		Mean
Developed countries (≥ 8)		
1	Middle East conflict	.323
2	North-South relations	.225
3	Energy problem	.186
4	Big-power relations	.127
6	Unequal distribution of wealth	.107
6	Nationalism	.107
6	East-West relations	.107
9	Economic conditions	.102
9	Food crisis	.102
9	Détente	.102
Developing countries (≤ 8)		
1	Middle East conflict	.282
2	Structural violence	.239
3	North-South relations	.230
4	State of the international economic system	.213
5	Domestic and regional interference	.170
6	Economic conditions	.145
7.5	Big-power relations	.128
7.5	Unequal distribution of wealth	.128
9.5	Racism and discrimination	.119
9.5	Food crisis	.119

Developed versus developing countries		t-value	p-value <
1	State of the international economic system	−4.09	.005
2	Structural violence	−2.79	.005
3	China-Soviet Union	−2.49	.01
4	Economic development	−2.43	.01
5	Cultural conflicts	−2.05	.02
6	Domestic and regional interference	−2.00	.02
7	East-West relations	−1.80	.05
8	Misperception	−1.73	.05

between the Soviet Union and the United States, especially as they affect the security of the world, the evolution of conflicts in their spheres of influence, and also the impact of these relations on the economic life of the Third World countries.

The four problems on the list of diplomats from developed countries that do not appear on the other list are energy, nationalism, East-West relations, and detente. With respect to these problems, the means for diplomats from developed countries are higher than those for their colleagues from developing countries. Only one, however, is significantly different: that for East-West relations ($t = 1.80$, $p < .05$). The only other problem on the list of diplomats from developing countries that does not appear on the list for the developed-country diplomats, and has not already been mentioned, is racism and discrimination. Some of the specific points mentioned by diplomats which were placed in this category are: the joke of ethnic superiority; the white domination in Africa; pockets of racism; the great unsolved problem of China and the Soviet Union, wherein racism is very strong; the education of antisemitism; and institutional racism. The mean for this problem is clearly higher for diplomats from developing countries, although it falls short of statistical significance. The trend suggests greater saliency of this problem for diplomats from those countries. Not included in either top-ten list, but significantly different for the two groups, are the China-Soviet Union, misperception, and cultural conflict problems. The first two problems are much more salient for diplomats from developed countries; the last problem, however, is more salient for developing countries. The meaning of the misperception problem can best be communicated by listing some of the points mentioned by the diplomats that were placed in this category: the misperceptions between the two superpowers and, in consequence, possible miscalculations; the yes-no thinking of the military; the distorted newsreels about the Middle East; the gap in international understanding; the tendency to infer, for example, from the knowledge of a person the character of a whole people; the frictions caused by cultural misconceptions; the existence of indoctrination; the myth of oil money; the tendency to deny real problems; and the misperception by leftists of communist dictatorships. The dissatisfaction of diplomats from developing countries with the state of cultural relations was indicated by their comments about issues such as cultural imperialism, cultural revolution, cultural rigidity, cultural gap, cultural condescension, and inability to communicate cross-culturally.

What do these results tell us about conditions antagonistic to peace? We have noticed a consensus among all diplomats about the antagonistic impact on peace of the Middle East conflict, the North-South polarization, big-power rivalry, unequal distribution of wealth, economic conditions, and the food crisis. In contrast to this consensus, we also find some significant differences, especially with respect to the conditions of the international economic system. For example, for developing countries, the dissatisfaction with their position in what we could call "the global dominance system" is very prominent. This and other differences in the perception of conditions antagonistic to international peace add support to the proposition that diplomatic thinking, here specifically with respect to the issue of peace, is a function of the international position of the diplomat's country.

4.5.4 Conditions for Peace

The third aspect of thinking about peace is the diplomat's peace profile. The questions and the conceptual framework for analysis were adapted from Hveem

(1968), who used them in a comparative study of foreign-policy thinking in Norway. The diplomats' peace profiles could be operationalized as their responses to a set of stimuli which represent concrete peace proposals. The set of proposals consisted of twenty-four items, sixteen of which were taken from Hveem. The list contains a representative sample of important and relevant peace proposals that have been put forward. In Table 4.5, the numbers at the left indicate the order in which the proposals were presented to the respondents; these numbers will later be used to represent the particular proposal.

Table 4.5 The optimal peace profile of the diplomats (percentage distribution of responses to 24 peace proposals, N = 200). *Source* Compiled by the author from his own research

Proposal item number	Proposal	Especially important to peace	Somewhat important to peace	Unimportant to peace	Against peace	Other	Mean
2	Abolish hunger and poverty in the world	76.6	19.9	2.5	0.0	1.0	4.4
13	General and complete disarmament must be realized	69.9	16.6	9.3	3.1	1.0	4.1
18	Western and Eastern countries must improve detente	64.6	28.8	4.5	0.5	1.5	4.2
9	More effective communication is necessary	64.3	31.2	1.5	2.0	2.0	4.1
1	The individual person must be educated to peace	64.0	26.0	7.5	1.0	1.5	4.1
19	We must strengthen the United Nations	62.9	28.4	7.6	0.5	0.5	4.0
12	Rich countries must give aid to the poor	60.7	34.7	3.1	0.0	1.5	4.1
23	States that naturally belong together must cooperate	50.6	42.1	5.1	1.1	1.1	3.8
22	Ideological disputes ought to be diminished	45.5	40.1	13.4	0.5	0.5	3.6

(continued)

Table 4.5   (continued)

Proposal item number	Proposal	Especially important to peace	Somewhat important to peace	Unimportant to peace	Against peace	Other	Mean
3	Relations between individuals must be more peaceful	45.5	33.3	19.2	0.0	2.0	3.5
8	Population growth must be controlled	43.1	37.1	16.2	2.0	1.5	3.4
16	There must be a military balance between states so that nobody dares to attack	40.8	36.7	9.2	12.8	0.5	3.2
6	We must strengthen regional organizations	40.1	45.7	12.2	0.0	2.0	3.5
5	The states must be more democratic	40.1	31.0	26.9	0.0	2.0	3.2
20	National boundaries ought to become more open	38.3	40.8	16.3	2.0	2.5	3.3
15	Free trade must be established between all countries	35.5	48.6	11.2	3.7	0.9	3.3
4	The small countries must have greater influence	33.7	45.1	16.6	3.1	1.5	3.2
24	Restrictions to migration should be gradually lifted	27.3	34.1	31.8	6.8	0.0	2.7
21	Each national group should be given its own country	19.3	27.3	23.0	29.4	1.1	2.0
10	The military alliances must be preserved	18.8	35.4	17.2	28.6	0.0	2.1
7	A world government must be established	18.4	23.8	46.5	10.3	1.0	2.1

(continued)

Table 4.5 (continued)

Proposal item number	Proposal	Especially important to peace	Somewhat important to peace	Unimportant to peace	Against peace	Other	Mean
14	Countries must become more socialistic	14.4	33.7	46.2	4.8	1.0	2.2
11	The nations must become more similar	12.7	27.5	48.1	11.1	0.5	1.9
17	One should create a world with small self-sufficient states	12.7	20.6	45.1	20.6	1.0	1.7

The proposals are ranked according to the degree of favourable responses elicited from the total sample of diplomats, that is, according to their assigned importance. This ranking gives us an idea of the ideal conditions for peace as seen by the diplomats, which will be called their optimal peace profile. To permit examination of differences in peace profiles as a function of the position of the diplomat's country, Table 4.6 presents the profiles of diplomats from developed and developing countries. Significant differences in the judgements of these two groups are listed in Table 4.7. Let us first look at the peace profile of the total sample of diplomats. It is fair to consider the proposals that received a mean of 4 or higher on a 0–5 scale as the main ingredients of an 'ideal' world. Thus, an ideal world is one without poverty, with general and complete disarmament, with reduced tension between East and West, and with a stronger United Nations, and, finally, one in which the individual is considered a relevant unit of concern in the building and preservation of peace. In contrast to these ingredients of an optimal international peace, we find at the bottom of the list a cluster of items which are perceived as not important to or even inhibiting the achievement of international peace in the short and long term. Seven items received a mean judgement below 3. Most of these items (17, 11, 14, 7, 10, 21, 24) reflect the perception of the diplomats that diminution of the nation-state's internal and external control would have minimal or even negative consequences for international peace.

Table 4.6 shows to what extent this general profile reflects the preferences of diplomats from developed and developing countries. If we focus on the top five proposals for each group, we find that both groups included items 2, 9, and 12 in this list. However, the importance of each of the three items was stressed more by diplomats from developing than from developed countries. As can be seen in Table 4.7, the means on these three items were significantly higher for diplomats from developing countries. These proposals refer to the alleviation of the poverty problem and the improvement of communication between countries. Two proposals rank among the top five for diplomats from developed countries only, but the means

Table 4.6 Conditions for peace, ranked by importance, as judged by diplomats from developed and developing countries (mean judgements of importance, ranked for each group). *Source* Compiled by the author from his own research

Proposal item number	Proposal		Developed countries		Developing countries
18	Western and Eastern countries must improve detente	(1)	4.25	(7)	4.09
2	Abolish hunger and poverty in the world	(2)	4.16	(1)	4.80
9	More effective communication is necessary	(3)	3.91	(5)	4.36
12	Rich countries must give aid to the poor	(4)	3.89	(4)	4.36
23	States that naturally belong together must cooperate	(5)	3.87	(8)	3.95
1	The individual person must be educated to peace	(6)	3.86	(6)	4.25
8	Population growth must be controlled	(7)	3.78	(13)	3.41
16	There must be a military balance between states, so that nobody dares to attack	(8)	3.63	(17)	3.06
13	General and complete disarmament must be realized	(9)	3.61	(2)	4.62
19	We must strengthen the United Nations	(10)	3.51	(3)	4.51
22	Ideological disputes ought to be diminished	(11)	3.43	(10)	3.81
20	National boundaries ought to become more open	(12)	3.36	(16)	3.39
6	We must strengthen regional organizations	(13)	3.29	(11)	3.74
15	Free trade must be established between all countries	(14)	3.22	(15)	3.39
24	Restrictions to migration should be gradually lifted	(15)	3.16	(18)	3.05
5	The states must be more democratic	(16)	3.13	(14)	3.40
3	Relations between individuals must become more peaceful	(17)	2.96	(9)	3.93
10	The military alliances must be preserved	(18)	2.80	(24)	1.67
4	The small countries must have greater influence	(19)	2.75	(12)	3.62
7	A world government must be established	(20)	1.92	(22)	2.34
14	Countries must become more socialistic	(21)	1.90	(19)	2.58
21	Each national group should be given its own country	(22)	1.81	(21)	2.27
11	The nations must become more similar	(23)	1.78	(20)	2.28
17	One should create a world with small self-sufficient states	(24)	1.06	(23)	2.06

Table 4.7 Significant differences between diplomats from developed and developing countries in their judgements of conditions for peace (t-tests between means). *Source* Compiled by the author from his own research

Proposal item number	Proposal	t-value	p-value <
19	We must strengthen the United Nations	−5.22	.005
13	General and complete disarmament must be realized	−4.77	.005
2	Abolish hunger and poverty in the world	−4.54	.005
3	Relations between individuals must become more peaceful	−4.37	.005
10	The military alliances must be preserved	4.15	.005
4	The small countries must have greater influence	−3.88	.005
17	One should create a world with small self-sufficient states	−3.12	.005
12	Rich countries must give aid to the poor	−2.79	.005
9	More effective communication is necessary	−2.44	.01
19	We must strengthen regional organizations	−2.19	.02
16	There must be a military balance	2.16	.02
11	The nations must become more similar	−2.08	.02
1	The individual person must be educated to peace	−1.93	.05
14	Countries must become more socialistic	−1.92	.05
22	Ideological disputes ought to be diminished	−1.67	.05

of the two groups are not significantly different. These items refer to the improvement of detente and cooperation. On the other hand, the two proposals that rank among the top five for only the diplomats from developing countries do show significant differences: strengthening the United Nations (t = −5.22, p < .005), and general and complete disarmament (t = −4.77, p < .005).

Let us now focus on the five lowest-ranked proposals, which the diplomats consider to have a minimal or negative effect on the realization of international peace. Diplomats of both groups agree in their negative appraisal of four items: 7, 21, 11, and 17. There is no significant difference between the two groups of diplomats with respect to their negative appraisals of the establishment of a world government or the idea of one-nation one-state. The idea that nations should become more similar and that the world should consist of small self-sufficient states received significantly more negative judgements from diplomats from developed countries (t = −2.08, p < .02 and t = −3.12, p < .005).

One proposal which appears among the five lowest items for only the diplomats from developed countries refers to the increase of socialism in the world. In contrast to diplomats from developed countries, their colleagues from developing countries, although they also rank this item quite low, consider it significantly more positive (t = −1.92, p < .05). One proposal which appears only in the bottom-five list for diplomats from developing countries refers to the preservation of military alliances.

In contrast to their colleagues from developed countries, they perceive the existence of military alliances or blocs as contributing significantly less to peace (t = 4.15, p < .005).

Finally, let us briefly look at some other proposals on which the two groups differ significantly. The diplomats from developing countries assign significantly greater importance to items 3, 4, 6, 1, and 22. An increase in the power of small countries (4) and the strengthening of regional organizations (6) are considered significantly more important by developing than developed countries. Further, items 3, 1, and 22 suggest that diplomats from developing countries are more inclined to approve "soft proposals" than their colleagues from developed countries. This conclusion is supported by the finding that developed countries are significantly higher on two typical hard proposals: 10 and 16. For developed countries the existence of a military balance (16) and the preservation of military alliances (10) are considered to be significantly more positive conditions (t = 2.16, p < .02, and t = 4.15, p < .005).

How can we interpret these findings on diplomats' optimal peace profiles? Although I will consider the optimal peace profile as a reflection of the real preferences of the diplomats, I do so with some reservation. Instead of expressing preferences that are really felt, their judgements may be part of the diplomats' effort at impression-management. As one diplomat strikingly pointed out: "Most people will favour democracy, but for some it barely covers their acceptance of the right to die of hunger, or their preference for the perfect peace as in Greece's slave democracy." Despite the possible intrusion of lip service, we had the feeling that most of the diplomats who made personal comments about the questionnaire or answered it in my presence expressed their real personal preferences. But even if the optimal peace profiles express real personal preferences, they are not necessarily indicative of real policy preferences. According to Hveem, proposals have to pass through a filtering process before becoming policy. "One such filter is the respondent's perception of how the real world meets the ideal proposals and makes them realistic, applicable or workable – or not" (Hveem 1968: 151).

To assess the effect of the real-world filter on the process, we asked the diplomats if they believed that the respective proposals for international peace were likely to be realized, moderately likely to be realized, or unlikely to be realized in the short or long term. The results for the total sample and our subsamples of diplomats from developing and developed countries are presented in Table 4.8. If we focus first on the bottom-five items, we see that the same items appear on the lists for the total sample and the two subsamples.

This means that there is general consensus that the following five proposals are most unlikely to be realized: "one should try to create a world with small self-sufficient states", "general and complete disarmament must be realized", "each national group should be given its own country", "the nations must become more similar", and "a world government must be established as soon as possible". Supporting the conclusion that there is consensus about the five unrealistic proposals is the fact that no significant differences were found between diplomats from developed and developing countries on these items.

Table 4.8 Conditions for peace, ranked by realism, as judged by diplomats from developed and developing countries (mean judgements of realism, ranked for each group on a 1–5 scale). *Source* Compiled by the author from his own research

Item No.	Proposal	Total		Developed countries		Developing countries
10	Military alliances must be preserved	3.5	(1)	3.7	(4)	3.4
18	Western and Eastern countries must improve detente	3.3	(4)	3.4	(5)	3.2
6	We must strengthen regional organizations	3.3		3.1	(3)	3.4
23	States that naturally belong together must cooperate	3.3	(5)	3.3	(2)	3.4
9	More effective communication is necessary	3.2	(3)	3.4	(1)	3.6
24	Restrictions to migration should be gradually lifted	3.1		4.3		2.6
12	Rich countries must give aid to the poor	3.0		3.2		2.9
1	The individual person must be educated to peace	2.9		2.2		2.3
16	There must be a military balance between states so that nobody dares to attack	2.7	(2)	3.5		2.3
19	We must strengthen the United Nations	2.6		2.4		2.7
14	Countries must become more socialistic	2.6		2.9		2.6
8	Population growth must be controlled	2.5		2.4		2.8
3	Relations between individuals must become more peaceful	2.4		2.2		2.7
15	Free trade must be established	2.4		2.8		2.3
4	The small countries must have greater influence	2.4		2.3		2.6
20	National boundaries ought to become more open	2.3		2.5		2.2
2	Abolish hunger and poverty in the world	2.2		2.9		3.0
5	The states must be more democratic	2.0		1.9		2.2
22	Ideological disputes ought to be diminished	1.9		1.9		2.0
17	One should create a world with small self-sufficient states	1.6	(20.5)	1.7	(20.5)	1.6
13	General and complete disarmament must be realized	1.5	(23)	1.3	(22)	1.5
21	Each national group should be given its own country	1.5	(22)	1.5	(20.5)	1.6
11	The nations must become more similar	1.5	(20.5)	1.7	(23)	1.4
7	A world government must be established	1.1	(24)	1.1	(24)	1.2
	Average			2.4		2.4

We also find a rather high consensus on the five most realistic proposals among the two groups. Four proposals appear among the top five for both (10, 23, 18, and 6). There are also no significant differences between the means on these four items. These results indicate that there is a high consensus that the following four proposals are realistic: "military alliances must be preserved", "Western and Eastern countries must improve detente", "states who naturally belong together must cooperate", and "more effective communication is necessary".

One proposal that ranks second in the list for developed countries and does not appear on the other list refers to the existence of a military balance as a precondition for peace. A strong significant difference is found between the means of the two groups on this proposal (t = 5.48, p < .005), indicating that diplomats from developed countries attribute greater realism to this proposal. The item that ranks third in the list for the developing countries, but does not appear on the other list, is concerned with strengthening regional organizations as a path towards peace. The mean of this item is somewhat greater for developing than developed countries, suggesting that they consider this proposal more likely to be realized than their colleagues from developed countries.

Let us now look briefly at the other proposals in which the two groups differ significantly (see Table 4.9). We find one additional item to which the diplomats from developed countries assign a greater likelihood of being realized: "national boundaries ought to be more open". On the other hand, diplomats from developing countries are significantly higher on items 2, 3, 5, and 8. For them, the abolition of hunger and poverty, the control of population growth, the realization of more peaceful relations between individuals, and the increased democratization of states are proposals whose realization they perceive as more probable than do their colleagues from developed countries.

Comparison of the results of the two groups of diplomats indicates clearly a relatively high degree of consensus among them in their assessments of the realism of the peace proposals. This conclusion is further supported by the greater

Table 4.9 Significant differences between diplomats from developed and developing countries in their attribution of realism to conditions for peace (t-tests between means). *Source* Compiled by the author from his own research

Proposal item number	Proposal	t-value <	p-value
16	The military balances must be preserved	5.48	.005
2	Abolish hunger and poverty in the world	−2.67	.01
3	Relations between individuals must become more peaceful	−2.50	.01
8	Population growth must be controlled	−2.11	.02
20	National boundaries ought to become more open	1.77	.05
5	The states must become more democratic	−1.72	.05

consensus in the attribution of realism than in the attribution of importance; whereas we found fifteen significant differences for the latter, there were only six for the former. After examining the two profiles separately, we combined them to construct the operational peace profile of diplomats. We did this by bringing together the two dimensions – importance and perception of realism – in one matrix, where the horizontal coordinate represents realism and the vertical coordinate the importance assigned to the proposals. We entered the means of each single proposal on each of the two coordinates and plotted one point for each proposal. An overview of the results for the total sample can be found in Fig. 4.1.

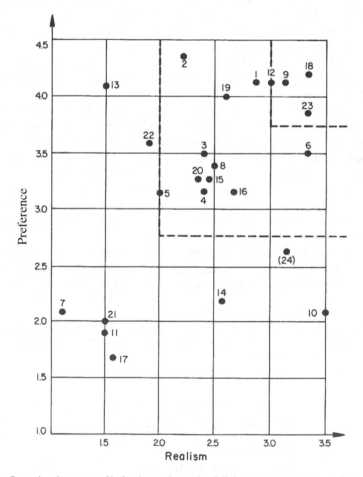

Fig. 4.1 Operational peace profile for the total sample of diplomats (means). *Source* Compiled by the author from his own research

To separate different clusters, we can distinguish three levels on each coordinate of the matrix: high, medium, and low. This gives us nine groups of proposals:

High/high	18, 9, 23
Medium/high	6
Medium/medium	20, 4, 3, 8, 15, 16
High/medium	2, 19, 1, 12
Low/high	24, 10
Low/medium	14
Low/low	7, 21, 11, 17
Medium/low	5, 22
High/low	13

The broken lines in the figure indicate the cuts between the three levels mentioned; the ranges of the levels are somewhat larger on the 'importance' coordinate because of the higher average means on that dimension. The cuts were made visually to correspond to clustering patterns. For comparative purposes, we also constructed two separate matrices, one for diplomats from developed countries and one for diplomats from developing countries (see Figs. 4.2 and 4.3).

Figures 4.1, 4.2 and 4.3 show that on several items the differences between the degree of importance and the degree of perceived realism are considerable. This finding answers the question about the relationship between realism and aspiration in the respondents' thinking.

Another test of this difference is the Pearson correlation coefficient. Scores on the importance and realism dimensions for the total sample yield an average correlation of .27, which supports the conclusion that the aspiration-realism difference is considerable.

An overview of the correlation coefficients, item by item, for the total sample and the subsamples of diplomats can be found in Table 4.10. If we compare the correlations of diplomats from developed and developing countries, we see that the latter are generally smaller. The average r for developed countries is .32; for the developing countries .22. This indicates that there is a greater convergence between aspirations and realistic expectations in the peace profiles of diplomats from developed countries than among their colleagues from developing countries. Several explanations could be given for this dissonance in the views about peace. One explanation is that the aspirations of these diplomats are, in fact, higher than the aspirations of their colleagues from developed countries. They see most of the proposals as more congruent with their national interests, as reflected in their optimal peace profiles, than do the diplomats from developed countries. Diplomats from developing countries assigned significantly greater importance than their colleagues from developed countries to thirteen of the proposals, while the latter assigned significantly greater importance to only two proposals (see Table 4.7). These data indicate that, given the higher aspirations, the same level of realism would yield larger discrepancies. Another possible explanation is that diplomats from developing countries compensate the power of word politics for their lower

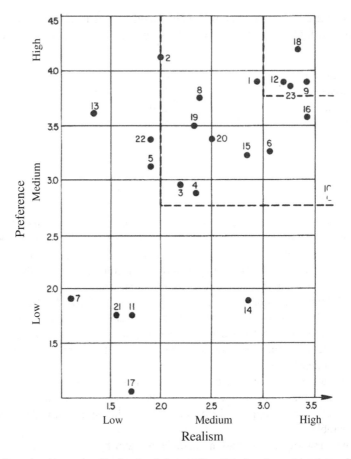

Fig. 4.2 Operational peace profile for the diplomats from developed countries (means). *Source* Compiled by the author from his own research

real-power position. In other words, they believe more than their colleagues that rhetoric can make a difference in international affairs. This greater belief in word politics, expressed in a tendency to accentuate aspirations more strongly, could account for the greater discrepancy found in developing countries.

Another expression of the discrepancy between aspirations and realistic expectations is the level of trust or distrust in the different proposals. The measure of trust is the mean difference between perceived realism and importance, as shown in Table 4.11, which presents the trust status of all proposals, ranked from most to least trusted, for the total sample and the two subsamples. A highly preferred proposal that is perceived as highly unrealistic represents an extremely distrusted proposal (reflected in a negative score). A highly realistic proposal which is less preferred might, on the contrary, be called a trusted proposal (reflected in a positive score).

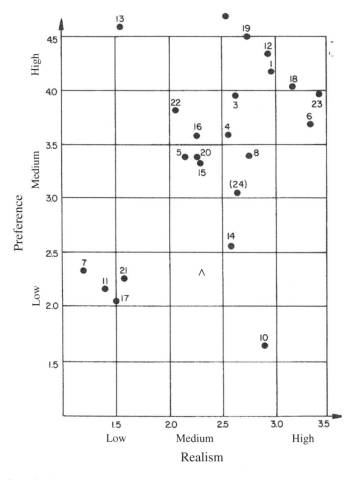

Fig. 4.3 Operational peace profile for the diplomats from developing countries (means). *Source* Compiled by the author from his own research

Let us first look at the most- and least-trusted peace proposals for the total sample. The 'migration' item calculations are based on a very small number of diplomats (n = 32) and will therefore not be discussed. We will also not comment on items 14 and 17 because of their low levels of importance. The remaining proposals with high-trust status for the total sample of diplomats are: "military alliances must be preserved", and "we must strengthen regional organizations". These two trusted proposals are low/high level and medium/high proposals.[2] A comparison of two arms policy proposals in Table 6.11 – "preserve alliances"

---

[2]In the terms "low/high" and "medium/high" the word before the slash refers to preference, and the second word refers to the realism level of the proposal.

Table 4.10 Pearson correlations between the importance and realism scores of each condition for peace for diplomats from developed and developing countries. *Source* Compiled by the author from his own research

Proposal item number	Proposal	Total	Developed countries	Developing countries
1	The individual person must be educated to peace	.21	.31	.06
19	We must strengthen the United Nations	.40	.58	.12
10	The military alliances must be preserved	.28	.29	.24
2	Abolish hunger and poverty in the world	.21	.24	.07
15	Free trade must be established between all countries	.33	.29	.39
21	Each national group should be given its own country	.47	.45	.50
22	Ideological disputes ought to be diminished	.16	.12	.19
8	Population growth must be controlled	.02	.05	.00
9	More effective communication is necessary	.19	.17	.19
7	A world government must be established	.34	.41	.30
11	The nations must become more similar	.21	.33	.15
12	Rich countries must give aid to the poor	.01	.14	.08
3	Relations between individuals must become more peaceful	.34	.51	.08
4	The small countries must have greater influence	.36	.43	.25
5	The states must be more democratic	.27	.33	.21
6	We must strengthen regional organizations	.40	.40	.40
13	General and complete disarmament must be realized	.17	.22	.10
14	Countries must become more socialistic	.25	.24	.31
16	There must be a military balance between states, so that nobody dares to attack	.32	.44	.13
17	One should create a world with small self-sufficient states	.46	.39	.52
18	Western and Eastern countries must improve detente	.27	.25	.27
20	National boundaries ought to become more open	.26	.29	.23
23	States that naturally belong together must cooperate	.21	.25	.17
24	Restrictions to migration should be gradually lifted	.16	.45	.25
	Average correlation	.27	.32	.22

and 'disarmament' – indicates that tough proposals like the former are trusted more than soft proposals like the latter. This conclusion is further supported by the greater trust status of "military balance" and 'detente'. The most distrusted proposals for the total sample are "general and complete disarmament must be realized", "abolish hunger and poverty in the world", "ideological disputes ought to be diminished", and "we must strengthen the United Nations". These four items are all located in the high/low and high/medium spaces of our operational peace profile (Table 4.11).

Table 4.11 Conditions for peace ranked according to the amount of difference between realism and importance attributions. *Source* Compiled by the author from his own research

Total sample		
10	Preservation of military alliances	+1.4
24	Lifting of migration restrictions	+0.4*
14	More socialism	+0.4
17	Self-sufficient states	−0.1
6	Strengthening regional organizations	−0.2
11	More similarity between nations	−0.4
23	Cooperation	−0.5
16	Military balance	−0.5
21	Each national group its country	−0.5
4	Small countries more power	−0.8
18	East-West detente	−0.9
9	More effective communication	−0.9
8	Control population growth	−0.9
15	Free trade between all countries	−0.9
20	More open boundaries	−1.0
7	Establishment of world government	−1.0
3	Peace between individuals	−1.1
12	Rich countries aid poor countries	−1.1
5	More democratic states	−1.2
1	Educate individuals to peace	−1.2
19	Strengthening of the United Nations	−1.4
22	Diminishment of ideological disputes	−1.7
2	Abolish hunger and poverty	−2.2
13	Disarmament	−2.6
Developed countries		
24	Lifting of migration restrictions	+1.1*
14	More socialism	+0.9
10	Preservation of military alliances	+0.9
17	Self-sufficient states	+0.6
11	More similarity between nations	−0.0
16	Military balance	−0.1
6	Strengthening regional organizations	−0.1

(continued)

Table 4.11 (continued)

21	Each national group its country	−0.2
4	Small countries more power	−0.4
15	Free trade between nations	−0.4
23	Cooperation	−0.5
9	More effective communication	−0.5
12	Rich countries aid poor	−0.7
3	Peace between individuals	−0.7
7	World government	−0.8
20	More open boundaries	−0.8
18	East-West detente	−0.9
1	Educate individuals to peace	−0.9
19	Strengthening of the United Nations	−1.1
5	More democratic states	−1.2
8	Population control	−1.4
22	Diminishment of ideology	−1.5
2	Abolish hunger and poverty	−2.1
13	Disarmament	−2.2
Developing countries		
10	Preservation of military alliances	+1.7
14	More socialism	−0.0
6	Strengthening regional organizations	−0.3
24	Lifting migration restrictions	−0.4*
17	Self-sufficient states	−0.4
23	Cooperation	−0.5
8	Control population growth	−0.6
21	Each national group its country	−0.7
9	More effective communication	−0.7
16	Military balance	−0.8
10	Preservation of military alliances	−0.8
18	East-West detente	−0.9
15	Free trade between all countries	−1.1
7	Establishment of world government	−1.1
20	More open boundaries	−1.2
5	More democratic states	−1.2
3	Peace between individuals	−1.2
1	Educate individuals to peace	−1.3
12	Rich countries aid poor countries	−1.4
22	Diminishment of ideology	−1.7
19	Strengthening of the United Nations	−1.8
2	Abolish hunger	−2.2
13	Disarmament	−3.0

*N = 32

Let us now find out if there are differences between diplomats from the two subgroups with respect to the trust status of their proposals. Excluding from both lists high-trust items with a low/low or low/medium operativeness, we are left with the following two lists of most trusted proposals: for the developed countries, "military alliances must be preserved", "there must be a military balance", and "we must strengthen regional organizations"; for developing countries, "military alliances must be preserved", and "we must strengthen regional organizations". Regional integration and military alliances have the highest trust status for both groups. The groups differ, however, with respect to the military balance proposal; the trust status of this item is clearly higher for diplomats from developed countries than for their colleagues from developing countries. These three proposals are all located in the medium/high or low/high spaces of the respective operational peace profiles.

The three most distrusted items that can be found in both lists are "general and complete disarmament must be realized", "abolish hunger and poverty in the world", and "ideological disputes ought to be diminished". The groups differ, however, with respect to the population control, foreign aid, and United Nations proposals. The diplomats from developed countries have a greater distrust of population control proposals than their colleagues. On the other hand, diplomats from developing countries show greater distrust of the foreign aid and United Nations proposals. These six proposals are located in the high/medium and high/low spaces of the respective operational peace profiles.

Looking back at the operational profiles, which were intended to categorize the proposals according to their degree of operativeness, the following qualification should be added. We feel it is fair to assume that diplomats would tend to opt for realistic policy proposals if they had to choose between optimal and realistic, or if they found that the highly preferred alternatives were not working overtime or turned out to be unrealistic. Thus we would suggest giving greater weight to the perceived realism side of the operational profile, which in turn means that the medium preference/high realism proposals emerge as relatively more operational. It also means that the regional middle-range proposals, which are highly trusted, are also the most operational (see Table 4.12). Regional peace policies are generally perceived as more operational than global, national, or subnational ones for the total sample and the two subsamples. This global-subnational dimension also raises the question whether and to what extent one or more specific levels in the international system are preferred for peace policies. Comparing the preferences of diplomats from developed and developing countries, we see that the former give the highest preference to the inter- and intraregional level, and the latter put more emphasis on the global and subnational levels. From all this we may conclude that the diplomats from developed countries do not go to extremes in their thinking about peace: they neither prefer nor perceive as very realistic the global approach or approaches that start with the individual or at the micro-sociological level. On the other hand, diplomats from developing countries emphasize the medium level as operational, while preferring the global and subnational as optimal.

Table 4.12 Preference, realism, and trust status of conditions for peace grouped according to system level for diplomats from developed and developing countries (means). *Source* Compiled by the author from his own research

	(0–5) Optimal profile average	(1–5) Realism perception average	Optimal–realistic difference average
Total sample			
Global	3.58	1.96	−1.62
Interregional	3.40	3.10	−0.30
Intraregional	3.65	3.30	−0.35
National	2.58	2.06	−0.52
Subnational/individual	3.80	2.65	−1.15
Developed countries			
Global	3.49	1.92	−1.58
Interregional	3.65	3.43	−0.22
Intraregional	3.63	3.22	−0.41
National	2.36	2.16	−0.20
Subnational/individual	3.41	2.53	−0.88
Developing countries			
Global	3.94	2.18	−1.76
Interregional	3.26	2.93	−0.33
Intraregional	3.89	3.40	−0.49
National	2.92	2.08	−0.84
Subnational/individual	4.10	2.81	−1.29

Global items 2, 7, 19, 13, 15
Interregional items 10, 12, 16, 18
Intraregional items 6, 23
National items 4, 5, 11, 20, 14, 17
Subnational/individual items 1, 3

Finally, let us look in somewhat greater detail at the relatively high discrepancy found between the aspirational and realism scores in the peace profiles. We can group the response tendencies of diplomats in four categories (see Table 4.13). A diplomat, most of whose responses are high on preference and low on realism, could be called an idealist or sophisticated realist, who also believes in the impact of verbal strategy and is able to live with this discrepancy.

Another position may be that of high preference/high realism, which is a more balanced state of thinking about peace. Realists or cynics, I assume, could be found in the high realism/low preference quadrant. Finally, a diplomat with a peace profile low on preference and low on realism may be called a defeatist. Although of interest, it is not within the scope of this study to test the validity of these tentative labels but rather to inquire into the differences between the diplomats of developing

Table 4.13 Aspirational-realism discrepancy in the peace profile of diplomats. *Source* Compiled by the author from his own research

Realism	Importance	
	High	Low
Total sample		
High	3.13	2.86
Low	9.51	2.94
Developed countries sample		
High	3.37	2.79
Low	8.46	3.35
Developing countries sample		
High	2.90	2.93
Low	10.57	2.54
Developed versus developing	t-value	p-value
High-high	.88	
High-low	−.46	
Low-high	−3.24	<.005
Low-low	2.49	<.01

Note The numbers represent the mean number of items that have the particular characteristics of the quadrant

and developed countries with respect to these four types. To that end, we classified all proposals scored by each diplomat as belonging to one of the four types. A proposal was classified as balanced if both the importance and realism scores were 5; as defeatist if both scores were 1; as idealist or sophisticated realistic if the realism score was lower than the importance score; and realistic or cynical if the realism score was higher than the importance score. The mean frequencies of each of these types, for the total sample and the subsamples, can be found in Table 4.13. The table clearly indicates that for both subsamples of diplomats, the idealistic or sophisticated-realistic pattern of low realism/high importance prevails. However, this pattern is significantly more frequent among diplomats from developing countries ($t = -3.24$, $p < .005$). The only other significant difference relates to the defeatist type of consonance. The data indicate that diplomats from developed countries tend to be higher on this type of consonance ($t = 2.49$, $p < .01$). This last finding should, however, be considered with some reservation because of the lower-than-possible validity of our measure of defeatism, which could have been made more valid if, instead of classifying a proposal as defeatist when both scores were 1, we had classified it as such when it had a preference score of 1 or 0, and a realism score of 1.

The relatively higher dissonance of the diplomats from developing countries seems to be due in large part to a greater percentage of idealists in that sample. This is indicated by Table 4.14, which contrasts the two groups of diplomats on some selected items. The diplomats from developing countries are shown as more soft

Table 4.14 Diplomats from developed and developing countries on some selected proposals (mean scores). *Source* Compiled by the author from his own research

	Important to peace		Realistic	
	Developed country	Developing country	Developed country	Developing country
Keep the balance	3.6	3.0	3.4	2.2
Preserve alliances	2.9	1.6	3.7	3.3
Educate individuals	3.8	4.2	2.1	2.2
World state	1.9	2.3	1.1	1.1

and more idealistic (believe more in the individual and the world state approach). Interestingly, while the two samples differ markedly on the importance they attach to the different proposals, there is much more agreement between them as to how realistic they deem the proposals to be. This explains much of the relatively higher dissonance of developing-country diplomats.

4.5.5   Preferred World Order

We turn, finally, to analysis of the diplomats' view of a preferred world order. Our primary focus is on the extent of agreement between diplomats representing developed and developing countries. We will also examine whether and to what extent the world-order values of the World Order Model Project (Mendlovitz 1975) are reflected in the diplomats' thinking about their preferred world order.

The data on which this analysis is based were gathered during the interviews in Washington and were evoked by the following question:

> How would the world you'd like to see for your children differ from that of today; or how would you describe your preferred world order?

We did not develop an a priori system for categorizing the answers but instead listed all the responses obtained and grouped them into meaningful categories. The list of the qualities of a preferred world order for the total sample can be found in Table 4.15, and the data on which we will base our description of the differences between diplomats from developed and developing countries are presented in Table 4.16.

As can be seen in Table 4.16, the two subsamples agree with respect to six qualities of a preferred world order: international cooperation, a viable United Nations, reduced polarity between the North and South, absence of domestic interference, economic welfare, and global approach. The global approach category includes all qualities that explicitly stressed the need for a global solution to a

Table 4.15 Preferred-world-order qualities for the total sample of diplomats (frequencies). *Source* Compiled by the author from his own research

International cooperation	44
A viable United Nations	33
Reduced polarity between North and South	28
No domestic interference	27
Open borders and migration	24
Absence of violence	22
Economic welfare	22
Mutual understanding	17
Equitable distribution of world resources	17
Arms control	16
Dignity and respect	16
International democracy	16
Global approach	16
Moral and humane international relations	15
New economic system	15
Disarmament	14
Communication improvement	14
Justice	14
Tolerance	14
Equality	11
Economic development	11
Detente	9
Mutuality of exchange	9
More meaningful interdependence	9
Population control	9
Absence of racism	8
Freedom	8
Regional integration	8
International legal order	7
Environmental improvement	6
World government	6

particular problem. With respect to these six qualities of a preferred world order, there are no significant differences between the two samples.

There are five qualities on the list for developed countries, which do not appear on the list for the developing countries: international democracy, improvement of communication, arms control, disarmament and arms reduction, and tolerance of differences. Of these five qualities, only one has a significantly higher mean for developed-country diplomats: improvement of communication. Although this category included all qualities relating to the improvement of international communications in general, most statements were specifically related to East-West

Table 4.16 Preferred-world-order qualities as judged by diplomats from developed and developing countries. *Source* Compiled by the author from his own research

		Means*	
Developed countries			
1	A viable United Nations	.127	
2	International cooperation	.107	
4.5	Economic welfare	.088	
4.5	No domestic interference	.068	
4.5	International democracy	.088	
4.5	Communication improvement	.088	
7	Reduced polarity North-South	.078	
8	Arms control	.068	
10	Global approach	.058	
10	Disarmament and arms reduction	.058	
10	Tolerance	.058	
Developing countries			
1	International cooperation	.196	
2.5	Reduced polarity North-South	.145	
2.5	Open borders and migration	.145	
5	A viable United Nations	.111	
5	Absence of violence	.111	
5	No domestic interference	.111	
7	Economic welfare	.094	
8	Equality	.076	
10	Equitable distribution of world resources	.068	
10	Global approach	.068	
10	Justice	.068	
Developed versus developing countries		t-value	p-value <
1	Open borders and migration	−3.05	.005
2	Communication improvement	−2.32	.01
3	Justice	−2.31	.05
4	Moral and humane international relations	−2.08	.05
5	Equality	−2.02	.05
6	Absence of violence	−1.94	.05
7	Regional Integration	−1.76	.05
8.5	Reduced polarity North-South	−1.58	
8.5	Dignity and respect	−1.58	

*Obtained by dividing the total number of responses provided by diplomats from each type of country by the number of diplomats in the subsample

relations. Regional integration did not emerge in the list of top-ranked qualities for developed-country diplomats, but it did have a significantly higher mean in this subsample than among developing-country diplomats (t = 1.76, p < .05).

It is not surprising that this category is high for developed countries, for most of the statements included referred specifically to the further integration of the European Common Market.

In the list of top-ranked qualities for the developing-country diplomats, we found five qualities which did not appear on the other list: open borders and migration, absence of violence, equality, equitable distribution of world resources, and justice. All these qualities, with the exception of equitable distribution of world resources, appear significantly more in the preferred world order of the diplomats from developing countries (see Table 4.16). The meaning of these qualities, with the exception of open borders, is self-evident. I will illustrate the meaning of this latter quality by quoting some statements which were placed in this category: a world with more open borders – political, economic, and cultural; travelling should be easier; a world without borders; easier immigration; greater freedom of movement of people around the world; opening of frontiers; no boundaries; and lifting restrictions and stimulating contact between peoples from East and West. This quality is uniquely characteristic of the preferred world order of developing-country diplomats; this is underscored by the fact that it occupies second place in their list, and also accounts for the most highly significant difference between preferred qualities for the two subsamples (t = −3.05, p < .005). Not included in the top list for developing-country diplomats, but appearing more frequently in the preferred world order of diplomats of developing countries, are moral and humane international relations, and dignity and respect.

Let us now see whether and to what extent the world-order values of the World Order Model Project (Mendlovitz 1975) are reflected in the preferred-world-order thinking of our diplomats. The World Order Model Project recognizes the absence of war, poverty, social injustice, ecological imbalance, and indignity as values having a global scope. In order to relate our data to these world-order values, we added together all categories that were related to each of these world-order values. The results can be seen in Table 4.17. For the diplomats from developed countries, we can divide the world-order values into two groups. Highly important and roughly equal are the absence of poverty and war, and markedly lower in importance are the absence of injustice, ecological instability, and indignity. The picture for diplomats from developing countries is more complex. The most salient value in their preferred world order is the absence of poverty; second in importance and more or less equal to each other are the absence of violence and injustice; third is the absence of indignity; and finally, the absence of ecological instability. The greatest differences are noticed in relation to the justice, dignity, and economic security values. These values are more salient in the preferred-world-order view of the developing-country diplomats. We also observe that all five values have higher scores among diplomats of developing countries. This is caused because our preferred-world-order question yielded a somewhat higher response rate from the diplomats of developing countries (1.8) than from diplomats of developed countries

Table 4.17 Qualities of preferred world order of the diplomats grouped according to six world-order values (mean frequencies). *Source* Compiled by the author from his own research

Quality	Developed countries	Developing countries
Absence of war		
Absence of violence	.039	.111
Arms control	.068	.042
Disarmament and arms reduction	.059	.059
Detente	.049	.034
Total	.215	.246
Absence of poverty		
Reduced polarity between North and South	.078	.145
Economic welfare	.088	.094
Economic distribution	.049	.068
Economic development	.029	.042
Total	.244	.349
Absence of justice		
Moral and humane international relations	.000	.059
Justice	.009	.068
Equality	.019	.076
Mutuality of exchange	.029	.042
Freedom	.029	.017
Total	.086	.262
Absence of ecological instability		
Environmental improvement	.029	.025
Population control	.029	.042
Total	.058	.067
Absence of indignity		
Dignity and respect	.019	.059
Absence of racism	.029	.034
Total	.048	.093

Note The six world-order values are those that are recognized by the World Order Model Project as being global in scope

(1.5). But even if we take into account this systematic bias, we see that the earlier presented interpretation would still be valid. From these results we can infer three conclusions: that the world-order values are represented in the diplomats' thinking about their preferred world orders; that some values are more salient than others; and that diplomats from developed and developing countries differ with respect to the priority ordering of these five values.

In summary, we may say that our investigation of the views of preferred world order has given us useful information about the future of international relations. Dreams of different future orders may not bring them about, but they may prevent the consolidation of the present order or enhance or inhibit the implementation of a particular future order.

4.6 Conclusion

These results give support to the propositions which we made explicit at the beginning of this chapter. First we observed the existence of significant differences among diplomats in their operationalization of peace and preferred world order. We also found that the gap between the present state of international peace and the ideal state is considerable. The greatest discrepancies between the aspirational and the realism levels were noticed by diplomats from developing countries. The larger discrepancy was accounted for by the stronger accentuation of their aspirations. Proposals at the inter- and intraregional levels were indicated as more operational than proposals at the global or national or subnational levels. The values espoused by the diplomat were also shown to be a function of the systemic position of the country he represented.

The impact of the systemic position was found to be greater on the appraisal of peace proposals in terms of their importance or preference than on the diplomat's assessment of the likelihood of peace proposals to be carried out. In other words, the systemic position tended to affect the accentuation of aspirations more than the appraisal of reality by diplomats. More generally, these data indicate that the meanings and weights associated with peace and preferred-world-order values generally reflect the particular interests of their respective proponents. Certain concerns and interests sensitize the diplomat to some indicators of peace and to particular international problems, and at the same time cause him to neglect a number of alternative indicators and problems that other diplomats deem important. This implies, for example, that if we want to create a greater consensus about international values, we have to cope not only with such problems as the improvement of international communication or empathy but also with a structural problem. In other words, universal consensus can be brought about only when we want to deal with the particularity of the systemic vantage points from which these terms are given meaning and are translated into action.

References

Aron, R. (1966). *Peace and War: A Theory of international relations*. (R. Howard, & A. Baker Fox, Trans.). New York: Doubleday. (Original work published 1962).

Bennett, G. (Ed.). (1953). *The Concept of Empire: Burke to Attlee, 1774–1947*. London: Black.

Berger, P. L., & Luckmann, T. (1966). *The social construction of reality*. New York: Doubleday.

Chomsky, N. (1974). *Peace in the Middle East: Reflections on Justice and Nationhood*. New York: Random House.

Claude, I. (1959). *Swords into Plowshares* (2nd edn.). New York: Random House.

Deutsch, K. (1972). *Peace Research: The Need, the Problems and the Prospects*. Middlebury, VT: Middlebury College Publications Department.

Dulles' Address on Peace in Washington on 11 April 1953 *New York Times* (12 April 1955): 6.

Fanon, F. (1968). *The Wretched of the Earth* (C. Farrington, Trans.). New York: Grove Press. (Original work published 1961).

Gadamer, H.-G. (1968). Notes on planning for the future. In S. Hoffmann (Ed.), *Conditions of World Order*. Boston, MA: Houghton Mifflin.

Galtung, J. (1967). *Peace Research: Science or Politics in Disguise*. Oslo, Norway: PRIO publication: 23–6. (Mimeographed).

Hoffmann, S. (1968). *Conditions of World Order*. Boston: Houghton Mifflin.

Howard, M. (1971). *Studies in War and Peace*. New York: Viking.

Hveem, H. (1968). Foreign Policy Thinking in the Elite and the General Population. *Journal of Peace Research, 2*, 146–170.

Hveem, H. (1972). Foreign Policy Opinion as a Function of International Position. *Cooperation and Conflict: Nordic Journal of International Politics, 2*, 65–86.

Hveem, H. (1973). Peace Research: Historical Development and Future Perspective. *IPRA Studies in Peace Research, 5*. (Mimeographed).

Janik, A., & Toulmin, S. (1973). *Wittgenstein's Vienna*. New York: Simon and Schuster.

Jaspers, K. (1971). *Man in the Modern Age*. New York: Doubleday.

Kara, K. (1968). On the Marxist Theory of War and Peace. *Journal of Peace Research, 1*, 1–27.

Kissinger, H. (1965). *The Troubled Partnership: A Reappraisal of the Atlantic Alliance*. New York: McGraw-Hill.

Kohn, H. (1942). *World Order in Historical Perspective*. Cambridge, MA: Harvard University Press.

Lagos, G. (1963). *International Stratification and Underdeveloped Countries*. Chapel Hill: The University of North Carolina Press.

Levi, W. (1964). On the Causes of Peace. *Journal of Conflict Resolution, 8*(1), 23–35.

Mannheim, K. (1936). *Ideology and Utopia*. New York: Harcourt, Brace and World.

Mazrui, A. A. (1967). *Towards a Pax Africana: A Study of Ideology and Ambition*. Chicago: The University of Chicago Press.

Mendlovitz, S. (1975). *On the Creation of a Just World Order: Preferred Worlds for the 1990s*. New York: The Free Press.

Northrop, F. S. C. (1946). *The Meeting of East and West*. New York: Collier Books.

Puchala, D. J. (1971). *International Politics Today*. New York: Dodd, Mead.

Rapoport, A. (1971). Various Conceptions of Peace Research. Paper presented at the Peace Research Society Conference, Ann Arbor.

Rawls, J. (1971). *A Theory of Justice*. Cambridge, MA: Harvard University Press.

Reychler, L. (1979). Values in Diplomatic Thinking: Peace and Preferred World Order. In L. Reychler, *Patterns of Diplomatic Thinking: A Cross-National Study of Structural and Social-Psychological Determinants* (pp. 112–162). New York: Praeger Publishers.

Rubin, Z. (1973). *Liking and Loving*. New York: Holt, Rinehart and Winston.

Schmid, H. (1968). Peace Research and Politics. *Journal of Peace Research, 3*, 219–232.

Scott, W. A. (1965). Effects of Cross-National Contact on National and International Images. In H. Kelman (Ed.), *International Behavior*. New York: Holt, Rinehart and Winston.

Sharp, G. (1973). *The Politics of Non-Violent Action*. Boston, MA: Porter Sargent.

t'Hart, W. A. (1955). Het psychologisch aspect van een wereldfederatie. In *Europa's geestelijke bijdrage tot wereld-integratie*. (n.p.): Broederschapsfederatie.

United Nations. (1945). *Charter of the United Nations and Statute of the International Court of Justice*. New York: United Nations.

Chapter 5
The Effectiveness of a Pacifist Strategy in Conflict Resolution

To find out to what extent the violence and exploitative behaviour of the opponent of a pacifist strategist could be controlled, the author tested within a laboratory context the impact of eight variables.[1] Together these variables accounted for more than 50% of the behaviour of the opponents. The predictions concerning the controlled variables were confirmed. A pacifist strategy tends to be most effective in reducing violence and exploitative behaviour when the human distance between the subject and opponent is small, the subject is well informed about the intentions of the pacifist, the subject is required to justify his behaviour post-facto, and when a partial third party is present. Male and female subjects did not differ with respect to the use of violence, but it was found that females had a higher propensity to stalemate, with a consequent reduction of their own and the pacifist's winnings. The strongest predictors of the effectiveness of the pacifist strategy were the demand characteristics and the image of the pacifist held by the opponent. A discussion of the controllability of these last two variables is included.

The search for alternatives to violence is not new. Many have proposed pacifist strategies for ethical reasons. Yet we still know very little about the effectiveness of pacifist bargaining behaviour – that is, the extent to which and the conditions under which those who bargain as pacifists succeed in resolving conflicts in a manner acceptable to them by non-violent means.

Since the mid-1960s, a group of experimental social psychologists have addressed themselves to the effectiveness questions: Solomon (1960), Lave (1965), Shure etal. (1965), Rapoport et al. (1965), Deutsch et al. (1967), Meeker/Shure (1969), Vincent/Tindell (1969) and Rapoport (1969).

Solomon (1969) found not only that his subjects exploited players who were unconditionally benevolent but also that the subjects were puzzled by such

---

[1]This text was first published as: Reychler, L. (1979, June). The Effectiveness of a Pacifist Strategy in Conflict Resolution. *Journal of Conflict Resolution* (June), *23*(2), 228–260. Reprinted with the permission of SAGE.

L. Reychler and A. Langer (eds.), *Luc Reychler: A Pioneer in Sustainable Peacebuilding Architecture*, Pioneers in Arts, Humanities, Science, Engineering, Practice 24, https://doi.org/10.1007/978-3-030-40208-2_5

unconditional behaviour. The output of Lave's (1965) experiment also shows the pacifist strategy to be ineffective. Subjects tended to exploit pacifist players. Shure et al. (1965) results corroborate the findings obtained by Solomon and Lave. The authors interpret their results as supporting the judgement that the pacifist strategy invites exploitation and aggression, even among those subjects who did not begin with such intentions. In a series of experiments, Deutsch et al. (1967) investigated the effect of different strategies on subjects' aggressive behaviour. Comparison of the effects of the display of a pacifist strategy with the effects elicited in a control situation reveals that the pacifist strategy elicited no more cooperative behaviour than did the strategy used in the control condition.

The results of the reported experiments suggest that pacifist strategies have little to recommend them as effective strategies for conflict resolution. There is, however, quite a lot of variance around this generally low 'effectiveness' mean, suggesting that the effectiveness of the pacifist strategy varies considerably from one type of situation to another. It is therefore important to identify the conditions under which a pacifist strategy is most effective.

As to the effectiveness of a pacifist strategy in a conflict situation, several variables are of particular relevance. They can be classified according to: (a) the social system in which the pacifist and his or her opponent operate, (b) their interaction structure, and (c) the decision-makers themselves.

5.1   Social Context

5.1.1   Third-Party and Constituency Influence

Meeker and Shure studied the effect of a neutral or uncommitted third party on the interaction between a subject and a pacifist. They conjectured that "any effects presumably would be pro-pacifist, since the effectiveness of pacifism is based on forcing the other player to decide between morality and self-interest; and the presence of an onlooker should serve to increase the salience of the situation's ethics" (Meeker/Shure 1969: 492). In a neutral third-party condition all subjects were told that their behaviour would be monitored by an observer who would interview them after the experiment. This condition produced significant results, such that more than twice as many of the subjects in the third-party condition planned at the outset to be cooperative; this is the point of greatest difference obtained between third-party and no third-party conditions. From this point the difference narrows until the end of the experiment, when the results are not significantly different. The same group of experimenters earlier investigated the effects of a constituency (also referred to as teammate pressure or cohort pressure) on the behaviour of the subject. When the subject had two simulated teammates, who in the initial phase of the experiment urged him to dominate, the teammate pressure seemed to have a significant effect on the interaction between the pacifist and the subject. Deutsch et al. (1967: 358) interpret Meeker and Shure's results as "un-doubtedly due to having fixed their subjects into a dominating strategy by group pressure under which the subjects were placed". Deutsch's explanation was later tested by Meeker/Shure (1969). They ran a no-constituency group for comparison, and found that the presence of constituency pressure does reduce the level of cooperation in the early trials, but that this initial pressure does not appear to make the subject more resistant to subsequent pacifist appeals than subjects who are initially uncooperative without exposure to cohort pressures.

5.1.2   Role Expectancy

Role expectancy refers to the demand characteristics of the situation or, in other words, to the subjects' answer to the question: "What do you think is expected of you?" There is ample evidence to support the contention that role expectation significantly affects the behaviour of an actor (cf. Orne 1962; Milgram 1965). According to Orne, demand characteristics of cues which convey an experimental hypothesis to the subject are significant determinants of the subject's behaviour, an effect which Orne attributes to a probable concern of subjects about the utility of their performance for a legitimate recognized authority. In an article reviewing the earlier experiments on the effectiveness of the pacifist strategy, Ofshe (1971: 264) suggests that the subject's knowledge of the experimenter's expectations for his behaviour will influence the subject to adopt a competitive strategy, and that this influence is present despite any other manipulations that are introduced. If we want to improve our ability to predict the effectiveness of the pacifist strategy, it is necessary to investigate the role of demand characteristics on the subject's beha-viour, and also the differential effect of the experimenter's expectations that are made quite explicit (as, e.g., "the opponent with the most points after fifteen games would be declared winner"; Vincent/Tindell 1969: 497) versus a situation with less defined demand characteristics.

5.1.3   Justification

Another variable which we assume to be of great importance is the existence of an institution requiring the actors to justify their behaviour. When a subject knows he must justify his behaviour, his freedom of action is limited by his desire to make certain justifying statements, or, as Snyder et al. conclude, "The decision to perform or not to perform a given act may be taken on the basis of the socially available answer to the question: 'What will be said?'" (Snyder et al. 1962: 146). This is, according to Goldman (1971), a procedure often followed in authoritative decision-making. Based on earlier experiments working with a standard prisoner's dilemma, Marlowe et al. concluded that "where anonymity was guaranteed, subjects were less cooperative than in any other condition and exploited their partners on an average of 30 trials" (Gergen 1965: 61). If, indeed, the perception of situational norms is a factor in the subject's decision-making process, then we predict that the subject will consider the positive and negative reactions of the other actors, and show a reduction in the use of violent responses and an increase in cooperative behaviour.

5.2 Interaction Structure Between the Pacifist and His Opponent

5.2.1 Pay-off Structure

This refers to the costs and benefits associated with alternative choices made by the different actors. Several studies show that variations in monetary incentives have a significant impact on the behaviour of the subjects. Gallo/McClintock (1965) refer to an earlier study in which they found that an important increase in the amount of money that could be earned by cooperative behaviour leads to a considerable increase in cooperative interactions between actors. Another positive effect of significance concerned manipulations of pay-offs. Meeker/Shure (1969) found moderate overall success for the pacifist strategy when they changed from a pay-off schedule where the subject earned more by dominating and the pacifist slightly more by cooperating to a pay-off schedule where cooperation produced outcomes that were strictly equal to the pacifist's over a number of trials.

5.2.2 Human Distance

This term refers to the physical (audio, visual, tactile) distance between actors. A common characteristic of earlier experiments is the audio-visual distance of the other actor. In Solomon's (1969) experiment the actors were placed in cubicles so that they were unable to determine the identity of their game partners. Also, the Deutsch et al. (1967) subjects were seated in separate cubicles; they did not see or speak to one another before or during the experiment. In the Shure et al. (1965) and Meeker/Shure (1969) studies, the subjects played against a computer. Vincent's subjects played against an audio-visually absent stooge. The audio-visual distance bias in the earlier experiments is certainly a factor that should be taken into account in the interpretation of the results. Social psychological studies support the contention that immediacy of the other is associated with a tendency to choose non-violent solutions. Milgram (1965: 63–64) suggests several alternative mechanisms explaining this change in behaviour:

- empathic cues, e.g., in a condition of audio-visual absence of the opponent, the pacifist's suffering possesses a more abstract, remote quality for the subject than when there is an audio-visual presence;
- denial and narrowing of the cognitive field: an audio-visual absence condition allows for a narrowing of the cognitive field so that the victim is out of mind. The subject considers the violent act as less relevant to moral judgement when it is no longer associated with the victim's suffering. When the victim is closer, it is more difficult to exclude him phenomenologically.
- phenomenal unity of act: in the remote condition, it is more difficult for the subject to gain a sense of relatedness between his own actions and the consequences of those actions for the victim.
- acquired behaviour dispositions: Milgram suggests that a shortcoming in education could be responsible for less restraint in the use of violence when the opponent is at a great distance, "Through a pattern of rewards and punishments he requires a disposition to avoid aggression at close quarters; a disposition which does not extend to harming others at a distance".

It is our contention that the audio-visual absence bias of the earlier experiments accounts for a significant amount in the unsuccessfulness of the pacifist strategy. We expect that a reduced human distance condition will decrease non-cooperation with the pacifist. This expectation is supported by the findings of Whichman (1970), who reported that visual isolation yields greater competition and visual contact produces greater cooperation in a prisoner's dilemma game, a relationship also suggested in the research of Exline (1963). Further support comes from an experimental demonstration of a positive relationship of visual contact with increased liking (Exline/Winters 1965; Machson/Wapner 1967) and also with decreased aggression (Milgram 1965).

5.2.3   Communication Facilities

The term *communication facilities* refers to the means of exchange of information. A distinction can be made between one-way, two-way, verbal and non-verbal, explicit, tacit, and so on. In Shure's (1965) experiment, written communications were exchanged after operating periods 5, 6, and 7. These written communications could be seen as reassurances concerning the pacifist's conciliatory intent, fair demands, refusal to use violence, and intention to force the other to use violence if the other was going to remain unfair. Shure found the change in effectiveness of the pacifist strategy in this condition, compared with a non-communication condition, to be statistically significant. From these findings we can conclude that the capacity of the pacifist to control and manage information that the subject obtains can be greatly enhanced by the use of multiple channels of communication, for it enhances the pacifist's power to influence, in his own favour, responses to the situation.

Of course, this is only true if the information conveyed by the different communication channels is not contradictory, but reinforces the projection of the desired image of the pacifist.

5.3   The Decision-Making Level

5.3.1   Perception of the Pacifist's Beliefs

This variable refers to how well one actor is informed about the other actor. The underlying assumption is that a more accurate perception of the pacifist by the other will make the pacifist strategy more effective. This was effected in Shure's (1965) experiment through an exchange of personal resumes, which indicated that the pacifist was a Quaker, normally committed to a position of non-violence. Shure's findings show that information about the pacifist's attitudes towards the use of violence is not enough to create a significant change in the behaviour of the subject. These findings are in line with what we would have expected. The rationale behind our expectation will be explained under "attribute structure".

5.3.2   Attribute Structure

This factor refers to the assignment of attributes and their configuration by one actor to another. The premises on which our predictions concerning the behaviour of the subject in relation to the pacifist are built are the following:

(a) An actor's response to another's behaviour is to a great extent influenced by his or her image of the other – in other words, by the configuration of the attributes assigned to the other. If an actor evaluates this image as positive, he will respond positively, and vice versa.

(b) The assignment of the attribute 'non-violent' to an actor does not per se evoke a positive evaluation. This contention is based on two observations. First, in our Western culture, violence is not always viewed negatively; e.g., if we look in the Bible, we find that God could use legitimate violence. In struggles for emancipation from structural violence, violence is often considered necessary. In a society where winning is highly valued, we find the implicit assumption that winning always implies some form of violence. Second, non-violence is only one of many attributes which people value, for example, intelligence, sense of humour, social concern, receptivity. Our contention is that a successful projection of the pacifist's image, stressing mainly his non-violence, is not a good predictor of the subject's behaviour. In order to improve our prediction power, we also need to know how the pacifist is perceived in relation to other attributes that are highly valued, and the weights attached to each attribute. We are quite sure that a lot of people would evaluate a sincere freedom fighter in Zimbabwe higher than a non-violent traitor.

5.3.3   Verbal Commitment

This term refers to the existence of oral or written commitments of an actor to behave in a certain way. Here the expectation is that subjects would be induced to cooperate better if given guarantees that the pacifist will not take advantage of them and retaliate. In his experiment, Shure (1965) introduced a condition in which the pacifist committed himself not to retaliate, namely, by an unconditional and uni-lateral disarmament pledge. Shure expected that this move would reduce the sub-ject's fear of the pacifist's retaliation and induce him to more co-operativeness. The expectations were not fulfilled; the guarantee against retaliation produced no sig-nificant differences in the pacifist's ineffectiveness.

5.3.4   Psychological Make-up of the Pacifist's Opponent

Intentions. Research findings indicate that the game plan or the start of intentions of the actors influences their behaviour. Vincent found that subjects who had decided to dominate tended to use more shocks (0.48 rho), make fewer draws (0.40 rho), win more games (0.52 rho), and score more game points (0.43 rho) than those who were either undecided on this matter or who entered the game with the decision to be cooperative.

Attitudinal orientation. This refers to an actor's place on certain attitude scales, for example, L. Ferguson's humanitarianism scale, K. Helfant's survey of opinions and beliefs about international relations, and T. W. Adorno's F-scale. In the Vincent and Tindell study (1969) the attitude scales showed scattered significant associa-tions with the subject's behaviour. In some cases, expectations were met (those who reject humanitarianism tend to make fewer reciprocations in respect to de-escalation requests (0.31); but in other cases, findings were completely contrary to expecta-tions (those who are more fascistic according to the F-scale tended to use fewer shocks, 0.45; make more draws, 0.45; win fewer games, 0.53; and win fewer game points, 0.48). According to Vincent, these mixed findings appear to be consistent with the inefficiency of such scales as predictors in connection with prisoner's dilemma games.

Sex. Sex has always been a variable that attracted the interest of experimenters. People expect differences in the behaviour of the sexes because of differential sex-role conditioning. The association between goal attainment, sense of competence, and esteem on the one hand and aggression on the other is more likely to exist for males in American society than for females. Thus, when frustrated, males are more likely than females to lower their anxiety by aggression. Hokanson/ Edelman (1966) had male subjects inflict shocks on another person on a pretext in the laboratory. Some of the shocked subjects were then given an opportunity to return shocks to the person who had previously shocked them. As a consequence of this aggressive action, their blood pressure dropped, indicating lower arousal. If the subjects were not given an opportunity to retaliate in this fashion, no such decrease in blood pressure was found. However, this drop in blood pressure did not occur among females who had a chance to return the shocks. Some other assumptions concerning sex differences are not clearly supported by empirical evidence – for example, the assumption that males might be more conditioned by our culture to be competitive and more resistant to shock than females. In relation to this assumption, Vincent found that women used fewer shocks, made more cooperative moves, and achieved fewer game points. However, the differences were not significant.

Intelligence. A question that tends to be raised among intellectuals is the following: "Is there a correlation between intelligence and moral behaviour – are intelligent people more or less violent than less intelligent people?" Vincent (1969) found no significant correlation between grade point average and the game behaviour. Of course, his sample was drawn from a quite homogeneous group in relation to intelligence (college sophomores). It might be interesting to see whether an extension of the variation of this variable, i.e., by drawing a sample from the more general population, would produce significant correlations.

The list of variables is, of course, not exhaustive, but it includes all the variables studied in the above-mentioned experiments. In relation to the improvement of our understanding and predictive power concerning the effectiveness of a pacifist strategy, I would like to make three suggestions:

(1) The experiments indicate that variables of the three system levels influence the degree of effectiveness of the pacifist strategy. While each level of analysis is necessary, they cannot by themselves sufficiently explain and predict the effectiveness of the pacifist strategy. Therefore, a three-levels-of-analysis approach should be used.

(2) In addition to studying the impact of the above-mentioned variables on the effectiveness of the pacifist strategy, we also need to investigate the *modes of interdependency of the variables.* The importance of understanding the cross-impact between variables can be illustrated by Meeker and Shure's finding that verbal communication in a non-third-party condition created an increase in cooperation, whereas verbal communication failed to have any effect on the third-party condition. These results refuted the expectations of the experimenters who conjectured that the third-party condition and the communication condition would make a stronger reinforcement together than separately. The ex post facto explanation of Meeker/Shure (1969: 493) is that "these results may be substantiation of the repeated admonition by Gandhi and other proponents of non-violent resistance, namely that the pacifist should be mindful

of his adversary's position. The embarrassment of socially admitting to a moral error may increase his intractability. Indeed in ensuring that the pacifist's position was clear and that his intent should not be misread – in our efforts to create a clear-cut pacifist programme – we may have limited his ability to forestall his adversary's guilt".

(3) A third proposal in relation to the improvement of our predictive power concerning the effectiveness of a pacifist strategy is to include in our cross-impact matrix several variables which were overlooked in earlier research but which we conjecture to be of vital importance.

In the experiments on the impact of a pacifist strategy, the impacts of nine variables were tested. Six of the variables were found to have a positive impact on the effectiveness of the pacifist strategy:

- the presence of a natural and uncommitted third party;
- the absence of a constituency, stimulating the subject to dominate;
- the use of multiple communication channels;
- the reception by the subject of information concerning the pacifist's intentions;
- a pay-off structure which rewards cooperative behaviour; and
- subjects with the intention to cooperate.

The expectations concerning the effects of the sex, attitudinal orientation, and intelligence on the pacifist's strategical effectiveness were not clearly supported by the research findings. We would like to include in the cross-impact matrix also the following four variables: role expectancy, human distance, public justification, and attribute structure. There is strong evidence that these variables account for a significant part of the outcome variance. The impact of these four and other variables were investigated in this experiment.

5.4   Description of the Experiment

The aim of the experiment was to measure the effects of a variety of potential influencing variables on the effectiveness of a pacifist strategy. In this experiment a "pacifist strategy" is defined as an approach to conflict resolution which (1) reduces or eliminates the threat of violence and the short- and long-term use of violence, and (2) will increase the likelihood of reaching a solution which is fair to both of the conflicting sides. A strategy conforming to this definition will be evaluated as an effective pacifist strategy. These definitions need some explanation in order to be reasonably clear.

For our purpose it is useful to conceive of conflict as occurring when two or more actors pursue incompatible or mutually exclusive values within the same interaction system. Two different kinds of conflict – substantive and emotional – can be distinguished. Substantive conflicts involve disagreements over policies and practices, competitive ends for the same resources, and differing conceptions of roles and role relationships. Emotional conflicts involve negative feelings between the parties (e.g., anger, distrust, scorn, resentment, fear, rejection). In our experiment we limited our attention to substantive conflicts.

When we use the term *violence* we are referring to the physical (e.g., deliverance of shocks) and psychological (e.g., communication of threats) hurting of another actor.

The effectiveness of the pacifist strategy under varying conditions has been investigated in a simulated conflict situation. In terms of procedural specifics, our basic conflict situation is a variant of Gerald Shure's (1965) bargaining game; formally, it is a two-person, mixed motive, non-zero-sum game.

5.4.1 Subjects and Physical Situations

Of all the subjects who signed up for the experiment, 17% refused to take part because of the possibility of administering or receiving an electric shock. We ended up with 64 subjects: 32 males and 32 females. Most of them were Harvard College undergraduates. The subjects were randomly assigned to one of the thirty-two conditions of the study: 2(V = human distance) $\times$ 2(J = public justification) $\times$ 2(I = information) $\pm$ 4(T = third party).

The experiment took place in two rooms divided by a one-way mirror through which the subject and third party in one room could observe the pacifist seated in the other room.

The physical equipment (panels and shock generator) was constructed in the electronics shop of the Department of Psychology and Social Relations at Harvard University. The panel (see Fig. 5.1) had all the capabilities of the Shure et al. (1965) game plus one additional feature, namely, the threat move.

5.4.2 Definition of the Situation

Instructions. Upon their arrival, subjects were informed of the use of electric shocks. Persons who chose not to participate received one dollar for their efforts and were dismissed. Subjects who chose to participate in the experiment were then told that they were going to perform as operators in a logistic system and that their task was to transmit energy. The following text briefed the subject concerning his role in the experiment.

You are going to play against another party who controls a panel identical to yours.

The task for you and the other party is to transport energy through a passage.

Both parties use the same passage. Only one party at a time can transport his energy through the passage. You have to transport your energy units from the right side to the left side of the panel, whereas the other party transports his energy units

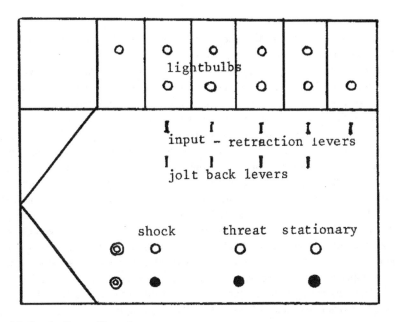

Fig. 5.1 Panel. *Source* The author

Fig. 5.2 Passage and storage rooms. *Source* The author

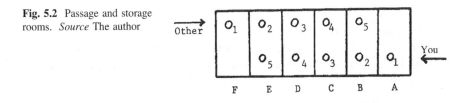

from the left to the right, as you see it. When you are able to store five energy units in the passage, you are permitted to pass through the passage, and you get paid 20 cents. The passage has six storage rooms (see Fig. 5.2): A, B, C, D, E, F.

In each storage room only one energy unit (symbolized by a light which you can turn on your panel, and by the letter 'O' in the diagram above) can be stored. Thus, either an energy unit of yours or one of the other party, but never both, can be stored at one time. Your energy units are located on the bottom row of the passage (O_5 O_4 O_3 O_2 O_1); the energy units of the other party are on the top row.

When it is your turn you can store an energy unit by pushing up a top-row level (try it). You have to start with energy unit O_1 and then, consecutively, O_2, O_3, O_4, O_5. Any other order is forbidden. Storing an energy unit is called an input move.

When it is your turn you have a choice between five, and, under certain conditions, six moves:

1. *Input Move*
2. *Stationary Move* (by pushing the push-button labelled Stationary): This means that you intend to stay where you were before,
3. *Retraction Move* (by pushing down one of the top row of levers that you have previously pushed up): This is the reverse of the input move. The meaning of this move is that you retract an energy unit that you have stored before in order to allow the other party to occupy the storage room.
4. *Shock Move* (by pushing the shock button): When you push this button, the other party receives a shock. This can be used as a reprisal or as a deterrent,
5. *Threat Move* (by pushing the threat button): When you push the threat button, you communicate to the other party that if he does not retract an energy unit, you will use a shock move or the jolt back move (if you possess the power to use it). On your next turn you are free to execute or not execute the threat.
6. *Jolt Back Move* (by pushing up one of the bottom row of levers); you can only use this move if you transported five energy units during the previous operating period. Otherwise you cannot use this move: When you use this jolt back move you empty one of the storage rooms that has been occupied by the other party and store an energy unit of yours. The use of this move is accompanied by the simultaneous deliverance of a shock to the other party.

There are sixteen operational periods. An operational period stops, and a new one starts when:

1. one of the parties has stored five energy units in the passage and consequently has passed through or,
2. there is a stalemate. We define an interaction as stalemated when sixteen moves have been made (sum of moves of both parties), and neither party has been able to transport five energy units through the passage.

The amount that you receive at the end of the experiment will be calculated as follows:

- you receive $1.50 guaranteed
- an additional 20 cents for each time you store five energy units in the passage (e.g., if you are able to store five units during each of the sixteen operating periods, you receive an additional 320 cents),
- however, each time you make a move you pay one cent for passage operating costs.

Example: If during the experiment you completed the passage fourteen times (i.e., fourteen times stored five energy units), and you used 85 moves, you will be paid:

$1.50 guaranteed
$2.80 for 14 transports through channel 0.85 operating costs $3.45
You can earn from $1.50 to $3.90.
If you have any further questions concerning the rules, ask the experimenter.
Do not talk to the third party present in the same room.

After the subjects had read the instructions, they were connected by means of two skin-electrodes attached to the left lower arm to a generator and given a sample shock. The subjects stayed connected until the end of the experiment. In the visual condition the pacifist was also connected by means of skin electrodes; however, his skin electrodes were not really connected to the generator. The pacifist used a little lamp that started flickering when he received a shock as a cue and expressed some facial irritation.

Pacifist. In the experiment the pacifist role was played by a programmed stooge (a male recently graduated from Harvard College). The pacifist-subject interaction was such that in the first operation period the pacifist allowed the subject to transport his energy. This move was to signal trust to the subject. In the next period, expecting a fair division of the earnings, the pacifist attempted to transport first. However, at that point the subject had the means of controlling transport. He could allow the pacifist to transport first, or he could use the jolt back move to force back the pacifist's stored units. The subject believed that the pacifist could choose to shock him or not shock him in retaliation. However, the pacifist stooge never used the jolt back or the shock move to retaliate. The pacifist continued to try to go first through the passage until his pay-off was equal to that of the subject.

Third party. In the same room with the pacifist, another stooge (a male senior of Harvard College) was presented as a third party. The job of the third party consisted of the delivery of a written communication to the subject before the experiment started and the registration of the moves made by the pacifist and the subject.

5.4.3   Controlled Variables and Predictions

In this experiment we were interested in several variables, of which four were controlled. The four controlled variables were concerned with human distance, justification, third-party effects, and information.

Human distance. In relation to this variable, we were interested in the comparison of the behaviour of the subject when confronted with an audio-visually absent pacifist (V) and a pacifist whose behaviour could be observed through a one-way mirror (V). We expected the subject to be more responsive to the pacifist's appeal in the visual condition than in a non-visual condition. More specifically, we predicted a considerable decrease in the use of violent responses in the visual condition. However, we did not expect any significant effect of the V condition on the cooperativeness of the subject. These predictions were based upon the following rationale: whereas in our culture the use of violence is strongly punished, successful competition and winning is effectively rewarded. Seeing the pacifist (without being seen by him) is a stimulus strong enough for triggering in the subject a moral restraint against the use of violence. Milgram (1965) has found evidence supporting the validity of this assumption. However, the same stimulus is not sufficient to evoke cooperative behaviour. Cooperative behaviour is more a function of the interpersonal exchange process than of the existence of cultural conditioning.

Therefore, we believe that if we want to stimulate cooperative behaviour, we need at least a visual exchange condition where both players can see each other. The findings of Whichman (1970), Exline (1963) and Hershel et al. (1973) support this last assumption.

Justification. Two conditions are distinguished: (1) justification condition (J), in which the subject receives the following message before the experiment starts: "When the experiment is over, you will meet the other two parties (party in the other room and third party present in your room), and you will be asked to explain your decision-making." (b) non-justification condition (J), in which the subject is notified that "When the experiment is over, you will not be able to meet the other two parties". We expected that the pacifist would receive less violent responses and score more winnings in a justification condition.

Third party. Earlier experiments investigated the impact of the existence of a neutral third party versus the absence of a third party. However, no studies investigated the impact of a partial and impartial third party. The term *partial* refers to partiality toward one of the parties (in our case toward the subject), whereas the term *impartial* refers to the third party's partiality towards a fair interaction system between the two parties. In our experiment we had four conditions:

(a) an impartial third party (Tl) – in this condition the third party was sitting in the same room as the subject and delivered to him, at the beginning of the experiment, a message saying that his function consisted of registering the moves made by the subject and other party and that he expected a fair interaction.

(b) a neutral third party (T2) – In this condition the third party informed the subject that his function consisted of registering the moves made by the subject and the other party,

(c) a partial third party (T3) – The subject received a message from the third party saying that his function consisted of the registering of the moves made by the two parties and that he hoped to see the subject cross the passage as often as possible, because the third party's pay-off would equal that of the subject.

(d) no third party (T4) – No third party was present.

Our predictions were that in a situation where the third party was present (Tl, T2, T3) there would be considerably less violence than in a situation with no third party (T4). When the third party was present, we expected most violence in T3, less in T2, and least in Tl.

Information. We distinguished between an information and a non-information condition. In the information condition (I) the subject and pacifist exchanged some personal information and a self-rating before the experiment started. The pacifist wrote the following standard information: "My name is Jib; I graduated in VES; I am interested in urban planning". On a semantic differential he rated himself as being active (rather than passive), non-violent (rather than violent), a dove (vs. hawk), a leader (vs. pawn), highly intelligent (vs. not very intelligent), and a sincere person (vs. an impression manager). The subject and pacifist also exchanged information before the fourth operating period. The standard pacifist message read as follows: "I am against every form of violence. So I won't use the shock or jolt back moves. However, I would expect that we share equally in the total pay-off". In the non-information condition (I) no personal information, self-ratings, or messages were exchanged.

We expected that the message would make the norm of reciprocity more salient and consequently reduce the amount of violent responses and increase the pacifist's winnings in the thirteen last operating period.

5.4.4 Non-controlled Variables

Sex. We did not expect sex to have a significant effect. However, if a difference existed, we expected females (F) to use less violence, in particular jolt backs and shocks. This prediction was made on the impression that females might be more conditioned by our society not to use physical violence in a conflict situation. In relation to cooperativeness, we did not expect females to differ from males.

Demand characteristics. After the experiment, we received from the subject an answer to the following question: "What do you think the experimenter expected from you?: (a) to cooperate with the other party (D1); (b) try to dominate the other party (D3); (c) did not care (D2)". We predicted the least violence and most fair interaction with a DI rating; most violence and the least fair interaction with a D3 rating. We expected the outcome of the D2 rating to lie between DI and D3 and nearer to D3.

Attribute configuration. After the experiment the subject was asked to describe his perception of the other party by checking the appropriate space on six scales of a semantic differential: active-passive; nonviolent-violent; highly intelligent-not very intelligent; sincere-impression manager; hawk-dove; obstinate-receptive. We expected everyone to rate the pacifist as non-violent and as a dove. Expecting no variation in the perception in relation to these two attributes, we decided not to use it as a predictor. We expected much more variation in relation to the other four attributes. Further, we predicted that a pacifist who was perceived as passive (Al), not very intelligent (A2), an impression manager (A3), and obstinate (A4) would receive far more violent responses and win less money than a pacifist who was perceived as active (Al), intelligent (A2), sincere (A3), and receptive (A4).

Monetary incentive. When the experiment was over, we asked the subject, "How important was the monetary incentive when you decided to take part in this experiment?" The subject could choose between (a) not important (Ml), (b) fairly important (M2), (c) very important (M3). We predicted the highest use of violence and the lowest number of pacifist winnings when the monetary incentive is very important. Additionally, we expected the results of M2 to fall between MI and M3.

5.5   Summary of Predictions

Our general expectation was that our data would corroborate earlier findings concerning the exploitative behaviour of a subject confronted with a pacifist in a conflict situation. However, we predicted that under particular conditions the exploitative behaviour would increase or decrease. In relation to the degree of exploitation measured by counting the amount of pacifist winnings (the more pacifist winning, the less exploitation, and vice versa), we predicted that:

J > £	Ml > M2 > M3
I > T*	DI > D2 > D3
T1 > T2 > T3 > T4	Al > Al
Fem = Male **	A2 > A2
	A3 > A3
	A4 > A4

An *effective pacifist conflict resolution* is characterized not only by a fair pay-off for both parties but also by the absence of the use of violence. In relation to the use of violent responses by the subject, we made the following predictions:

V < V	Ml < M2 < M3
J < T	DI < D2 < D3
I < I*	Al < Al
T1 < T2 < T3 < T4	A2 < A2
F < M **	A3 < A3
	A4 < A4

5.6   Results

Overviewing the results, we quickly discover how ineffective the pacifist strategy is in bringing about a pacifist conflict resolution. In our experiment approximately 37% of the subjects did not use any move involving the delivery of an electric shock. Though this percentage is considerably higher than the 12% found by Shure and the 14% in Vincent's study, we still find 63% of the subjects deciding to use violence in the resolution of conflicts. Apart from using violence, most subjects took advantage of the pacifist. A comparison of the average winnings (number of energy transmissions) and earnings (money received) of the subject and the pacifist clearly indicates a tendency to exploit the pacifist strategy (see Table 5.1).

Table 5.1 Comparison between winnings and earnings of pacifists and subjects

Winnings (0–16)		Earnings (0.00–3.90)	
Pacifist	4.118	Pacifist	$1.40
Subject	9.687	Subject	$2.47
t = 7.43 (p < .001)		t = 7.043 (p < .001)	

5.6.1 Controlled Variables

Table 5.2 indicates that approximately 60% of the effectiveness variation of the pacifist strategy is accounted for by our controlled variables and the interactions among the variables.

Human distance. A comparison of the means of violent response under a visual and nonvisual condition (12.75 and 21.50 respectively) indicates that when the subject can observe the pacifist without being seen by the pacifist he tends to reduce his aggressive behaviour. As predicted, the visual presence or absence variable has a significant effect ($p < .05$) on the total amount of violent responses. It also accounts for 5.27% of the total sum of squares of the violent responses.

If we look at the means of the pacifist winnings (4.06 and 3.34), we see more winnings under the visual than non-visual condition. However, as predicted, the effect is not significant, and the human distance condition accounts for only 1.23% of the variance.

Justification. A comparison of the means of violent responses and pacifist winnings under a justification and a non-justification condition (13.47 and 20.78;

Table 5.2 Percentage of the total "violent responses" and "pacifist winnings" variation explained by each controlled variable and the interactions among the variables

Controlled variables	Violent responses	Pacifist winnings
V	5.27 (p < .05)	1.23
J	3.68 (p < .10)	4.30 (p < .10)
I	4.68 (p < .10)	6.54 (p < .05)
T	5.48	4.96
V × J	0.03	0.19
V × I	1.55	2.24
J × I	0.00	0.28
V × T	5.34	2.37
J × T	9.50 (p < .10)	2.80
1 × T	1.04	9.69 (p < .10)
V × J × I	0.00	0.00
V × J × T	11.11 (p < .05)	16.48 (p < .01)
V × I × T	4.68	2.33
J × t × T	3.94	1.91
V × J × I × T	3.48	5.28
	59.79	60.60

4.16 and 3.03) indicates that when subjects are notified before the interaction that they will be required to justify their behaviour to other parties, they tend to reduce their aggression and allow the pacifist more winnings. The justification condition accounts for 3.68% of the violence variance (p < .10) and 4.3% of the pacifist winnings (p < .10).

Information. As predicted, the exchange of information seems to evoke a benevolent cycle of reciprocation, reducing the use of violence and increasing the pacifist winnings. A comparison of the means (13 and 21.25; 4.53 and 2.88) shows the effect clearly. The information condition accounts for 4.68% of the variance of violent responses (p < .10) and 6.54% of the pacifist winnings variance (p < .05).

Third party. The third-party condition accounts for 5.48% of the violent response variance and 4.96% of the pacifist winnings variance. The overall third-party effect is not significant. However, if we break the variance down, we find a borderline significant effect (p < .10) when we compare the importance of the presence versus absence of the third party on the total count of violent responses. This comparison does not give a significant effect in relation to the winnings of the pacifist. It seems that the presence of the third party is more of an inducement towards a reduction in violence than a stimulus towards cooperativeness for the subject. Further, we predicted a linear relationship between the four levels of the third-party condition and our two dependent variables. In relation to total violence and pacifist winnings, our predictions were corroborated. In both cases the linear comparison was significant at p < .10. The means the violent responses and pacifist winnings under impartial, neutral, partial, and absent third-party conditions were, respectively, 11.31, 15.81, 17.62, 23.75 and 4.65, 4.06, 3.43, 2.68.

In the Anova of the total number of violent responses, we found a two-way (justification × third party) interaction with a significance level of p < .10, accounting for 9.5% of the total variance; and a three-way interaction (visual × justification × third party) with a p < .05 significance, accounting for 11.11% of the variance. Breaking down the variance of the two-way interaction, we found a significant p < .05) when we looked at the interaction between T1 and T3 versus T2 and J versus J. This interaction comparison explained 80% of the two-way interaction variance.

Studying Fig. 5.3, we find that the effects induced by T3 and T1 under both conditions are as expected. However, totally unexpected are the results created by T2. In a situation where no justification is required, the neutral third party induces less violence than the impartial third party. In the justification condition, we find the opposite to be true; namely, a neutral third party induces more violence than a third party expressing his biases (T1 and T3). A possible explanation of this significant interaction is that, in a situation where no justification is required, unsolicited interference (here in the form of moral advice given by T1) is perceived as unsuitable and evokes some kind of embarrassment. The embarrassment of socially admitting moral incompetence may increase the intractability of the subject and evoke aggression. There is historical evidence that the induction of guilt feelings in an amoral situation can lead to the use of more violence. Opton's (1971) study of My Lai ends with the admonition that efforts to induce consciousness of guilt in

Fig. 5.3 Two-way interaction between justification and third-party condition. *Source* Compiled by the author from his own research

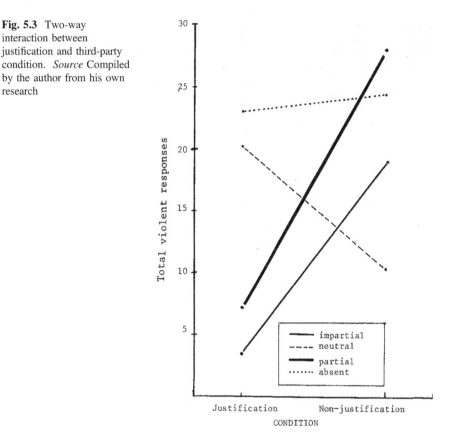

people who lack inner strength to bear it can backfire, evoking behaviour that relieves queasiness by demonstrating that what is feared can be done, even more and worse, without catastrophic consequences to oneself. In our experiment the subgroup that is most susceptible to this effect consists of the subjects who said that they initially intended to dominate. For them the moral appeal of the third party to cooperate creates most cognitive dissonance. In comparison to the other subjects, who either intended to cooperate or had not made up their minds, the ones who had at the outset decided to dominate used more violence when the third party appealed for fairness in the non-justification condition.

The fact that neutrality induces more violence in a justification condition could be explained as follows: a third party who will be present when the subject has to justify his behaviour and seems to be totally unconcerned reinforces the amorality of the definition of the situation to the subject, thus reducing the impact of the justification condition.

Breaking down the variance of the three-way interaction (Figs. 5.4 and 5.5), we found a significant interaction comparison between T1 and T2, J and J, and V and V

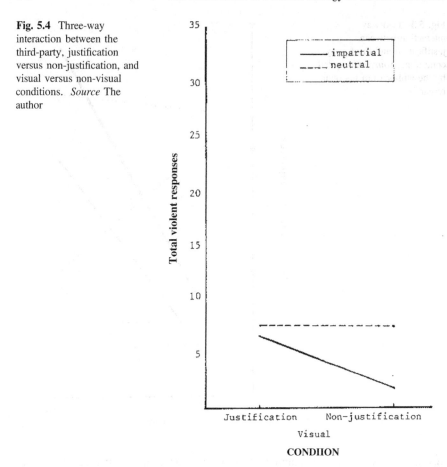

Fig. 5.4 Three-way interaction between the third-party, justification versus non-justification, and visual versus non-visual conditions. *Source* The author

at p < 0.05. This comparison accounts for approximately two-thirds of the three-way interaction. From the figures we can infer that in order to achieve a more accurate prediction of the effect accounted for by the three-way interaction we should be informed about the human distance condition. The effects of the two-way interaction are clearly present in the non-visual condition and practically absent in the visual condition.

In the Anova of the pacifist winnings, we have one significant three-way interaction (visual × justification × third party) at p < .01, which accounts for 16.48% of the total variance. Breaking down the interaction, we found an interaction comparison (T1 vs. T2, J vs. J, and V vs. V) at the p < .01 level, accounting for approximately 70% of the three-way interaction variance.

If we look at Figs. 5.6 and 5.7, we observe a similarity in the three-way interaction of the violence Anova. This similarity is quite obvious after turning the pacifist winnings scale upside down. The three-way interaction of the violence and winnings Anovas gives some support to two important hypotheses. First, the

Fig. 5.5 Three-way interaction between the third-party, justification versus non-justification, and visual versus non-visual conditions. *Source* The author

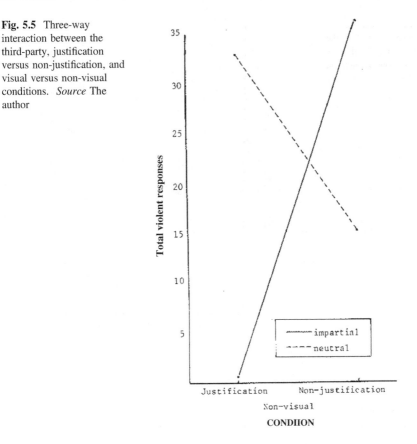

existence of neutrality in a social system where justification is institutionalized tends to be counterproductive to the inducement of more cooperative and less violent behaviour. Second, this trend seems to be especially true in a context where human distance is great.

5.6.2   Non-controlled Variables

Sex. In our experiment we had 32 females and 32 males. From the correlation matrix we could infer the non-existence of significant differences between sexes in relation to the winnings of the pacifist and the use or non-use of violent responses by the subject. The correlation between the total amount of violent responses and sex is -0.030; the correlation between sex and pacifist winnings is somewhat higher, namely 0.178. Male subjects tend to be more cooperative than women.

As a predictor of the number of pacifist violent responses, sex is useless. However, for predicting the number of pacifist winnings, sex is a predictor (see Table 5.3).

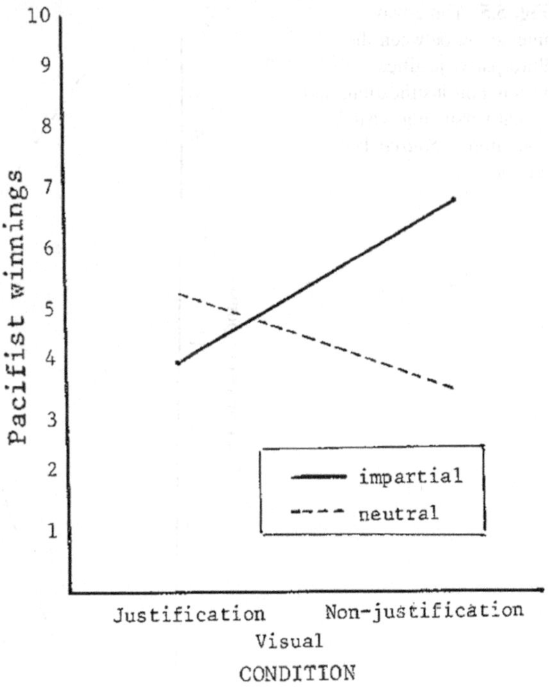

Fig. 5.6 Three-way
interaction between the
third-party, justification
versus non-justification and
visual versus non-visual
conditions

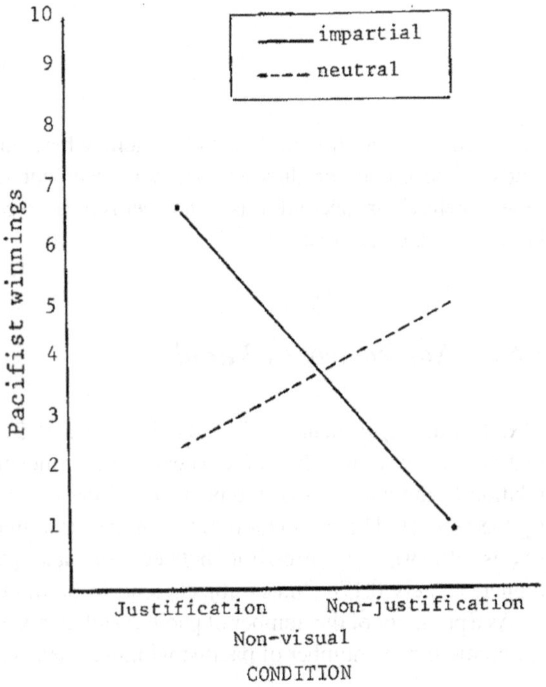

Fig. 5.7 Three-way
interaction between the
third-party, justification
versus non-justification and
visual versus non-visual
conditions

Table 5.3 Sex differences

	Female	Male
Total violence	17.5	17.3
Pacifist winnings	3.84	4.93

Note The independent variables are entered according to the highest partial correlation between the dependent variable and the independent variable, controlling for independent variables already entered into the equation.

In a multiple regression analysis using our controlled and uncontrolled variables as independent and the number of the pacifist winnings as dependent variables, we found that sex accounts for 6.6% of the variance and its coefficient is statistically significant (<0.01). A pacifist tends to win more when confronted with a male than when interacting with a female. This could be explained by the only significant correlation (-0.288) found between sex and one of the dependent variables, namely stalemates. This correlation indicates a greater tendency for females to stalemate in a conflict situation than males – with a consequent reduction in their own and the pacifist's winnings.

Demand characteristics. Demand characteristics have a correlation of -0.180 with the total number of violent responses of the subject, and -0.204 with the number of pacifist winnings. For a comparison of the means, see Table 5.4.

According to the multiple regression analysis for pacifist winnings, demand characteristics account for approximately 11% of the total variance. In the multiple regression analysis with the total violence as a dependent variable, it accounts for 9% of the variance. This variable is an important predictor for the two dependent variables.

Attribute configuration. For the distribution of the assignments of attributes to the pacifists, see Table 5.5. The fact that the numbers do not add up to 64 is caused by some subjects forgetting to check one or more of the attribute dimensions.

As expected, no subject perceived his opponent as violent or as a hawk. That being equal for all subjects, we predicted that the variation in the behaviour of the subjects would be caused by the differential assignment of the other attributes. The

Table 5.4 Demand characteristics. *Source* Compiled by the author from his own research

	Pacifist winnings	Total violence
Cooperate (14)[a]	7.6	3.5
Did not care (25)	3.84	17.0
Dominate (25)	3.0	24.6

[a]Number of subjects

Table 5.5 Distribution of the assignment of personality attributes to the pacifists. *Source* Compiled by the author from his own research

Non-violent	62	0	Violent
Dove	59	0	Hawk
Active	20	3	Passive
Highly intelligent	49	1	Low intelligence
Sincere	49	1	Impression manager
Receptive	43	1	Obstinate

Table 5.6 Correlation of personality attributes with pacifist winnings and violent responses. *Source* Compiled by the author from his own research

	Pacifist winnings	Violent responses
Active	.191	−.309 (0.05)
Intelligent	.369 (0.01)	−.345 (0.05)
Sincere	.345 (0.01)	−.319 (0.05)
Receptive	.451 (0.001)	− .371 (0.01)

four other attributes have highly significant correlations with the pacifist winnings and the total amount of violent responses of the subjects (see Table 5.6).

The attribute that accounts for the most variance in the pacifist winnings and the total number of violent responses is *receptivity versus obstinacy.* In the multiple regression analysis (MR) for pacifist winning, it accounts for 23% of the variance. In the MR for total violent responses, receptivity accounts for 12.8% of the variance. Of all the variables, the attribution of receptivity accounts for the most variance and seems to be a very good predictor of the effectiveness of the pacifist strategy.

The attribution of *intelligence* which is not correlated with receptivity, but which has a high correlation with the attribution of sincerity [0.507 (0.001)], accounts for 5% of the variance in the MR of violence and for 2% of the pacifist winnings variance. Only in the MR of the pacifist winnings is the coefficient significant ($p < .05$). We conclude that the attribution of intelligence is a fairly good predictor of the total number of violent responses but a weak predictor of the amount of pacifist winnings.

A third attribute which accounts for 5.5% of the variance of the total amount of violent responses is the attribution of *activeness versus passiveness.* The coefficient of this attribute is significant ($p < .05$). If we want to achieve more accurate predictions concerning violent responses, we should also take this variable into account. However, in relation to the pacifist winnings, this attribution has no effect at all.

The attribution of sincerity has highly significant correlations with the number of pacifist winnings and the violent responses received by the pacifist. Thus, if the pacifist is perceived as an impression manager, he receives a significantly higher dose of violent responses than when sincerity is attributed to him. The same relation in the reverse direction, however, is valid in relation to the number of pacifist winnings. Because of the high correlation of this attribute with intelligence, we left it out of the multiple regression analysis. However, we could say that the sincerity attribution can be substituted for the intelligence attribution as a predictor. This contention is supported not only by the high correlation between them but also by the results of a factor analysis of all our independent variables. In this factor analysis, we found that both intelligence and sincerity load very high on the first independent factor (0.710 and 0.830).

To summarize, we can say that information concerning the assignment of attributes other than non-violence or dove is more important for the prediction of

Table 5.7 Correlations between the four attribute and information variables. *Source* Compiled by the author from his own research		Active	Intel	Sincere	Recept	Inform
	Active	1.000				
	Intel	.005	1.000			
	Sincere	.297	.507	1.000		
	Recept	.063	.044	.151	1.000	
	Inform	.259	.418	.447	.011	1.000

the subject's behaviour. In relation to his or her violent responses, they account for 23% of the variance, and for 25% of the pacifist winnings variance.

Having established the importance of the pacifist image, it would be interesting to know to what extent it is possible to influence the image of the pacifist held by his opponent. The correlation matrix (Table 5.7) indicates a positive impact of the controlled information variable on three attribute dimensions: active versus passive (0.259), high versus low intelligence (0.418), and sincere versus impression manager (0.447). No such impact exists, however, with respect to the receptive versus obstinate dimension (0.011). Thus, we seem to have no control over the attribute variable receptive-obstinate which accounts for most of the pacifist winnings or violence variance.

Money. Differences in monetary motivations do not correlate either with the total number of violent responses (0.090) or with the number of pacifist winnings (−0.051). In our multiple regression analysis (MR) with total violent responses as dependent variables, the monetary incentive variable accounts for no variance (0.000). This is also the case in the MR with the number of pacifist winnings as dependent variables (0.006). However, we ought to mention one significant correlation (−0.270) between the monetary incentive variable and the number of retraction moves. There seems to be a tendency for people to whom the monetary incentive is very important to make less use of the retraction move than for people to whom the money is not very important. Persons for whom the monetary incentive is high seem to be very resolute in their pursuit to win during an operation period once they have decided to do so.

5.7 Conclusion

We have studied the impact of different conditions on the effectiveness of a pacifist strategy. Predictions were made concerning the effects of four controlled and four uncontrolled variables on the cooperativeness (pacifist winnings) and the amount of violent responses of the subjects. Together, seven of the eight studied variables (monetary incentive excepted) account for 61% of the pacifist winnings and 51% of the violent responses variance. (Corrected R^2 = 0.54 for pacifist winnings and 0.45 for the violent responses.) All our predictions concerning the controlled variables were confirmed. We found that the pacifist strategy tends to be most effective when the human distance between the subject is reduced (V), the subject is well informed about the intentions of the pacifist (I), the subject is required to justify his behaviour post facto (J) and a third party is present.

The data also show that an impartial third party evokes more positive responsiveness in the subject toward the pacifist than a neutral party; the least positive responsiveness is induced by a partial third party. Though the relationships described above are valid in most situations, they are invalid in some; e.g., a third party's plea for a fair interaction (impartiality) in a non-justification condition tends to evoke more negative responsiveness (in the form of violent responses and less cooperation) than a third party's neutrality. Under certain conditions, the absence of third-party mediation contributes more to the effectiveness of a pacifist strategy than a well-intended but unsuitable intervention. Thus, successful mediation depends not only on the level of mediation chosen, but also on the third party's thorough understanding of the interaction between different scope conditions and the levels of mediation.

Apart from the four above-mentioned variables for the prediction of the effectiveness of the pacifist strategy, sex should also be taken into account. Although sex makes no difference in relation to violence, it accounts for 6.6% variance of the pacifist winnings.

A sixth variable which has a considerable effect on the dependent variables, demand characteristics, accounts for approximately 10% of the variance of each of the dependent variables. This variable is uncorrelated with our other independent variables and can be considered a strong independent predictor.

Differences in the degree of monetary motivation for taking part in the experiment had no effect whatsoever on our dependent variables. This finding was rather unexpected. The relative value difference between the pay-offs for cooperation ($2.30) and competition ($3.90) was probably too low to induce subjects to compromise their basic attitudes towards conflict resolution.

The strongest predictor of the effectiveness of the pacifist strategy is the image of the pacifist held by the subjects. There was no variation between the subjects in relation to the attribution of non-violent or dove to the pacifist; each subject perceived the pacifist as non-violent and as a dove. However, there was more variation between the subjects in relation to the assignment of the four other attributes (active, intelligent, sincere, receptive). Treating these four attributes as independent variables, they account for 25% of the pacifist winnings and for 23% of the violence variance. From these findings, we can infer that it is not enough for the pacifist to project an image stressing only his non-violent nature. In addition to a successful communication of his non-violent intentions, the pacifist should also try to promote an image of other attributes which are highly valued by the subject. A pacifist perceived as receptive, intelligent, active and sincere has a much better chance of evoking positive responses from the subject than a pacifist perceived as obstinate, not very intelligent, passive, and as an impression manager. The extent to which we could influence the opponent's perception of the pacifist was fair with respect to the active, intelligent, and sincere attributions, but we had no control whatsoever over the receptive versus obstinate attribution which accounts for most of the variance of the effectiveness of the pacifist strategy.

With the exception of the information and three attribution variables, the correlation coefficients between our studied variables are low. This indicates that our variables can be considered to be quite independent pieces of information, and consequently cannot substitute for each other.

In this study, demand characteristics and attributions were treated as independent variables in our model for predicting the effectiveness of a pacifist strategy. However, from the correlations between our dependent variables and the demand characteristics and attribution data which we collected when the experiment was finished, we cannot infer the direction of the causal relationship; and therefore cannot be sure about their role as independent variables. We could easily conjecture that they are dependent variables and that they fulfil a cognitive adjustment function for the opponent. It is psychologically more comfortable to an exploiting opponent to attribute some responsibility for his behaviour to the experimenter or to picture the pacifist as an impression-manager and a not-too-intelligent, passive, and obstinate person. Such a definition of the situation could perform a dissonance reduction function for the exploiting opponent.

In conclusion, we will answer two questions: (a) what did we learn from this study? and (b) can we extrapolate these results to other system levels? We learned that instead of saying that people are exploitative we should better conclude that people are more exploitative under some conditions and less under other conditions. Our answer to the second question is one of caution. As with any basic laboratory finding, one cannot extrapolate conclusions to particular real-life situations. Our position is that a study on an analogue level should be treated as a heuristic device. At most, one can expect from this experiment hints on what sort of variables may be pertinent. The findings do suggest directions we should explore to check out real-world possibilities. Such an exploration remains to be carried out for our pacifist strategy. To illustrate its relevance, let us assume that our experimental findings are also valid in international military situations. Our prediction would be that unilateral renunciation of the use of military force as an instrument of foreign policy has a high probability of leading to exploitation by another.

Therefore, if a foreign policy planner believes that another government is controlled by a non-pacifist elite, he should advise against a pacifist strategy. But since the outcome of a strategic interaction depends on the presence of several scope conditions, he should also advise the government which prefers non-violence to create conditions which inhibit the exploitative behaviour and enhance the cooperative behaviour of the opponent.

References

Aldous, J., & Tallman, I. (1972). Immediacy of Situations and Conventionality as Influences on Attitudes Toward War. *Sociology and Social Research, 56*, 356–367.

Case, C. M. (1923). *Non-Violent Coercion: A Study in Methods of Social Pressure*. New York: Century.

Cooley, C. N. (1922). *Human Nature and Social Order*. New York: Charles Scribner's.

Deutsch, M., Epstein, Y., Canavan, D., & Gumpert, P. (1967). Strategies of Inducing Cooperation: An Experimental Study. *Journal of Conflict Resolution, 11*(3), 345–360.

Exline, R. (1963). Explorations in the Process of Person Perception: Visual Interaction in Relation to Competition, Sex, and Need for Affiliation. *Journal of Personality, 31*(1), 1–20.

Exline, R., & Winters, L. C. (1965). Affective Relations and Mutual Glances in Dyads. In S. Tomkins & C. Izard (Eds.), *Affect, Cognition and Personality* (pp. 319–350). New York: Springer.

Gallo, P. S., & Mcclintock, C. (1965). Cooperative and Competitive Behavior in Mixed Motive Games. *Journal of Conflict Resolution, 9*(1), 68–78.

Galtung, J. (1957). Pacifism from a Sociological Point of View. *Journal of Conflict Resolution, 3*(1), 67–84.

Gergen, K. (1965). *The Psychology of Behavior Exchange*. Reading, MA: Addison-Wesley.

Goldman, K. (1971). *International Norms and War between States*. Stockholm: Laromedelsforlagen.

Gouldner, A. (1960). The Norm of Reciprocity: A Preliminary Statement. *American Sociological Review, 25*(2), 161–178.

Hershel, G., Kaplan, K., Firestone, I., & Cowan, G. (1973). Proxemic Effect on Cooperation, Attitude, and Approach-Avoidance in a Prisoner's Dilemma Game. *Journal of Personality and Social Psychology, 27*(1), 13–18

Hokanson, J., & Edelman R. (1966). Effects of Three Social Responses on Vascular Processes. *Journal of Personality and Social Psychology, 3*(4), 442–447.

Holmes, D. (1966). Effect of Overt Aggression on the Level of Psychological Arousal. *Journal of Personality and Social Psychology, 4*(2), 189–194.

Katz, D. (1960). The Functional Approach to the Study of Attitudes. *Public Opinion Quarterly, 24*(2), 163–177.

Kelley, H. (1972) Attribution in Social Interaction. In E. Jones et al. (eds.) *Attribution in Perceiving the Causes of Behavior*. Morristown, NJ: General Learning Press.

Lave, L. (1965). Factors Affecting Cooperation in the Prisoner's Dilemma. *Behavioral Science, 10* (1), 26–38.

Marlowe, D., Gergen, K. J., & Doob, A. N. (1966). Opponent's Personality, Expectation of Social Interaction and Interpersonal Bargaining. *Journal of Personality and Social Psychology, 3*(2), 206–213.

Meeker, R., & Shure, G. (1969). Pacifist Bargaining Tactics: Some Outside Influences. *Journal of Conflict Resolution, 13*(4), 487–493.

Milgram, S. (1965). Some Conditions of Obedience and Disobedience of Authority. *Human Relations, 18*(1), 57–76.

Morrison, B. J., Enzle, M., Henry, T., Dunaway, D., Griffin, M., Kneisel, K., & Gimperling, J. (1971). The Effects of Electrical Shock and Warning on Cooperation in a Non-Zero-Sum-Game. *Journal of Conflict Resolution, 15*(1), 105–108.

Machson, I., & Wapner, S. (1967). Effect of Eye Contact and Physiognomy on Perceived Location of Other Person. *Journal of Personality and Social Psychology, 7*(1), 82–89.

Ofshe, R. (1971). The Effectiveness of Pacifist Strategies: A Theoretical Approach. *Journal of Conflict Resolution, 15*(2), 261–269.

Opton, E. (1971). It Never Happened and Besides They Deserve It. In N. Sanford (Ed.), *Sanctions for Evil*. San Francisco: Jossey-Bass.

Orne, M. (1962). On the Social-Psychology of the Psychological Experiment: With Particular Reference to Demand Characteristics and Their Implications. *American Psychologist, 17,* 776–783.

Pilisuk, M., Potter, P., Rapoport, A., & Winter, A. (1965). War Hawks and Peace Doves: Alternate Resolutions of Experimental Conflicts. *Journal of Conflict Resolution, 9*(4), 491–508.

Rapoport, A. (1969). Games as Tools for Psychological Research. In I. Buehler & H. Nitini (Eds.), *Game Theory in the Behavioral Sciences.* Pittsburg: University of Pittsburgh Press.

Rapoport, A. (1963) Formal Games as Probing Tools for Investigating Behavior Motivated by Trust and Suspicion. *Journal of Conflict Resolution, 7*(3), 570–579.

Rapoport, A., Shubik, M., & Thrall, R. (1965). Editorial Comment. *Journal of Conflict Resolution, 9,* 66–67.

Sharp, G. (1951). The Meanings of Non-Violence: A Typology. *Journal of Conflict Resolution, 3*(1), 41–66.

Shaw, M., & Wright, J. (1967). *Scales for the Measurement of Attitudes.* New York: McGraw-Hill.

Shure, G., Meeker, R., & Hansford, E. (1965). The Effectiveness of Pacifist Strategies in Bargaining Games. *Journal of Conflict Resolution, 9*(1), 106–117.

Sibley, M. (1968). Pacifism. In D. L. Sills (Ed.), *International Encyclopedia of Social Sciences* (353–357). New York: Macmillan.

Snyder, R., Bruck, H., & Sapin, B. (1962). *Foreign Policy Decision-Making.* New York: Free Press of Glencoe.

Solomon, L. (1960). The Influence of Some Type of Power Relationships and Game Strategies Upon the Development of Interpersonal Trust. *Journal of Abnormal and Social Psychology, 61*(3), 223–230.

Vincent, J., & Tindell, J. (1969). Alternative Cooperation Strategies in a Bargaining Game. *Journal of Conflict Resolution, 13*(4), 494–510.

Walton, R. (1969). *Interpersonal Peacemaking: Confrontations and Third Party Consultation.* Reading, MA: Addison-Wesley

Whichman, H. (1970). The Effects of Isolation and Communication on Cooperation in a Two-Person Game. *Journal of Personality and Social Psychology, 16*(1), 114–120.

Wilson, W. (1969) Cooperation and Cooperativeness of the Other Player. *Journal of Conflict Resolution, 13*(1), 110–117.

Chapter 6
The Art of Conflict Prevention: Theory and Practice

At different international organizations, such as the UN, the CSCE, NATO, and the EC-WEU, efforts are being made to improve their conflict prevention facilities.[1] In academic circles, research data about crisis prevention and management, preventive diplomacy and other studies related to the prevention of unwanted escalation of conflicts are being reviewed and new research projects set up. Conflict prevention (CP) is in again. This proactive approach to conflict has been enhanced by recent positive and negative experiences.

The Gulf War and the Yugoslavian civil war have destroyed the initial peace-euphoria of the Post-Cold-War period. Instead, we have found ourselves with an increasing number of conflicts. Many have been unforeseen; they have escalated faster than expected and rapidly became internationalized. Furthermore we have not been equipped to handle them effectively. Euphoria has been replaced by the expectation that large-scale violence will be part of the European scene because internal instability is likely to persist in Central and Eastern Europe for at least the next generation, and probably even longer (Goodby 1992: 154). The *Sarajevo* effect, or the fear of becoming entangled in ethnic and nationalist disputes, has rekindled interest in the prevention of destructive conflicts.

Equally important has been the high cost of peacekeeping and peace enforce-ment after conflicts have escalated. The Gulf War, measured in direct outlays for military expenditure, cost fifty billion dollars. The budget for two days of Operation Desert Storm would have paid for all UN peacekeeping operations worldwide for a year. The former has not been challenged; the latter remains a chronic problem.

Another criticism is that peacekeeping without peacemaking frequently leads to a 'peaceful perpetuation' of disputes (Ryan 1990: 141). The presence of the Blue

---

[1]This text was first published as: The Art of Conflict Prevention: Theory and Practice. In W. Bauwens and Luc Reychler (Eds.): *The Art of Conflict Prevention* (London: Brassey's, 1994): 1–21. Reprinted with permission.

L. Reychler and A. Langer (eds.), *Luc Reychler: A Pioneer in Sustainable Peacebuilding Architecture*, Pioneers in Arts, Humanities, Science, Engineering, Practice 24, https://doi.org/10.1007/978-3-030-40208-2_6

Helmets and the resulting diffusion of the crisis may ease the sense of urgency felt by the parties to resolve the dispute. They may find it in their interests to perpetuate the peacekeeping presence. Finally, there is a growing awareness of the fact that the management of conflicts after they have escalated is much more difficult. Feelings of hatred and frustration tend to protract them and require increased effort to bring them to a satisfactory conclusion (Azar 1990).

The discussion of conflict prevention is part of a search for a more cost-effective security organization. A major lesson of recent experience is that "one ounce of prevention is far better than a ton of delayed intervention". Another lesson, following from the great transformation of 1989–1990, is that preventive conflict management has become a feasible option.

6.1   A Sensible Concept in Post-Cold War Diplomacy

Conflict prevention has become a key concept in the discussion of Europe's new security and peace organization. The semi-permanent battleground of Yugoslavia has taught us that Europe and the United States should be better prepared next time. As long as Central and Eastern Europe remain relatively unstable, incidents may occur that will become threats to European peace and security (Goodby 1992: 155). Despite the prevailing view that there is an urgent need for better conflict prevention, the concept remains a sensitive one and has not been spared from criticism. The reservations expressed are not addressed at the aims, but at the ways and means of implementation.

One source of criticism is from those people who tend to associate conflict prevention with the creation of a *collective security system*. They are concerned with not only its feasibility but also its desirability. There is the fear of being dragged into war whenever and wherever aggression occurs. There is also a concern about the loss of national sovereignty. The latter, however, could be alleviated by a looser collective security system in which no automatic and binding commitments would be made, and cooperation would be judged on a case by case basis. These people additionally do not expect the preconditions to be realized that must be present if a collective security organization is to take shape and function effectively. They believe that whilst the international moral-political climate has improved, it is as yet far from being sufficiently developed for a collective security system to function successfully. National self-interests are still far from being equated with the concept of the welfare and stability of the international community.

A second source of criticism lies in those who associate conflict prevention with *domestic interference and intervention*, including military intervention for humanitarian or security reasons. These people protest at the undermining of the principles of non-intervention and sovereignty. Many regimes from Third World countries, for example, believe that they are most vulnerable to collective sanctions and fear that the major powers would misuse intervention to further their own selfish interests (Buzan 1991). The last source of criticism is those people who associate the concept of conflict prevention with the possible *use of military force* to keep or restore peace. They fear a military interpretation of the idea of conflict prevention.

All these reservations and criticisms make it clear that conflict prevention is one of those things that is easier said than done.

6.2   A Conceptual Framework of Analysis Conflict Management and Prevention

In confronting the prospect of increased demand for conflict prevention, it is useful to define the terms and to outline a typology of instruments for preventing unwanted escalation. This is necessary because the field of study is characterized by a great deal of conceptual confusion. One should, of course, keep in mind that the following definitions and typologies are intellectual creations to highlight policy issues rather than neat depictions of reality. The term *conflict prevention* refers to a particular kind of conflict management, which can be distinguished from *conflict avoidance* and *conflict resolution* (Mitchell 1981). Efforts to avoid the development of contentious issues and the incompatibility of goals are called *conflict avoidance*; measures which contribute to the prevention of undesirable conflict behaviour once some situation involving goal incompatibility has arisen fall under the heading of *conflict prevention;* and any strategy activated at the stage when a conflict involves incompatible goals, hostile attitudes and disruptive behaviour is categorized as *conflict settlement* or, in special cases, *conflict resolution*. The concept of 'conflict prevention' refers to two types of effort: those which prevent behaviour defined within the relevant international system as undesirable, and those which attempt to confine conflict behaviour within clearly defined limits of permissible activity. The concept is not new. Older terms such as "preventive diplomacy", "crisis-prevention" and "deterrent diplomacy" all refer to similar efforts.

6.2.1   A Typology of Conflict Prevention Measures

The development of a typology of conflict prevention measures requires an understanding of the causes of war and peace. These causes can be clustered into three groups of factors, the first related to the conflict itself, the second to opportunity structure and the third to the decision-making process (L. Reychler) (Table 6.1).

Table 6.1 Typology of conflict prevention measures. *Source* The Author

Dealing with conflict	Peacemaking	Imposing a solution through coercion Judicial dispute settlement Negotiation Traditional methods
	Peacebuilding	Attitudinal measures Socio-economic measures Political monitoring of elections Strengthening of the democratization process and human rights
Dealing with opportunity structure	Non-military measures	Political-diplomatic Legal Economic Informative-educational Normative-ideological
	Military measures	Nuclear deterrence Collective security system Collective defence system Arms control CSBMs Peace enforcement Peacekeeping
Dealing with decision making		Reducing misperceptions Creating less cumbersome decision-making Searching for better CP strategies

6.2.1.1   Measures Dealing with the Conflict

To prevent a conflict from escalating, one can deal with the conflict itself and take measures either to avoid the development of contentious issues and goal incompatibilities or address them after they have arisen. One can take measures to remove the source of the conflict, through some form of settlement or by resolution of the conflict itself. Each type of conflict requires a different remedy. If it is a facts-based conflict or a disagreement over "what is", it could be solved by an exchange of information, an increase in military transparency and a more effective set of verification agreements. If it is an interest-based conflict or a disagreement over who will get what in the distribution of scarce resources (e.g., power, territory, economic benefits or respect), it could be managed constructively through peacemaking and peacebuilding efforts.

Peacemaking is concerned with the search for a negotiated resolution of the perceived conflicts of interest between the parties. Three groups of peacemaking methods can be identified (Ryan 1990). The first method is to try to impose a solution through either violence or power. The second method relies on the judicial settlement of disputes. The problems with the use of violence and power and the inappropriateness of this recourse to the legal instrument, however, have led many to regard negotiation as the best method of conflict settlement or resolution.

Negotiation can be subdivided into two broad categories. The first contains traditional diplomatic instruments, such as negotiation, good office, mediation, enquiry, conciliation and arbitration (Riggs/Plano 1988: 186–187). Criticism of conventional methods led to the development of a whole series of alternative methods for dispute resolution. These methods focus a great deal of attention on the process of interaction rather than the content of the negotiated positions. Exemplary in this field is the work of Roger Fisher, who works with the Harvard Negotiations Project, and the workshop approach of Burton (1969), further developed by Azar (1990) and Kelman (1989). These all believe that conflict managers should not wait until a conflict is ripe for negotiation. Instead they hold that a Track Two non-governmental type of diplomacy should be utilized, designed to establish a pre-negotiation stage in which analytical breakthroughs by the conflicting parties themselves are encouraged. Azar (1990) found that this process tended to create conditions enabling the introduction of other components of conflict management, such as strategies for economic development and political institution-building. This brings us to 'peacebuilding'. This term refers to a strategy which tries to avoid or resolve conflicts through measures of an attitudinal, socio-economic and political nature. Whereas peacekeeping is about building barriers between the armed groups, peacebuilding strategies involve greater inter-party contact. Peacebuilding tries to build bridges between people. It tries to avoid or reduce conflict escalation through contact plus forgiveness, contact plus the pursuit of superordinate goals, contact plus confidence-building, and contact plus education for mutual understanding (Ryan 1990). Peacebuilding also advocates development diplomacy. The latter aims to reduce the socio-economic deprivation and build responsive political institutions. It requires an investment in the human rights of individuals and minorities, political democratization and socio-economic development.

6.2.1.2   Measures Dealing with the Opportunity Structure

To prevent a conflict from escalating, one can also deal with the power available to the conflicting parties and the factors which constrain or allow its use. A first distinction can be made between military and non-military measures. The latter involve measures of a political-diplomatic, legal, economic, informative-educational and moral-ideological nature. For example, a country or group of countries could threaten to close diplomatic posts, the Security Council could delegitimize certain kinds of behaviour, the International Court of Justice could decide against a party, an economic boycott could be implemented, violations of international law could be exposed to world opinion, a moral authority could promote a liberation theology and thereby disapprove of the neglect of human rights. Among the military measures used to prevent conflicts one can distinguish: nuclear deterrence, balance of power, collective security systems, arms control, Confidence and Security Building Measures (CSBMs), peace enforcement and peacekeeping. With respect to the recent violence in Europe, a great deal of attention has been given to peacekeeping and peace enforcement. *Peacekeeping* forces are used in situations where a cease-fire has been established and when the main parties to the conflict want help in preserving an unstable peace. They carry light arms and their mission can be compared to that of police forces. *Peace enforcement* is used in situations in which violence is still raging. The purpose is to impose peace. For such an operation, well-equipped units are necessary and their mission is comparable to that of a military campaign. To cope with the conflict in Yugoslavia, the international community has made use of peacekeeping operations. Peacekeeping forces could be used for different missions: to carry out humanitarian functions, such as organizing shipments of food and medicine under hazardous conditions, to observe a situation that contains some risk of conflict, to patrol borders or other sensitive areas, to establish a buffer zone between adversarial military forces, and to protect enclaves of ethnic minorities (Goodby 1992: 166). The preconditions for using peacekeeping are very strict. Urquhart (1990) notes that over the years, agreement has been established on the following seven interconnected preconditions for successful peacekeeping missions:

- the consent of the parties involved;
- the continuing and strong support by the mandating authority, the Security Council;
- a clear and practicable mandate;
- the maintenance of strict neutrality;
- the non-use of force except in the last resort of self-defence, including resistance to attempts by forceful means to prevent the peacekeepers from carrying out their duties;
- the willingness of troop-contributing countries to provide adequate numbers of capable military personnel and to accept the degree of risk which the mandate and the situation demand;
- the will of the member states, and especially of the permanent members of the Security Council, to make available the necessary financial and logistical support.

The reluctance of some EU nations to authorize the use of peacekeeping forces in Yugoslavia under conditions of an unstable peace considerably inhibits the effective use of peacekeeping. This "eighth precondition" allows the most irresponsible forces in the conflict to determine the use or non-use of peacekeeping forces. This seems to be a very irresponsible way of managing Post-Cold-War crises in Europe.

6.2.1.3   Measures Dealing with Decision-Making

These measures help the decision-makers manage conflicts more constructively by:

- Reducing misperceptions. A great deal of research has been done on the role of misperception in conflict behaviour. For example, the studies of decision-making under crisis (e.g., Frei 1982; Hopple et al. 1989; Janis 1972) are very useful. The decision-makers could use these findings to improve their conflict management skills. Also important is the availability of specialized and real time information and early warning facilities. Recent history has taught us that we are not very good at forecasting conflict behaviour. To a great extent this can be accounted for by a lack of institutionalized forecasting (Laulan 1991) and an inadequate understanding of conflict dynamics. Realistic means should be devised to help governments, with respect to Yugoslavia, for example, to assess the risks of a limited use of force compared to the risks involved in either a major escalation of force or not using force at all (Goodby 1992: 165).

- Providing them with a less cumbersome decision-making system which makes it possible to respond more effectively. The CSCE "unanimity minus one" voting rule is an improvement but leaves the organization structurally impaired as a credible crisis manager. Some interesting suggestions to recast the CSCE and to balance the need to reflect power realities with the need to foster consensus have been made by Charles and Kupchan (1991). Equally interesting are the proposals made by Boudreau (1984) regarding the establishment of an adequate global communication system and the use of satellite diplomacy.
- By searching for better strategies to prevent unwanted conflict behaviour. Conflict managers should be provided with a wider set of options, including assessments of anticipated costs and benefits.
- The need for a comprehensive conflict prevention strategy.

The inter-relationships among the three above-mentioned conflict-prevention strategies are very important. All three strategies tend to be essential for effective conflict resolution. Ineffective peacekeeping or enforcement can seriously hamper the peacemaking or peacebuilding process. Any group wishing to sabotage peace efforts will find it easier to provoke armed clashes, increase the number of martyrs for the next generation, and provide all the requirements for a conflict to become a protracted one. If peacemaking and peacekeeping efforts are made without adequate peacebuilding, people will be left out of the peace process. If decision-makers exceed the limits of people's tolerance, they may lose their support and be forced to abandon their peace efforts. Peacekeeping can discourage violence but cannot by itself build the foundations of an enduring peace. Peacekeeping can buy time but it cannot settle or resolve anything. Conflict prevention requires a comprehensive strategy. Reducing conflict prevention to mere peacekeeping missions or new negotiation efforts will not do. What is needed is a package of conflict prevention measures and much better coordination between the conflict-managing organizations.

6.3   Evaluation

During the past few years, at various levels (UNO, CSCE, NATO, EC-WEU), we have noticed serious efforts on the part of these organizations to evaluate and adapt their respective security strategies and organizations to the new strategic environment. These efforts, however, have not been translated into a successful handling of some of the new challenges. The historic opportunity to replace the old bipolar security organization by a Security Community stretching from Vancouver to Vladivostok is slipping through our fingers. Several ethnic and nationalist conflicts have turned into bloody civil wars. Of course, there have also been successes, for example, those scored by the new generation of peacekeeping operations in Central America, Namibia, Cambodia and the Western Sahara. But on the whole, much more will have to be done to prevent international conflicts more effectively.

The state of the art of conflict prevention could be rated on a set of scales of factors which are expected either to inhibit or enhance conflict prevention.

6.3.1 Lack of Interest or Common Interest

The interests of a majority of CSCE countries seem to be limited to containing the Yugoslav conflict within its boundaries and to preventing a possible spill-over into neighbouring regions. For many people the conflict is "far away from their bed". For them, the European House does not extend from San Francisco to Vladivostok.

The 'we-ness' feeling of many Western Europeans does not seem to reach far beyond their Little Europe. They do not consider peace to be indivisible beyond these subjective boundaries. The other Europeans have their hands full with their own domestic problems. Detracting attention from this conflict are the chronically overloaded diplomatic agendas. Equally responsible is the weakly developed strategic culture in Europe. The past forty years, characterized by an unusual stability and dependence on the Superpowers, seem to have retarded the development of a European strategic culture. In some small countries this underdevelopment has been reinforced by the existence of "small power fatalism" and a strong moral-legal approach to international politics. This weak strategic culture embraces several illusions: for example, the illusion that peace has broken out and that the peace dividend can be used for domestic purposes; or the illusion of distance, or the feeling that one should not be too concerned with faraway conflicts; and, finally, the illusion of time, or the belief that important decisions can be postponed to the distant future.

6.3.2 Lack of Foresight or Early Warning

Many of the recent upheavals in international relations have been political surprises. Think of the "velvet revolutions" in Eastern Europe, the Persian Gulf War, the rapid escalation of the ethnic conflicts in Yugoslavia, and the growing tensions between nations and states. It is not only the lack of foresight that is problematic, but also the complacency of the professional diplomats, the diplomatic correspondents and the academics about these matters. The model diplomat tends to be more interested in facts than probabilities. Few State departments have effective planning offices that do anything more than writing speeches for the Minister of Foreign Affairs. Political forecasting and contingency planning is far from being institutionalized. All this is unacceptable for two reasons. First, because the future will inevitably produce less attractive scenarios and confront us with new crises and threats and historical opportunities could also be more easily overlooked. Second, because such an attitude will not enhance the development of preventive diplomacy or more effective conflict prevention. Several explanations could be given for the failure to anticipate major developments in international relations (Reychler 1992). The traditional explanations attribute the lack of foresight in international politics to the state of the art of the science of international relations, including conflict dynamics, and to the lack of good intelligence. A comprehensive explanation, however, should also include: the general lack of interest; cognitive biases; societal and epistemic pressures; and neglect of the power of public opinion. Among the cognitive biases there is, for example, the propensity to treat the future as an extension of the past and as a gradual process. This is psychologically more comforting than living with an uncertain future. Under the heading of societal and epistemic pressures, one can distinguish: (1) the existence of myths and taboos; (2) the short-term attention span of democracies; and (3) the disinformation spread by pressure groups. During the Cold War, there were myths about "the irreversibility of communism", the "domino effect", and the "convergence between capitalism and communism", etc. One of the major causes of the many surprises in international affairs, however, is the relative neglect of the opinions of the common people. In Eastern Europe, for example, not enough attention was given to the development of the post-totalitarian mind. According to Goldfarb (1991), the development of these changes in the public opinion preceded the political and institutional changes. To anticipate developments in the Arab world, research relating to the development of the post-fundamentalist mind should now be undertaken. In addition, a better understanding of the revolutionary thresholds of the people would give us better insight into the dynamics of conflict.

6.3.3 Propensity to React or Propensity to Proact

International organizations are often reactive institutions. Getting their attention seems to require a certain quantity of blood. This attracts cameras, and then journalists and researchers wake up when the media appeal to their "instant expertise". In most universities practically no attention is being given to the development of future-orientated thinking. Most national and international governmental organizations seem to suffer from bureaucraticization of the art of diplomacy. Encountering unfamiliar problems, such as the fighting in Yugoslavia, they shrink from taking resolute action. With respect to this slow reactive approach, Winston Churchill made the following commentary: "We shall see how the counsels of prudence and restraint may become the prime agents of mortal danger; how the middle course adopted from desires for safety and a quiet life may be found to lead directly to the bull's-eye of disaster. We shall see how absolute is the need of a broad path of international action pursued by many states in common across the years, irrespective of the ebb and flow of national politics" (cited by Goodby 1992: 165). Somewhat related to this strong propensity to evade risks is the general inclination to wait for disputes to ripen before attempting a resolution (Haas 1990). Reactive conflict management is enhanced by the competitive political environment of democratic systems. The use of peacekeeping or peace enforcement creates openings for the opposition factions. Hence, prior to risking the domestic costs of people being killed in peace operations, democratic elites require more than the assumption of a potential future threat. Driven by the will to stay in power, incumbents tend to insist on evidence of a clear and present danger and usually require several provocations. The low-cost tolerance of the public also contributes to the political exhaustion and isolationism that often afflict democratic states (Schweller 1992). "Europeans, and Americans too, repeatedly made the tactical mistake of advertising their reluctance to engage in military action. Their purpose presumably was to assure the public that no adventures and no heavy costs were in store" (Rosenfeld 1994).

6.3.4 Traditional Diplomacy or New Diplomacy

Several characteristics of traditional diplomacy tend to inhibit effective prevention of the escalation of ethnic and nationalist disputes. First, traditional diplomacy draws a sharp distinction between civil wars, in which no external power is supposed to interfere, and international conflicts, which are the concern of all states. Civil war will only pose a threat when it spills over the boundaries or when major powers become involved. For them, the right approach is to isolate or quarantine the war zone. This is the logic of realpolitik. Related to this is the overriding concern with the principles of sovereignty and non-intervention. There is insistence on their "sovereign equality" and on the "one-state one-vote" principle, or the fiction that all states are equal and each state should have one vote. The new diplomatic thinking, on the other hand, is well aware of the complex interaction between internal and external conflict management and considers certain kinds of intervention necessary for realizing a stable international peace (Henrikson 1991). Second, traditional diplomacy tends to reduce conflict prevention to the employment of the familiar methods of peacemaking and/or peacekeeping and enforcement.

In contrast to this reductionism, the new diplomacy stresses the importance of a comprehensive peace strategy, including peacekeeping, peacemaking and peace-building measures. Third, traditionalists are not acquainted with the battery of new conflict management techniques.

6.3.5   Lack of Consensus or Strong Consensus

The prevention of the nationalist conflicts in Yugoslavia was seriously hampered by a lack of consensus among the major third-party powers involved. The negotiation efforts of the European Community were weakened by contradictory signals sent by the member states. In contrast to the other eleven member states, Germany opted for early recognition of Slovenia and Croatia. The German pledge that unsuccessful negotiations would result in recognition stimulated the parties who wanted independence to ensure failure. The US assumed a very low profile and warned the twelve members of the European Community that premature and selective recognition would damage prospects of peace and lead to greater bloodshed. Russia opposed the expulsion of Yugoslavia from the Conference on Cooperation and Security in Europe.

6.3.6   Cumbersome Decision-Making or Effective Decision-Making

The prevention of the Yugoslav conflict was inhibited not only by the overloaded domestic and diplomatic agenda of third parties, but also by the cumbersome methods for reaching decisions within the international organizations involved. The CSCE mechanism for the peaceful settlement of disputes has not so far proved to be fit for its task. Changes in the methods of reaching decisions are needed, probably including a qualified majority vote under clearly defined conditions or the creation of a European security council. The CSCE "unanimity minus one" principle is still too cumbersome. Only a smaller body which is more genuinely representative of the order of power within the international system can exercise a greater influence on the actions of states. The UN debate on the Yugoslav conflict in November 1991 showed that non-European members are prepared to obstruct peacekeeping operations for reasons having nothing to do with Europe. This is certainly one sound reason for developing regional peacekeeping machinery in Europe (Goodby 1992: 168). A second problem hindering international organizations from being more effective in fast-moving crisis situations is the element of competition between international organizations and the lack of a clear division of operational responsibilities between them.

6.3.7   Lack of Infrastructure or Adequate Infrastructure

Effective conflict management is impaired not only by the cumbersome decision-making procedures of the international organizations involved, but also by the near absence of contingency planning. Contingency planning within and between the UN, NATO, EC-WEU, and the CSCE Conflict Prevention Centre should be authorized so that governments have a better grasp of the issues to be decided if they, for example, are ever to authorize military force in crisis situations (Goodby 1992: 166).

Equally important is the availability, at short notice, of adequate peacemaking, peacekeeping and peacebuilding facilities.

An asset for Europe would be the constitution of a permanent Rapid-Response Peace Force that could be used for peacekeeping or peace enforcement. Urquhart's (1990, 1991) suggestion of equipping peacekeeping operations or other forms of preventive action with a trip-wire clause is also to be recommended. If the peaceful settlement of dispute measures provided under Chapter VI of the UN Charter proves to be of no avail, then under certain circumstances an automatic transition to the type of action outlined in Chapter VII of the Charter would be allowed. With such a clause, preventive action and deterrence might begin to have more effect (Urquhart 1990).

6.4 Suggestions

Scanning the literature, one finds a plethora of conflict prevention suggestions. To improve the state of the art in Europe, measures could be taken to advance.

6.4.1 Common Security Thinking

A strengthening of "common security" thinking requires the development of a European strategic culture and an extension of a 'We-ness' feeling throughout the whole of Europe. A better understanding of the impact of geopolitical trends on Europe's security would serve to strengthen Pan-European identity (Buzan 1991). A clear strategic insight would make people understand that the creation of a Pan-European Security Community would provide them, from the military point of view, with the most cost-effective security organization. They would also realize that this kind of security requires serious investments of a military and, even more, a non-military nature (Reychler 1991).

6.4.2 Early Warning

The installation of early warning systems stands high on the agenda of conflict prevention. Many concrete suggestions are also found here. To enable the UN Secretary General to exercise his preventive role effectively, Boudreau (1984) requests the installation of adequate global communications. Johansen (1991) pleads for an international monitoring agency, under the auspices of the UN, to detect violations of arms control agreements. Jonah (1990) stresses a strengthening of the UN Office for Research and the Collection of Information (ORCI). But, in any case, a successful early warning system will require more "intelligence sharing". Efforts will also have to be made to eliminate the factors which impede political forecasting. The institutionalization of forecasting in the foreign policy-making process and fundamental research into international conflict dynamics are also recommended.

6.4.3 Proactive Policy-Making

Proactive policy-making could be enhanced by the authorization of contingency planning. High-powered analysts would plan not only for a single, but also for a range of possible futures (Starling 1988). They would focus not only on short-term but also on long-term developments and remedies. The education of the democratic public opinion is of great importance. Efforts will need to be undertaken to convince the people that often the price of not looking ahead is at best a surprise and sometimes a catastrophe.

6.4.4 New Diplomacy

Conflict prevention would be enhanced by:

- the development of comprehensive conflict prevention strategies, including peacebuilding, peacekeeping and peacemaking;
- the extension of the rule of law to permit international intervention, such as humanitarian, security and environmental intervention (Henrikson 1991);
- bringing to bear all the conflict management resources available. The skills of non-governmental organizations or unofficial diplomats, largely ignored in diplomacy, could play major roles in peacemaking and peacebuilding (Berman/ Johnson 1977).

6.4.5 Consensus Building

The lack of consensus between the major external parties in Yugoslavia is to be accounted for not only in terms of differences of interests, but also in terms of different assessments of the options considered. Realistic means should be devised to help governments assess the risks of different options under consideration. The availability of a thorough assessment of policy options would help the governments involved to take resolute action and would probably enhance consensus-building.

6.4.6   Effective Decision-Making

To overcome decision-making problems, measures should be taken to cope more effectively with a chronically overloaded diplomatic agenda in order to facilitate swift and effective decision-making and to improve cooperation between international organizations. To cope simultaneously with several conflicts, these conflicts could be assigned to a department or an ad hoc interdepartmental or interagency team. The burden of conflict prevention could be shared or divided between several countries. A country or group of countries could *adopt a conflict* and assume responsibility for it. The structural impairment of the decision-making in international organizations such as the CSCE could be reduced through the introduction of qualified majority voting under clearly defined conditions or by the creation of a European Security Council (Kupchan/Kupchan 1991; Eyskens 1992). Better cooperation and division of labour among the UN, CSCE, NATO and the EC-WEU would also help.

6.4.7   An Adequate Infrastructure

The impact of declaratory conflict prevention, i.e., delineating in advance what actions will not be tolerated by the world community, is to a great extent dependent on the availability of an adequate infrastructure for peacemaking, peacekeeping and peacebuilding. This entails the further development of conciliation and mediation facilities. It necessitates new ways and means for peacebuilding in order to strengthen respect for human rights, the democratization process and sustainable socio-economic development. Some concrete proposals include the creation of an International Court for Human Rights or the more frequent use of election monitoring units. With respect to efforts to prevent or stop violence, one finds a great many suggestions concerning arms control and confidence-building measures. There is also a trend in the direction of developing a new generation of peacekeeping operations, including, for example, naval peacekeeping operations, peacekeeping in support of emergency humanitarian relief, and peacekeeping methods for the control of drugs (Diehl/Kumar 1991; Weiss/Campbell 1991). According to Norton-Taylor (2011) and Weiss/Campbell (1991), there is also an undercurrent of thought for the development of new and more forceful intervention techniques. Urquhart (1990) advises that peacekeeping forces could also be used as a tripwire which, after suitable warnings, would set in motion preplanned enforcement action under Chapter VII of the UN Charter. To realize those ideas, Henrikson (1991) suggests the creation of a UN Peace Force consisting of (1) a Standing-Reserve Peace Force composed of predesignated national units; (2) a smaller and more tightly organized Rapid-Response Force; and (3) a Permanent Peacekeeping Force.

A truly effective system of conflict prevention will require major investment. Hopefully the new interest in conflict prevention reflects a megatrend in security thinking and a determination to relieve the peoples of the world of unnecessary conflicts, excessive armament and the constant threat of war. Despite the slow pace of security reforms and the existing reservations vis-à-vis certain measures, investment in conflict prevention would be prudent in the long run.

References

Azar, E. (1990). *The Management of Protracted Social Conflict*. Dartmouth: Dartmouth Publishing Company.

Berman, M., & Johnson, J. (1977). *Unofficial Diplomats*. New York: Columbia University Press.

Boudreau, T. E. (1984). *The Secretary General and Satellite Diplomacy: An Analysis of the Present and Potential Role of the United Nations Secretary-General in the Maintenance of International Peace and Security*. New York: Council on Religion and International Affairs.

Buzan, B. (1991). New Patterns of Global Security in the Twenty-First Century. *International Affairs, 67*(3), 431–451.

Celente, G. (1991). *Trend Tracking*. New York: Warner Books.

Chee, C. H. (1991). The UN: From Peace-Keeping to Peace-Making? *Adelphi Papers, 32*(265), 30–40.

Diehl, P., & Kumar, C. (1991). Mutual Benefits from International Intervention: New Roles for United Nations Peace-Keeping Forces. *Bulletin of Peace Proposals, 22*(4), 369–375.

Eyskens, M. (1992). *Diplomatie Préventive*. Texte distribue par la Belgique & ses partenaires des CE lors du Conseil. Lisbon (17 February).

Fisher, R., & Ury, W. (1983). *Getting to Yes*. London: Penguin Books.

Frei, D. (Ed.). (1982). *Managing International Crises*. Beverly Hills: Sage Publications.

Goldfarb, J. (1989; 1991). *Beyond Glasnost: The Post-Totalitarian Mind*. Chicago: University of Chicago Press.

Goodby, J. (1992). Peacekeeping in the New Europe. *The Washington Quarterly, 15*(2), 153–171.

Haas, R. (1990). *Conflicts Unending*. New Haven: Yale University Press.

Henrikson, A. (1991). *Defining a New World Order* (Fletcher School of Law and Diplomacy Discussion Paper). Cambridge: The Fletcher School of Law and Diplomacy.

Holst, J. (1990). Enhancing Peace-Keeping Operations. *Survival, 32*(3), 264–276.

Hopple, G., Andriole, S., & Freedy A. (Eds.). (1989). *National Security Crisis Forecasting and Management*. Boulder: Westview Press.

Inman, B., Nye, J., Perry, W., & Smith, R. (1991). Lessons from the Gulf War. *The Washington Quarterly, 15*(1), 57–66.

Janis, I. (1972). *Victims of Groupthink*. Boston: Houghton Mifflin.

Joffe, J. (1992). Collective Security and the Future of Europe. *Survival, 34*(1), 36–50.

Johansen, R. (1991). A Policy Framework for World Security. In M. T. Klare, & D. C. Thomas (Eds.), *World Security: Trends and Challenges at Century's End* (pp. 441–444). New York: St Martin's Press.

Jonah, J. O. C. (1990). Office for Research and the Collection of Information (ORCI). In H. Chestnut, P. Kopacek, & T. Vamos (Eds.), *International Conflict Resolution Using System Engineering* (pp. 69–72). Oxford: Pergamon Press.

Kupchan, C. H., & Kupchan, C. I. (1991). Concerts, Collective Security and the Future of Europe. *International Security, 16*(1), 114–160.

Laulan, Y. M. (1991). *La Planète Balkanisde*. Paris: Economica.

Luard, E. (1990). *The Globalization of Politics*. London: Macmillan.

Mitchell, C. R. (1981). *The Structure of International Conflict.* New York: St Martin's Press.

Norton-Taylor, Richard (2011). An Attack on Iran Would Be Disastrous. *The Guardian Weekly* (4 November): 11.

Powaski, R. (1991a). United Nations and the Future of Global Peace. *USA Today, 120,* 19–20.

Reychler, L. (1991b). Querela Pacis. In J. Nobel (Ed.), *The Coming of Age of Peace Research* (pp. 89–96). Groningen: Styx.

Reychler, L. (1991). A Pan-European Security Community: Utopia or Realistic Perspective? *Disarmament: A Periodic Review by the United Nations, 14*(1), 42–52.

Reychler, L. (1992). The Price is a Surprise on Preventive Diplomacy. *Studia Diplomatica, 45*(5), 71–82.

Riggs, R., & Plano, J. (1988). *The United Nations.* Chicago: The Dorsey Press.

Rosenfeld, S. (1994). Peacekeeping: The Doctor is Indisposed. *Washington Post* (14 January).

Ryan, S. (1990). *Ethnic Conflicts and International Relations.* Aldershot: Dartmouth Publishing.

Schweller, R. (1992). Domestic Structure and Preventive War: Are Democracies More Pacific? *World Politics, 44*(2), 235–269.

Starling, G. (1988). *Strategies for Policy Making.* Chicago: The Dorsey Press.

Urquhart, B. (1990). Beyond the 'Sheriff's Posse'. *Survival, 32*(3), 196–205.

Urquhart, B. (1991). The UN: From Peace Keeping to a Collective System? *Adelphi Papers, 32* (265), 18–29.

Weiss, T., & Campbell, K. (1991). Military Humanitarianism. *Survival, 33*(5), 451–465.

Chapter 7
Religion and Conflict

7.1 Introduction: Towards a Religion of World Politics?

The New World Order cannot be understood without accounting for the role of religion and religious organizations.[1] During the Cold War, not much attention was paid to the phenomenon of nationalism and religion. Marxists, Liberals, nation-builders and integration specialists treated it as a marginal variable. In the Western political systems a frontier has been drawn between man's inner life and his public actions, between religion and politics. The West is characterized by a desecular-ization of politics and a depoliticization of religion. Part of the elite Western opinion views religion as irrational and premodern; "a throw-back to the dark centuries before the Enlightenment taught the virtues of rationality and decency, and bent human energies to constructive, rather than destructive purposes" (Weigel 1991: 27). In the Communist block, religion was officially stigmatized as the opium of the people and repressed. In theories of integration and modernization, secu-larization was considered a *sine qua non* for progress. Consequently, the explosion of nationalist and ethnic conflicts was a great surprise.

What about religious conflicts? Are we in for surprises too? The answer is: yes, probably. As late as August 1978, a US Central Intelligence Agency (CIA) Paper asserted confidently that "Iran is not in a revolutionary or even a pre-Revolutionary situation". Williamson (1990) considers this the most glaring example of Western incomprehension and misconception of Modern Islam. The fundamental mistake of Western observers, he argues, is the assumption that since Christianity plays little direct role in shaping policy in Western nations, the separation of religion and political decision-making can be assumed in the Middle East as well. Since the fall of the Shah, research about the role of religions in conflict dynamics has increased.

[1]This text was first published as: Religion and conflict (1997). *International Journal of Peace Studies* (January), 2(1): 21–38. Reprinted with permission.

L. Reychler and A. Langer (eds.), *Luc Reychler: A Pioneer in Sustainable Peacebuilding Architecture*, Pioneers in Arts, Humanities, Science, Engineering, Practice 24, https://doi.org/10.1007/978-3-030-40208-2_7

The amount of research, however, lags considerably behind the boom of studies of ethnic and nationalistic conflicts.

The attention for the role of religion in conflicts has been stimulated by positive and negative developments, including the desecularization of the world and the rise of religious conflicts. In most Strategic Surveys, attention is now paid to the militant forms of religious fundamentalism as a threat to peace. Also important has been the phenomenon of realignment or the cross-denominational cooperation between the progressives and traditionalists with respect to certain specific issues (Hunter 1991). An example of this is the view of the Catholic Church and the Islam Fundamentalists vis à vis the Report by the United Nations Population Fund on population growth.

Attention has also been drawn by the increased engagement of churches and church communities in the search for detente or the constructive management of conflicts. Think of the voice of American bishops in the nuclear debate in the Eighties; the role of churches in the democratic emancipation of Central and Eastern Europe; and the impact of church leaders on the conflict dynamics in several African conflicts. All have attracted considerable attention, not only in South Africa with Desmond Tutu and Allan Boesak, but also, for example, in Sudan (Assefa 1990; Badal 1990), Mozambique and Zaire. Mgr. Jaime Gonpalves, the archbisop of Beira, played an important role in the realization of a peace agreement in Mozambique on 4 October 1992. It ended a gory war in which a million lives were wasted and half of the population was on the run for safety. In Zaire, Monseigneur Laurent Monsengwo was elected chairman of the High Council of the Republic, and played a central role in the difficult negotiations between President Mobutu and his opponents. The Burundian Catholic bishops, representing half the population, are now mediating towards the development of a more collegial government to prevent further violence. Finally, we should also mention the role of the Church in empowering people in the Third World with Liberation theology and many recent efforts to provide peace services, including field diplomacy, in conflict areas.

To get a better grasp of what religions and religious organizations could do to promote a constructive conflict dynamic, one could start by systematically investigating which positive or negative roles they play now. Consequently, suggestions would be made about how to reduce the negative and strengthen the positive impact. Religious organizations can act as conflicting parties, as bystanders, as peacemakers and peacebuilders (see Table 7.1).

7.2 Religious Wars

Since the awakening of religion, wars have been fought in the name of different gods and goddesses. Today most violent conflicts still contain religious elements linked with ethno-national, inter-state, economic, territorial, cultural and other issues. Threatening the meaning of life, conflicts based on religion tend to become dogged, tenacious and brutal types of war. When conflicts are couched in religious

Table 7.1 Religious organizations in conflict dynamics. *Source* The author

Conflicting parties
Religious wars
Low-intensity cultural violence
Structural violence
Violence
Bystanders
Peacebuilders and peacemakers
Peacebuilding
Empowering people
Influencing the moral political climate
Development cooperation – humanitarian aid
Peacemaking
Traditional diplomatic efforts
Track II peacemaking
Field diplomacy

terms, they become transformed into value conflicts. Unlike other issues, such as resource conflicts which can be resolved by pragmatic and distributive means, value conflicts have a tendency to become mutually conclusive or zero-sum issues. They entail strong judgements of what is right and wrong, and parties believe that there cannot be a common ground to resolve their differences. "Since the North-South conflicts in the Sudan have been cast in religious terms, they developed the semblance of deep value conflicts which appear unresolvable except by force or separation" (Assefa 1990). Religious conviction is, as it has ever been, a source of conflict within and between communities. It should, however, be remembered that it was not religion that has made the twentieth the most bloody century. Lenin, Stalin, Hitler, Mao Tse-tung, Pol Pot and their apprentices in Rwanda maimed and murdered millions of people on an unprecedented scale, in the name of a policy which rejected religious or other transcendent reference points for judging its purposes and practices (Weigel 1991: 39). Those policies were based on an ideology possessing the same characteristics as a religion.

In a world where many governments and international organizations are suffering from a legitimacy deficit, one can expect a growing impact of religious discourses on international politics. Religion is a major source of soft power. It will, to a greater extent, be used or misused by religions and governmental organizations to pursue their interests. It is therefore important to develop a more profound understanding of the basic assumption underlying the different religions and the ways in which people adhering to them see their interests. It would also be very useful to identify elements of communality between the major religions.

The major challenge of religious organizations remains to end existing and prevent new religious conflicts. In December 1992, twenty-four wars with a religious background were counted (adjusted AKUF-database on wars). Most of them

Table 7.2 War with a religious dimension. *Source* Gantzel et al. (1993)

1. Myanmar/Burma	1948	Buddhists versus Christians
2. Israel/Palestine	1968	Jews versus Arabs (Muslims-Christians)
3. Northern Ireland	1969	Catholic versus Protestants
4. Philippines (Mindanao)	1970	Versus Christians (Catholics)
5. Bangladesh	1973	Buddhists versus Christians
6. Lebanon	1975	Shiites supported by Syria (Amal) versus Shiites supported by Iran (Hezbollah)
7. Ethiopia (Oromo)	1976	Muslims versus Central government
8. India (Punjab)	1982	Sikhs versus Central government
9. Sudan	1983	Muslims versus Native religions
10. Mali-Tuareg Nomads	1990	Muslims versus Central government
11. Azerbeijan	1990	Muslims versus Christian Armenians
12. India (Kashmir)	1990	Muslims versus Central government (Hindu)
13. Indonesia (Aceh)	1990	Muslims versus Central government (Muslim)
14. Iraq	1991	Sunnites versus Shiites
15. Yugoslavia (Croatia)	1991	Serbian Orthodox Christians versus Roman Catholic Christians
16. Yugoslavia (Bosnia)	1991	Orthodox Christians versus Catholics versus Muslims
17. Afghanistan	1992	Fundamentalist Muslims versus Moderate Muslims
18. Tajikistan	1992	Muslims versus Orthodox Christians
19. Egypt	1977	Muslims versus Central government (Muslim); Muslims versus Coptic Christians
20. Tunisia	1978	Muslims versus Central government (Muslim)
21. Algeria	1988	Muslims versus Central government
22. Uzbekisgtan	1989	Sunnite Uzbeks versus Shiite Meschetes
23. India (Uttar Pradesh)	1992	Hindus versus Muslims
24. Sri Lanka	1983	Hindus versus Muslims

were situated in Northern Africa, the Middle East, the ex-USSR and Asia. In Europe there were only two: Yugoslavia and Northern Ireland. No religious wars were registered in the Americas (see Table 7.2).

These wars could be further classified by distinguishing violent conflicts within and between religions and between religious organizations and the central government. In Europe during the 1990s, Bosnian Muslims were brutally harried by Serbs who called themselves Christians. On the border between Europe and Asia, Christian Armenians have thumped Muslim Azzeris, and Muslims and Jews still shoot each other in Palestine. Further east, Muslims complain of the Indian army's

Fig. 7.1 A cross-impact analysis of conflicts in which religion is involved. *Source* The author

brutality to them in Kashmir, and of Indian Hindus' destruction of the Ayodhya mosque in 1992. Islam, as Samuel Huntington has put it, has bloody borders (Huntington 1993). It was Huntington who recently provided the intellectual framework to pay more attention to the coming clash of civilizations. Civilizations are differentiated from each other by history, language, culture, tradition and, most importantly, religion.

He expects more conflicts along the cultural-religious fault lines because (1) those differences have always generated the most prolonged and the most violent conflicts; (2) because the world is becoming a smaller place, and the increasing interactions will intensify the civilization consciousness of people, which in turn invigorates differences and animosities stretching or thought to stretch back deep in history; (3) because of the weakening of the nation-state as a source of identity and the desecularization of the world with the revival of religion as a basis of identity and commitment that transcends national boundaries and unites civilizations; (4) because of the dual role of the West. On the one hand, the West is at the peak of its power. At the same time, it is confronted with an increasing desire by elites in other parts of the world to shape the world in non-Western ways; (5) because cultural characteristics and differences are less mutable and hence less easily compromised and resolved than political and economic ones; (6) finally, because increasing economic regionalism will reinforce civilization-consciousness.

Of course there are no **pure** religious conflicts. It is the correlation with other integrating or disintegrating pressures which will determine the dynamics of a conflict. There is a need for a more sophisticated typology.

For each conflict in which religion is involved, a cross-impact analysis is necessary of at least six variables which together could reinforce a constructive or a destructive conflict dynamic (see Fig. 7.1).

7.3 Low-Intensity Violence

To further their interests, religious organizations also make use of low-scale violence, political repression and terrorism. The British-Indian writer Salman Rushdie and the writer, physician and human rights activist Taslima Nasrin in Bangladesh were forced into hiding from Muslim fundamentalists who wanted to punish them with death. Each religion has its fanatical religious fundamentalists. The Kach Party, which was headed by Rabbi Meir Kahane until his death in November 1990, used tactics of abusing and physically attacking Palestinians. Kahane believed in a perpetual war and preached intolerance against the Arabs. Christian fundamentalists in the US embrace a "Manifest Theology", a fundamentally Manichean world-view in which 'we' are right, and all civil and aggressive intentions are projected on to 'them' (Galtung 1987).

"Because they are evil and aggressive forces of chaos in the world, we then have to be strongly armed, but do not perceive ourselves as aggressive even when attacking other countries" (Williamson 1992: 11). Intolerance is also spawned by a minority of Islamic organizations, like Egypt's Gama'at al-Islamiya, Lebanon's Hezbollah and Algeria's Islamic fundamentalists. All pursue a policy of violent confrontation, based on the convention that armed struggle or jihad is a necessary and appropriate response to the enemies of God, despotic rulers and their Western allies.

7.3.1   Structural Violence

Several religious organizations also support structural violence by endorsing a centralized and authoritarian decision-making structure and the repression of egalitarian forces. Churches have sympathized with authoritarian governments. The concord of the Vatican with Portugal in 1940, the agreement with Franco in 1941, and the support of authoritarian regimes in Latin America were clear statements. More recently, the Vatican disapproved of the candidacy of Aristide for President in Haiti. On the contrary, it recognized the military regime.

7.3.2   Cultural Violence

One of the major contributions of Johan Galtung to the understanding of violence is his exposure of cultural violence, or the ways and means to approve or legitimize direct and indirect violence. Cultural violence could take the form of distinguishing the chosen from the unchosen, or the upper classes being closer to God than the lower classes and possessing special rights. John Paul II, opening the Santo Domingo meetings, warned the Latin American bishops to defend the faithful from the "rapacious wolves" of Protestant sects. His language dealt a blow to twenty years of ecumenical efforts (Stewart-Gambino 1994: 132). Cultural violence declares certain wars as just and others as unjust, as holy or unholy wars.

The peace prize given to Radovan Karadžić, the Serbian leader in Bosnia, by the Greek Orthodox church, for his contribution to world peace could easily be labelled cultural violence. In July 1994, Kurt Waldheim was awarded a papal knighthood of the Ordine Piano for safeguarding human rights when he served with the United Nations. His services in the Balkans for the Nazis were seemingly forgiven. Both were made religious role models.

It is clear that the causes of religious wars and other religion-related violence have not disappeared from the face of the earth. Some expect an increase of it. Efforts to make the world safe from religious conflicts should then also be high on the agenda. Religious actors should abstain from any cultural and structural violence within their respective organizations and handle inter-religious or denominational conflict in a non-violent and constructive way. This would imply several practical steps, such as a verifiable agreement not to use or threaten violence to settle religious disputes. It must be possible to evaluate religious organizations objectively with respect to their use of physical, structural or cultural violence. A yearly overall report could be published. Another step would be furthering the *depoliticization* of religion. Power also corrupts religious organizations. In addition, the depoliticization of religion is a major precondition for the political integration of communities with different religions.

Very important is the creation of an environment where a genuine debate is possible. Extremist rhetoric flourishes best in an environment not conducive to rational deliberation. Needless to say, extremist rhetoric is very difficult to maintain in a discursive environment in which positions taken or accusations made can be challenged directly by rebuttal, counter-propositions, cross-examinations and the presentation of evidence. Without a change in the environments of public discourses within and between religious organizations, demagogy and rhetorical intolerance will prevail. In his latest book, *Projekt Weltethik* [Project World Ethics]. Küng (1990) rightly concludes that world peace is impossible without religious peace, and that the latter requires religious dialogue.

7.3.3 Religious Bystanders

Religious organizations can also influence the conflict dynamics by abstaining from intervention. As most conflicts are *asymmetrical,* this attitude is partial in its consequences. It is implicitly reinforcing the *might is right* principle. During the Second World War, the Vatican adopted a neutral stand. It didn't publicly disapprove of the German atrocities in Poland or in the concentration camps. To secure its diplomatic interests, Rome opted for this prudence and not for an evangelical disapproval. The role of bystanders, those members of the society who are neither perpetrators nor victims, is very important. Their support, opposition, or indifference based on moral or other grounds shapes the course of events. An expression of sympathy or antipathy by the head of the Citta del Vaticano, Pius XII, representing approximately 500 million Catholics, could have prevented a great deal of the violence. The mobilization of the internal and external bystanders in the face of the mistreatment of individuals or communities is a major challenge to religious organizations. To realize this, children and adults, in the long run, must develop certain personal characteristics, such as a prosocial value orientation and empathy. Religious organizations have a major responsibility in creating a worldview in which individual needs would not be met at the expense of others and genuine conflicts would not be resolved through aggression (Fein 1992).

7.4   Peacebuilding and Peacemaking

Religious organizations are a rich source of peace services. They can function as a powerful warrant for social tolerance, for democratic pluralism, and for constructive conflict-management. They are peacebuilders and peacemakers.

7.4.1   Peacebuilding

Religions contribute to peacebuilding by empowering the weak, by influencing the moral-political climate, by developing cooperation and by providing humanitarian aid.

7.4.1.1   Empowering People

In the last quarter of this century, religious actors have been a major force for social justice in the Third World and a movement for peace in the industrial countries in the Global North.

People can be empowered by offering support to protest movements, for instance, the "God Against the Bomb" action in North America and Europe. In both East and West, churches issued a declaration in the 1980s supporting the goals of the peace movement. The ecumenical peace engagement was particularly important in creating a mass constituency for peace. The pastoral letter "The Challenge of Peace is God's Promise and our Response", issued by American Catholic bishops in May 1983, challenged the very foundation of US nuclear policy and opposed key elements of the Reagan administration's military build-up.

People can also be empowered by providing them with theological support against injustice. In the Third World, many varieties of theology have been developed which are critical of structural violence. The best known are the Liberation theology in Latin America and the black theology in South Africa. These theologies speak for putting an end to suffering caused by physical, structural, psychological and cultural violence. The existence of a Christianity of the poor is a powerful social force, especially for those confronted with repression and exploitation. Hundreds of church workers, catechists, priests and bishops have undergone death threats, been tortured or murdered while working on the abolition of poverty and injustice (Lernoux 1982).

A wide variety of measures were recently taken to protect people from violence. For instance, *Peaceworkers* went into conflict areas to accompany people whose lives were in danger. In one case, from the Philippines, twenty-five volunteers came at short notice to be with 650 refugees in a church surrounded by death squads who were threatening to kill the refugees. The presence of the international group, which was holding a press conference, prevented a potential massacre (Jeger 1993). Other examples include the activities of the **Cry for Justice** organizations, which were, according to Father Nangle, a response to a call from Haiti "for as many internationals as possible for as long as possible to go into the most violent places in Haiti's countryside to be a protective presence, to protect human rights abuses, and to foster a climate for free and open dialogue and assembly."

7.4.1.2   Influencing the Moral-Political Climate

The major variable which religious organizations can influence is the moral-political climate. The moral-political climate at the international or domestic level can be defined in terms of the perceived moral-political qualities of the environment in which the conflicting parties operate. Some climates tend to be destructive, but others enhance conditions for constructive conflict-management. Religious organizations influence the moral-political climate by justifying war or peace, tolerance or intolerance, conservatism or progressivism (see Fig. 7.2).

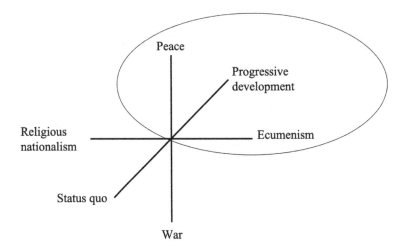

Fig. 7.2 Dimensions of the moral-political climate. *Source* The author

7.5 War Versus Peace

With respect to war and peace, the religious approaches could be divided into two main categories: pacifism and just war doctrine (Life and Peace Report 1990). Several varieties of pacifism can be distinguished: optimistic, mainstream and pessimistic (Ceadel 1987). For some, pacifism means an unconditional rejection of participation in armed struggle; for others, it refers to active engagement in peacemaking. Quakers have traditionally devoted themselves to the dismantling of enemy-images and reconciliation. With Gandhi as a source of inspiration, others have developed non-violent peacemaking strategies.

Christianity is contributing to arms control and non-violent conflict resolution through the evolution of the just war tradition and the acceptance of it as its mainstream normative framework for reflecting on problems of war and peace. The *ius ad bellum* principles (which determine when resort to armed force is morally justified) and the *ius in bello* principles (which determine what conduct within war is morally acceptable) have significantly influenced current international law. Some analysts consider liberation theology to be a recent but radical variant of this doctrine, even though certain liberation theologians tend to be advocates of non-violence and some are considering it necessary to develop just revolution principles. In Latin America, some theologians of liberation pondered on the use of *revolutionary* violence, and in South Africa revolutionary **second violence** was endorsed against the *first violence* of the Apartheid system. A just war or just revolution discourse is also alive in the Muslim world. Modernist authors have argued that the doctrine of **jihad** can also be regarded as a theory of *bellum justum* (Peters 1979). However, the Islamic-Christian National Dialogue Committee in Lebanon declared in 1989 that their respective religions could not be used to justify violence.

7.5.1 Religious Nationalism Versus Ecumenism

Religious organizations also make efforts to overcome religious-intolerance, sectarianism and nationalism, and to develop an ecumenical climate. Hans Küng urges, as a first step, the development of an ecumenical and concrete theology for peace between Christians, Jews and Muslims (Küng 1990). Systematic analysis of their divergences and convergences, and their potential for fostering either conflict or cooperation, would be a helpful step forwards.

7.5.2   Status Quo Versus Progressive Development

Religious organizations have also played an important role in clearing the social space for pluralism, thereby enhancing the potential for establishing an environment characterized by persuasion and consent rather than coercion. The impact of religious conviction and religious actors on the revolution in 1989 in Central and Eastern Europe began to be documented soon after the events took place. In his book *Beyond Glasnost: The Post-Totalitarian Mind* (1989), Goldfarb demonstrated that the political revolution was preceded by a moral and cultural revolution. Garton Ash refers, for example, to the impact of John Paul II on his homeland, Poland, in March 1979. "For nine days the state virtually ceased to exist, except as a censor doctoring the television coverage. Everyone saw that Poland was not a communist state. John Paul II left thousands of human beings with a new self-respect and renewed faith, a nation with rekindled pride, and a society with a new consciousness of its essential unity" (Garton Ash 1985). Religious organizations also played a crucial role in the peaceful revolution in South Africa. There is the steadfastness of the condemnation of violence by black religious leaders. Of great significance was also the role of the Dutch Reformed Church, which distanced itself from Apartheid and condemned it as biblically unwarranted. This helped to deprive the racial radicals of the moral legitimacy of violence.

7.6   Development, Cooperation, and Humanitarian Aid

A great number of INGOs, engaged in all kinds of development projects, have a religious base. I do not know of any study assessing the efforts of religious INGOs, but scattered data suggest that these efforts are considerable. In 1974 Belgium had 6,283 Catholic missionaries in the world. In 1992 this number decreased to 2,766: 1,664 in Africa, 633 in America, 464 in Asia, and 5 in Oceania. In the same year Caritas Catholica Belgium spent 145 million Belgian francs on emergency aid, 290 million Belgian francs on food aid, 22 million Belgian francs in Yugoslavia, 17 million Belgian francs on micro-projects in other parts of the world, and 21 million Belgian francs to help refugees and migrants in Belgium.

7.6.1   Peacemaking

Several religious organizations distinguish themselves through peacemaking efforts. Those efforts could be of a traditional diplomatic nature or be categorized as Track II or field diplomacy.

7.6.1.1 Traditional Diplomatic Efforts

The Vatican has been involved in several cases as a mediator. Its Secretariat of State has 150 members divided across eight language desks. It is represented in 122 countries. Its highest ranking ambassador is called a *nuncio*, who represents the Pope to the Heads of State and to the local church. The Vatican's Secretariat gains much information about local affairs from the 150,000 parish priests who preside over 800 million Catholics worldwide. To the extent that the Vatican lacks traditional state interests and maintains a sense of *objectivity*, it makes it a good candidate for international mediation. During 1978–1984 Pope John Paul II mediated successfully between Argentina and Chile over a few islands at the tip of South America. Both countries narrowly averted war by turning to the Vatican as a mediator. However, the Vatican was successful only at keeping the parties at bay, not at reaching a settlement. In the end, that proved quite enough once domestic factors, especially those in Argentina after the Maldives/Falklands war, changed.

According to Thomas Princen (1992), the Papacy has special resources that few world leaders share. Six resources, which appear to be common to other international actors, stand out.

(a) *Moral legitimacy:* The Pope has a legitimate stake in issues such as peace-making or human rights with a spiritual or moral component. During the Beagle Channel mediation, the Pope appealed to the moral duty to do all necessary to achieve peace between the two countries.

(b) *Neutrality:* In the dispute mentioned above, there was no question that the Vatican had any interests in the disputed islands.

(c) *The ability to advance the political standing of others:* A papal audience, a papal visit or involvement confers political advantage on state leaders. This advantage can be used at key junctures in a mediation: gaining access; deciding on agenda and other procedures; delivering proposals. The démarches made by the President of the United States, Jimmy Carter, to resolve the issue peacefully were distrusted by the two military governments who found his stand on human rights annoying (Princen 1992).

(d) *The ability to reach the (world) public opinion:* The Pope can command the attention of the media. This was especially true for the perigrinating John Paul II.

(e) *Network of information and contacts:* The information and communication network of the Catholic Church is extensive. For a localized dispute, communication channels outside conventional diplomatic channels can be significant.

(f) *Secrecy:* Confidentiality is a major asset for mediation. As an organization with no claim to democratic procedures or open government, the Holy See is known to be able to keep a secret. Maintaining confidentiality is a standard operating procedure in the Vatican.

Princen (1992) concludes his analysis of the mediation by the Vatican by observing that when pay-offs are not the primary obstacle, when the interaction between disputes is inadequate, when face-to-face talks and face-saving devices are in short supply, a powerless transnational actor can influence disputants in subtle ways. He also notes that the mediation effort in the dispute mentioned above turned out to be a terrible headache for the mediation team and the Pope. What started out as a six-month enterprise turned out to be a six-year ordeal. He further observes that, on the whole, however, the Vatican remains a reactive player, for whom power politics continues the dominant paradigm.

7.6.1.2 Track II Peacemaking

The peacemaking activities of NGOs, be they of a religious or non-religious nature, are getting more attention. A great deal of research is, however, needed to obtain a proper insight into the potential of the rich amount and variety of peace services.

Traditionally, a lot of peace work has been delivered by the Quakers. Urging that humanity should "deny all outward wars, strife and fightings with all outward weapons for any end", various efforts have been directed toward conciliation. Under the heading 'conciliation' come especially efforts to promote better communication and understanding by bringing people together in seminars and efforts to work with the conflicting parties. Adam Curie and Kenneth Boulding are the two most well-known academic spoksmen of this approach. The definition given by Adam Curie for conciliation describes the Quakers assumptions: "Activity aimed at bringing about an alteration of perception (the other is not so bad as we imagined; we have misinterpreted their actions, etc.) that will lead to an alteration of attitude and eventually to an alteration of behaviour" (Yarrow 1977).

This kind of conciliation is the most appropriate if conflicts primarily arise over a different definition of the situation. For other conflicts related to gross injustices or unequal power, the Quakers use methods of witness or advocacy. Essential for effective conciliation is the establishment of confidence, impartiality and independence. Yarrow (1977) describes the kind of impartiality which tends to promote *balanced partiality*, that is, listening sympathetically to each side, trying to put themselves in the others party's place. Another characteristic of balanced partiality is the Quakers' concern for all people involved in a situation. The Quakers' teams – Quakers have tended to entrust mediatory work to at least two Friends – emphasize the need to maximize the gains that might accrue to both sides through a settlement. Also, several other religious organizations are increasingly engaged in peacemaking efforts. An important role was played by the Catholic community of San Egidio in Rome to reach a Peace Agreement in Mozambique in October 1992 (Sauer 1992).

Non-governmental peace-makers tend to approach conflicts from a different perspective than that shared by traditional diplomacy. The new approach, carrying different names, such as Track 13, parallel, multitrack, supplemental, unofficial, citizen diplomacy, or *interactive problem-solving diplomacy*, reflect a new conflict resolution culture. This new conflict resolution culture differs from the traditional one with respect to four points.

(a) *Goals*: Non-governmental diplomats tend to make a distinction between conflict settlement (by authoritarian and legal processes) and conflict resolution (by alternative dispute resolution skills). Conflict resolution aims at an outcome that is self-supporting and stable because it transforms the problem to the long-term satisfaction of all the parties (Burton 1984).

(b) *Attitude vis-à-vis the conflicting parties*: Track II diplomacy assumes that the motivations and intentions of the opposing sides are benign; this contrasts strongly with the conflict culture of the traditional diplomacy, in which distrust and a more negative perception of men prevails. Track II peace-makers further believe that only the conflicting parties can arrive at a solution; in other words, their task consists mainly of facilitating the process. They also try to foster understanding of so-called **irrational behaviour** that is disapproved of by dominant social norms. From the point of view of the decision-maker, it could be perceived as the best they would do given what they know about the intentions of the other parties and the perceived options. They believe that not only the government, but also different layers of the respective society, should have a say in the peacemaking process. A stable peace ought to be embedded in a democratic environment.

(c) *Towards a multi-level and comprehensive approach*: Track II peace-makers see their efforts as complementary to the official diplomatic efforts. They believe that peace has to be a multi-level effort and that governmental as well as non-governmental actors should be involved. The latter could be private persons or organizations, and national or transnational institutions of a secular or religious nature. They also believe that a sustainable peace requires a comprehensive approach in which the necessary diplomatic, political, military, economic, cultural and psychological conditions are created.

(d) *Peace, a learning process*: Track II peace-makers assume that, in many cases, violence and war are the consequences of a wrongful assessment of the consequences of war or of a lack of know-how about how to manage conflicts in a more constructive way. They also believe that warlike or peaceful behaviour is learned behaviour, and that what is learned could be unlearned through peace research and peace education.

Track II diplomacy involves a series of activities such as (1) the establishment of channels of communication between the main protagonists to facilitate exploratory discussions in private, without commitment, in all matters that have or could cause tensions; (2) setting up an organization which can offer problem-solving services for parties engaged in conflicts within and between nations; (3) the establishment of a centre to educate people undertaking such work; and (4) the creation of a research centre or network in which know-how and techniques are developed to support the above-mentioned tasks.

7.6.2 Field Diplomacy

In several parts of the world there have been new initiatives to develop what could be called *field diplomacy*. This new creative energy has been jolted by three factors. First of all, there is the failure of the traditional diplomacy of governmental and intergovernmental organizations to prevent conflicts (Bauwens/Reychler 1994).

Also important is the explosion of peacekeeping and humanitarian relief efforts. These efforts, which absorb huge budgets, do not solve conflicts, and could have been used at an earlier phase of the conflict to prevent violence escalation. Peacekeepers and humanitarians seem to be doomed to Sisyphean efforts, an endless pushing up of peacebuilding stones against the mountain of human suffering. When the task seems over, the stones role back and they have to start again.

A third factor stimulating *field diplomacy* is the growing awareness that case studies and especially practical experience in the field would enhance the research work and the training of professional conflict-managers. Compared with other professions, such as lawyers, economists, and physicians, who have respectively their law courts, business and patients or dead corpses to try out their theories, most peace researchers have no practical experience. In addition, certain kinds of the information necessary to understand the dynamic of a conflict require the analyst to be in the field.

The confluence of these factors stimulated the development of a third generation of peacemaking approaches: *field-diplomacy.* This refers to sending non-governmental teams to conflict areas for an extended period to stimulate and support local initiatives to prevent conflict, and entails the creation of a network of persons based on trust. Such a network or trustbank is necessary to observe the conflict dynamics, provide early warnings, assess needs and take timely measures. Such measures could consist of keeping the communication channels between the conflicting parties open, creating a favourable climate for exploring solutions, developing a constructive conflict culture, keeping account of the total costs and benefits of the conflict, giving advice and evaluating official peace proposals or agreements. An effective contribution to conflict prevention requires a credible presence in the conflict area of professional volunteers who empathize with the concerns, needs and preferences of the communities in which they operate. To be effective, they need to earn the respect and trust of local opinion-leaders. The organization, the personnel and the methods for this kind of peacemaking and peacebuilding are still in an embryonic state.

Among the pioneering organizations, we could mention, for example, *Witness for Peace*, which operated in Nicaragua, and the Swedish *Peace Monitoring in South Africa* (PEMSA). Some organizations specialize in one task. The *Peace Brigade International* takes care of the security of threatened activists. *International Alert,* an INGO situated in London, tries to improve conflict prevention by networking with humanitarian organizations, research institutes, peace movements, conflict resolution networks, human rights organizations and the media of each country, and by organizing training sessions in constructive conflict management.

An NGO with ample experience of constructive conflict management is *Search for Common Ground*, based in Washington (1993 Report). It began in 1982 and focused originally on Soviet-American relations. Now it works in the Russian Federation, the Middle East, South Africa, Macedonia and the United States. They develop what they call a common ground approach which draws from techniques of conflict resolution, negotiation, collaborative problem-solving and facilitation. The aim is to discover not the lowest but the highest denominator.

In Belgium, a similar NGO called *International Dialogue* went on to be created. Meanwhile, at the American University of Beirut, "Training for trainers in conflict resolution, human rights and peace democracy" was organized by the International Peace Research Association (IPRA), International Alert and UNESCO.

An initiative which could be referred to as a model for field diplomacy is the *Centre for Peace, Non-Violence and Human Rights* in Osijek, Croatia, very close to the Serbian border. The Peace Centre was founded by a small group of people in May 1991. In the early 1990s there were twenty altogether, with a core group of five, including Croats, Serbs and Muslims. Under the Chairman, Katarina Kruhonja, several initiatives were undertaken to help and protect people against threats. The members of the Centre would sit in apartments with Serbs so that they could confront the soldiers who came with orders to evict them. The members of the Centre promote human rights, teach methods and strategies of active non-violence, assist in the resettlement of refugees, and mediate in conflict situations, etc. The members of the Centre were frequently threatened and have been accused by the authorities of being unpatriotic traitors. During the conflict, one of the heads of the local government said that the Centre would be destroyed and members would lose their jobs if they continued their activities, but today it continues to flourish.

Several organizations are addressing violence in their own countries. The activist Carl Upchurch was focusing on violence in Los Angeles, where 857 young men and women were killed in group-related violence in 1992 alone. Field diplomacy projects are being developed not only at non-governmental but also at governmental levels. The United Nations Volunteers undertake projects to support the peace process in several conflict areas in the world. Their project in Burundi endeavours to promote peace at a community level through grass-roots confidence-building measures aimed at enabling the emergence, return and social reintegration of persons in hiding, internally displaced people and refugees. Part of the project's efforts is to shift the dynamic of inter-group relationships from animosity and confrontation to mutual esteem and cooperation. Training in conflict resolution supports the efforts of community leaders to achieve reconciliation, and has laid the foundations of a national capacity to detect, pre-empt and defuse future tensions and latent conflict situations. Ongoing peace education and the promotion of respect for human rights are also essential. The project is also helping in building up a fabric of local and international NGO support for the peace process. It promotes complementarity between peacebuilding efforts and humanitarian relief, and also provides practical support in advancing the agenda towards sustainable social recovery and human development. To further the cooperation between all these scattered peace services in the field, in May 1994, Elise Boulding organized a mini-seminar in Stensnas (Sweden) to prepare the way for the establishment of a *Global Alliance of Peace Services* (GAPS).

Although, to a great extent, inspired by and using techniques and methods of Track II peacemaking, field diplomacy distinguishes itself in several ways.

First, field diplomacy requires *a credible presence in the field*. One has to be in the field to help transform the conflict effectively. A credible presence in the field is needed to build a trust bank or network of people who can rely on each other. This is necessary to get a better insight into the concerns of the people and the conflict dynamics, and to take timely measures to prevent destructive action.

Second, *serious engagement* is necessary. Just as the adoption of a child cannot be for a week or a couple of months, so field diplomacy is a long-term commitment. Facilitating a reconciliation process could be depicted as a long and difficult journey or expedition.

Third, field diplomacy favours a *multi-level approach* of the conflict. The actors in the conflict could be located at three different levels: the top leadership, the middle-level leaders and the representatives of the people at the local level. A sustainable peace needs the support of the people. Since they have a major stake in peace, they should be stake holders in the peacemaking, peacekeeping and peacebuilding process.

Fourth, field diplomats believe that peace and the peace process cannot be prescribed from the outside. They *favour the elicitive approach*. One of the most important tasks of field diplomacy is to identify the peace-making potential in the field. The role of field diplomats is to catalyse and facilitate the peace process. Any peace process should be seen as a learning experience for all the people concerned.

Fifth, field diplomats have a *broad time perspective,* both forward and backward. A sustainable peace demands not only a mutually satisfying resolution of a specific conflict but also a reconciliation of the past and a constructive engagement towards the future.

Sixth, field diplomacy also focuses attention on the deeper layers of the conflict – *the deep conflict*. Most peace efforts focus on the upper layers. They are concerned with international and national peace conferences and peace agreements signed with pomp. A lasting peace needs to take care of the deeper layers of the conflict: the psychological wounds; the mental walls; and the emotional and spiritual levels. The latter refers to the transformation of despair into hope; distrust into trust; hatred into love. Our understanding of these soft dimensions is very limited. We have a long way to go.

Seventh, another characteristic of field diplomacy is the *recognition of the complex interdependence of apparently different conflicts*. Field diplomats do not only reject the artificial distinction between internal and external conflicts, but pay attention to the interdependence of different conflicts in space and time. Many Third World conflicts do not only have roots within the country or the region, but also in the Global North. The conflicts in Rwanda and Burundi cannot be uprooted if not enough attention is paid to the behaviour of Belgium or France in the past and present. There is also some fieldwork to be done here and now.

Eighth, field diplomats stress the importance of a more *integrative approach of the peace process.*

7.7 Religious Peacemaking: Strengths and Weaknesses

7.7.1 Strengths

Several factors endow religions and religious organizations with a great and under-utilized potential for constructive conflict management.

First, more than *two-thirds of the world population belongs to a religion.* In 1992, 29.2% of the religious constituency was Christian; 17.9% Muslim; 13% Hindu; 5.7% Buddhist/Shintoist; 0.7% Confucianism/Taoist. Together, all those religious organizations have a huge infrastructure with a communication network which reaches all corners of the world. They have great responsibility, and leadership is expected from them.

Second, religious organizations have the *capacity to mobilize people and to cultivate attitudes of forgiveness and conciliation.* They can do a great deal to prevent dehumanization. They have the *capacity to motivate and mobilize people* to forge a more peaceful world. Religious and humanitarian values are one of the main roots of voluntarism in all countries: doing something for someone else without expecting to be paid for it. They are problem-solvers. They do not seek conflict. But when a need is seen, they want to do something about it. They are a force to be reckoned with (Hoekendijk 1990).

Third, religious organizations can rely on a set of *soft power sources* to influence the peace process. Raven and Rubin (1983) developed a useful taxonomy for understanding the different bases of power. It asserts that six different sources of power exist for influencing another's behaviour: reward, coercion, expertise, legitimacy, reference, and information.

Reward power is used when the influencer offers some positive benefits (of a tangible or intangible nature) in exchange for compliance. If reward power relies on the use of promises, coercive power relies on the language of threat. Expert power relies for its effectiveness on the influencers' ability to create the impression of being in possession of information or expertise that justifies a particular request. Legitimate power requires the influencer to persuade others on the basis of having the right to make a request. Referent power builds on the relationships that exist between the influencer and recipient. The influencer counts on the fact that the recipient, in some ways, values his or her relationship with the source of influence. Finally, informational power works because of the content of the information conveyed.

To mediate, religious organizations can rely on several sources of power. There could be the referent power that stems from the mediation position of a large and influential religious family. Closely related could be legitimate power or the claim to moral rectitude, the right to assert its views about the appropriateness and acceptability of behaviour. Religious leaders could refer to their *spiritual power* and speak in the name of God. Also important could be the informational power derived through non-governmental channels; groups like the Quakers could use expertise power on the basis of their reputation of fine mediators.

Fourth, religious organizations could also use *hard sources of power*. Some religious organizations have reward power, not only in terms of promising economic aid, but, for example, by granting personal audiences. Use could also be made of coercive power by mobilizing people to protest against certain policies. Think of Bishop James McHugh, warning President Clinton of an electoral backlash for the administration's support of abortion rights at the United Nations population conference in Cairo. Integrative power, or power of **love** (Boulding 1990), is based on such relationships as respect, affection, love, community and identity.

Fifth, there is a *growing need for non-governmental peace services*. Non-governmental actors can fulfil tasks for which the traditional diplomacy is not well equipped. They would provide information not readily available to traditional diplomats; they could create an environment in which parties could meet without measuring their bargaining positions, without attracting charges of appeasement, without committing themselves, and without making it look as if they were seeking peaceful solutions at the expense of important interests. They could monitor the conflict dynamics, involve the people at all levels, and assess the legitimacy of peace proposals and agreements.

Sixth, most can make use of their *transnational organization* to provide peace services. Finally, there is the fact that *religious organizations are in the field* and could fulfil several of the above peace services.

7.7.2 Weaknesses

Several weaknesses limit the impact of religious organizations in building a world safe from conflict. Several religious organizations are still *perpetrators of different kinds of violence*. In many of today's conflicts they remain primary or secondary actors or behave as passive bystanders.

Also inhibiting religious peacemaking efforts is the fact that, as third parties, religious organizations tend to be *reactive players*. They seem to respond better to humanitarian relief efforts after a conflict has escalated than to potential violence. A third weakness is the *lack of effective cooperation* between religious organizations. Most of the peacemaking or peacebuilding efforts are uncoordinated. Finally, there is a *need for more professional expertise in conflict analysis and management*.

7.8 Conclusion

Religious organizations have a major impact on inter-communal and international conflicts. During the Cold War, religious as well as ethnic and nationalist conflicts were relatively neglected in the study of international relations and peace research. After the implosion of the communist block, the escalation of nationalist violence was a surprise. Some expect an escalation of religious conflicts as well. Despite an increase in the attention to the religious dimension of conflicts, it remains an under-researched field. There is no useful typology of religious conflicts; no serious study of the impact of religious organizations on conflict behaviour; no comparative research into the peacemaking and peacebuilding efforts of different religious organizations.

The world cannot survive without a new global ethic, and religions play a major role, as parties in violent conflicts, as passive bystanders and as active peacemakers and peacebuilders. Küng's (1990) thesis that there cannot be world peace without a religious peace is right. Representing two-thirds of the world population, religions have a major responsibility to help create a constructive conflict culture. They will have to end conflicts fuelled by religion, stop being passive bystanders and organize themselves to provide more effective peace services. Religions and religious organizations have an untapped and under-used integrative power potential. Assessing this potential and understanding which factors enhance or inhibit joint peace ventures between the Christian religions, but also between the prophetic religions (Judaism, Christianity, Islam), the Indian religions (Hinduism and Buddhism) and the Chinese wisdom religions, are urgent research challenges.

References

Assefa, H. (1990). Religion in the Sudan: Exacerbating Conflict or Facilitating Reconciliation. *Bulletin of Peace Proposals, 21*(3), 255–262.

Badal, R. K. (1990). Religion and Conflict in the Sudan: A Perspective. *Bulletin of Peace Proposals, 21*(3): 263–272.

Bauwens, W., & Reychler, L. (Eds.). (1994). *The Art of Conflict Prevention.* London: Brassey's.

Boulding, K. (1990). *Three Faces of Power.* Newbury Park: Sage Publications.

Burton, J. (1984). *Global Conflicts: The Domestic Sources of International Crises.* Maryland: Wheatsheaf Books.

Ceadel, M. (1987). *Thinking about Peace and War.* Oxford: Oxford University Press.

Fein, H. (Ed.). (1992). *Genocide Watch.* New Haven: Yale University Press.

Galtung, J. (1997–98). Religions, Hard and Soft. *Cross Currents* (Winter), *47*(4). At: http://www.crosscurrents.org/galtung.htm.

Gantzel, K., Schwinghammer, T., & Siegelberg, J. (1993). *Kriege der Welt. Ein systematischer Register kriegerischen Konflikte 1985 bis 1992.* Bonn: Stiftung Entwicklung und Frieden.

Garton Ash, T. (1985). *The Polish Revolution: Solidarity, 1980–82.* New York: Scribner, 1984).

Goldfarb, J. (1989; 1991). *Beyond Glasnost: The Post-Totalitarian Mind.* Chicago: University of Chicago Press.

Hoekendijk, L. (1990). Cultural Roots of Voluntary Action in Different Countries. *Associations Transnationales,* 6.

Hunter, J. D. (1991). *Culture Wars: The Struggle to Define America.* New York: Basic Books.

Huntington, S. (1993). *The Clash of Civilizations?* New York: Foreign Affairs.

Küng, H. (1990). *Mondiale Verantwoordelijkheid.* Averbode: Kok-Kampen Altioria.

Lernoux, P. (1982), *Cry for the People.* London: Penguin.

Princen, T. (1992). *Intermediaries in International Conflict.* Princeton University Press, Princeton.

Raven, B. H., & Rubin, J. Z. (1983). *Social Psychology.* Hoboken: Wiley.

Reychler, L. (1979). *Patterns of Diplomatic Thinking.* New York: Praeger Publishers.

Sauer, T. (1992). *The Mozambique Peace Process* (Working Paper). Bologna: John Hopkins University, Bologna Center.

Search for Common Ground. (1993). *Report.* Washington, D.C.: Search for Common Ground.

Shenk, G. (1993). *God with Us? The Roles of Religion in Conflicts in the Former Yugoslavia* (Research Report of the Life and Peace Institute). Uppsala: Life and Peace Institute.

Stewart-Gambino, H. (1994). Church and State in Latin America. *Current History, 93,* 129–133.

Weigel, G. (1991). Religion and Peace: An Argument Complexified. *The Washington Quarterly, 14*(2), 27–42.

Williamson, R. (1990). Why Is Religion Still a Factor in Armed Conflict? *Bulletin of Peace Proposals, 21*(3), 243–253.

Williamson, R. (1992). *Religious Fundamentalism as a Threat to Peace: Two Studies* (LPI Working Paper No. 18). Uppsala: Life and Peace Institute.

Yarrow, M. (1977). Quaker Efforts Towards Reconciliation in the India-Pakistan War of 1965, in: M. R. Berman and J. E. Johnson (Eds.), *Unofficial Diplomats.* New York: Columbia University Press.

Chapter 8
Lessons Learned from Recent Democratization Efforts

Our democratic creed is predicated on the possibility of improving the organization and the ability of citizens to achieve their purposes and to better their lot.[1] Efforts by the international community, however, do not always result in successes; there have been failures (March/Olsen 1995). These failures indicate that although we have acquired a lot of insight into the processes of change, we still do not know enough to be confident that the effects we produce will be intelligent ones. Some lessons from recent experiences and concrete proposals for coping with the problems are discussed in this part of the report.

8.1 Early Warning: The Devil Is in the Transition

The creation of an instant democracy is now widely accepted as an illusion, and elections are not to be confused with the installation and consolidation of democracy. Democracy as a form of human existence is possible in an environment characterized by an accumulation of concrete conditions. It takes time and effort to install and consolidate a democracy (Linz 1997).[2] Additional research is needed (a) to monitor different types of transition processes more effectively; (b) to identify the crucial factors which influence successful and unsuccessful transitions; and (c) to develop effective ways and means to facilitate the transition process.

---

[1] This text was first published as: Reychler (1999). Reprinted with permission.

[2] Bad democracies are better than authoritarian rule or political chaos since they contain the potential for gradual, peaceful improvements.

© The Editor(s) (if applicable) and The Author(s), under exclusive license to
Springer Nature Switzerland AG 2020
L. Reychler and A. Langer (eds.), *Luc Reychler: A Pioneer in Sustainable Peacebuilding Architecture*, Pioneers in Arts, Humanities, Science, Engineering, Practice 24, https://doi.org/10.1007/978-3-030-40208-2_8

8.1.1 Diagnosis First, Policy Second

The development of a more effective peacebuilding policy has been hampered by diagnostic problems. Therefore, an effective Conflict Impact Assessment System (CIAS) should be developed and incorporated in the conflict prevention decision-making at the macro and micro project level. This includes a democratic need assessment by in- and outsiders. The analytic framework focusing on ten democratic building blocks and several internal and external support systems is a useful tool for identifying, analysing and coping with the problems associated with democratic transition (Reychler 1997b).

8.1.2 Democratic Peacebuilding Architecture

Although the emergence of democracy may have been, in most cases, an unintended consequence of structural changes in a country or region, it has always been possible to further or obstruct democratization by conscious strategies of democratization. The less favourable the structural conditions are for democracy, the more the success or failure of democratization depends on the skills and strategies of political leadership. A distinction can be made between three major democratization strategies: direct, indirect and mixed. The direct democratization strategy aims at the establishment of political institutions that make it possible to share power democratically between competing groups. The indirect democratization strategy aims at the transformation of the social structures affecting the distribution of economic and intellectual power resources. A mixed strategy focuses on both aims simultaneously. "In Eastern Europe and the Soviet Union, political democratization preceded, in most cases, the decentralization of economic power through privatization, whereas in China the decentralization of economic power resources preceded political democratization that has not yet started in many European and Latin American countries, as well as in India, Japan, South Korea, Taiwan, and in some other Asian countries, land reforms and educational reforms preceded democratization or were started with democratization in order to consolidate new democratic structures" (Vanhanen 1997: 169–175).

To enhance the chances of democratization, more efforts and creativity are needed to develop a democratic peace architecture (DPA). The aim of a democratic peace architecture is to build political institutions that provide circumstances for meaningful political participation, facilitate power-sharing among competing groups, and prevent unnecessary violence. In several cases, such as in asymmetric conflicts between two ethnic groups, many obstacles need to be overcome to build democratic peace. Some ingredients, such as federalism, proportional representation, and parliamentarism, are strongly recommended to satisfy the demands of equality and reciprocity. The establishment and consolidation of democracy requires more than "political engineering" focussed solely on political institutions.

Democratic peace architecture is, and remains, an art involving different sectors, levels, time frames, and layers of operation. More means should be provided to develop the art or practice of designing and building democratic peace.

8.2   Timing

In democratic peace architecture, timing is of the utmost importance (Di Palma 1990).[3] This is true because the democratic peacebuilding process takes time. The time horizons for measuring progress are half-decades and decades. The external actors promoting or assisting the democratization process must opt for programmes designed for the long haul, design such programmes accordingly, commit the appropriate personnel to the Field, and, above all, be patient (Barkan 1997: 389–390). This, however, should not imply that democratization should be seen as a prolonged affair. Everything suggests that matters get prolonged not because democracy needs time but because democratization has met some hefty stumbling blocks. Attention should be paid to the identification and the dismantling of these obstacles. The democratization process could be locked in a kind of halfway house, called dictablanda or democradure. Dictablandas are non-competitive dictatorships in which a degree of bland liberalization has been used to justify the status quo (e.g., Chile during the period in which Pinochet prepared the regime for a referendum in 1988). Democraduras go a step further and allow a competitive system, but the competition is limited in three ways by fairly explicit or formal pacts: participation is restricted to conservative forces; these forces share government offices according to consociational arrangements fairly independent of electoral verdicts; and they bypass touchy issues of the political agenda (Di Palma 1990: 153–156).

A second remark relates to short-term versus long-term planning. The management of the democratic peacebuilding process is, to a great extent, the management of political time in a complex and volatile environment. There is always tension between the pressure to achieve results in response to the immediate circumstances and the time needed to build a sustainable outcome with long-term stability. Bloomfield and Reilly make a distinction between a slow-fast and a fast-slow approach. A *slow-fast* approach to managing conflicts is one where the initial stage of reaching the agreement is done as slowly as necessary, to ensure that the agreement, when reached, is as comprehensive and detailed as possible. A good example is the South African constitutional negotiations of the early 1990s. By contrast, many agreements are reached in a *fast-slow* mode: pressure to get results encourages the parties to rush through the negotiation phase and reach a less-than-optimum agreement, so that problems remain which slow down, or

---

[3]Time is a tactical resource. Regime opponents have good reasons to wish for a speedy transition, and frequently so do secessionists.

altogether obstruct, the implementation phase. This approach is exemplified by the Dayton negotiations of November 1995 (Harris/Reilly 1998: 25–26).

A third remark about time relates to the scheduling of different efforts. There is, for example, the ongoing debate about the simultaneity of economic and political reform. Some argue that political and economic gains must not only be pursued, but have to occur simultaneously. They support the so-called "tightly-coupled hypothesis". Others, like Linz/Stepan (1997: 28–29), believe that democratic policies can, and must, be installed and legitimized by a variety of other appeals before the possible benefits of the market economy fully materialize. Their more "loosely-coupled hypothesis" assumes that the relationship is not necessarily one-to-one. At least in a medium-range time horizon, they believe people can make independent and opposing assessments about political and economic trends. They further assert that when people's assessments of politics are positive, this provides a valuable cushion for painful economic restructuring. Evidence gathered in six East Central European countries tends to support their thesis. However, they do not believe that political-economic incongruence can last forever.

Another example illustrating the importance of good scheduling concerns *security pacts*. When elections went wrong in highly conflictual situations, for example in Angola in November 1992 or Algeria in 1991, a rather majoritarian electoral system was in place, and the parties entered the contest without a pre-election security pact (Sisk 1998: 109).

An additional observation about time relates to the so-called "post election blues". Over the past six years, analysts have pointed to enfeebled political institutions and the virtual absence of effective mechanisms of governmental accountability in many Latin American countries. Elections may now be routine, but democratic practice (reflected in accountability, participation, representation, and the rule of law) is anaemic at best (Hakim/Shifter 1977: 8–11). As a consequence, the political "relative deprivation" has increased considerably. Comparative public opinion studies in 1995 and 1996 show that the majority of respondents are not satisfied with the actual functioning of democracy. The free, periodic elections which today characterize nearly every Latin American country have produced illiberal democracies and widespread discontent. Without the addition of solid democratic institutions – legislatures, political parties, labour unions, judicial systems, and civic organizations – political regimes have inadequate peace-inducing qualities.

8.3   External Democratization Efforts Make a Significant Difference

Although domestic factors play a predominant role in the transition of democracy, the impact of the external environment should not be underestimated (Di Palma 1990: 183–199). Democracy does not happen in an international vacuum. Three types of impact can be distinguished: (a) *democracy by diffusion* or by demonstration effects: The affluence of liberal democracies in the world has a showcase effect on non-democratic and less affluent countries. Diffuse demonstration effects often go together with a more active promotion of democracy; (b) *democracy by trespassing*: The democratization of a country can be implanted forcefully, for example in Germany and Japan after the Second World War, or a hegemony, such as the former Soviet Union, could remove its veto to democratization; (c) *democracy facilitation*: Most of the international democratization efforts today intend to support indigenous democratization efforts.

Among the external influences on non-democratic regimes discussed earlier are: the security of the international environment; the international moral-political climate; foreign intervention; and transnational civil society. The interdependence between internal and external (international) democratization is very complex. More research is needed for a better understanding of the direct and/or indirect impact of changes in the international environment on the democratization chances of a particular country or region. The promotion and/or facilitation of democracy should remain high on the foreign policy agenda of the International Community and the European Union. Sustained and synergetic involvement at different levels is needed to increase the chances of installing and consolidating democracies.

8.4 Psychology of Democratization

To analyse and transform political systems, more attention needs to be paid to political-psychological variables. It concerns, for example, patience, dissatisfaction, and the existence of "sentimental walls"[4] which inhibit the democratic transition process and constructive conflict transformation. Impatience is politically significant because its consequences can be immediately manifest. Political patience encourages people to be more supportive of the current regime and to retain expectations of it becoming better in the future. Impatience, on the other hand, can generate public protest and is a real problem insofar as impatient people are unwilling to support the new regime or agree on an authoritarian alternative to replace it. Another variable to monitor is the popular dissatisfaction with the effects of economic and political transformation. Popular dissatisfaction with the effects of political and economic changes is perhaps the largest challenge to prescriptions for political and economic transformation. The five most frequent reactions can be categorised as opportunistic, materialistic, defensive, victim-like, and mobilization. It is important to consider the impact of the distribution of these different responses to the basic conflict inherent in transitional societies. Finally, attention should be paid to the identification and dismantling of the mental and sentimental obstacles on the way to democratic peacebuilding. These mental and sentimental obstacles

---

[4]For detailed explanations of this term see Chaps. 9 and 10.

inhibit not only adequate analysis of the conflict dynamics, but also the will to do something about it and the development of know-how. The detection, exposure and dismantling of those obstacles is a must, if we want to create a more effective conflict prevention policy. A preliminary scan gave the following list of mental obstacles which inhibit the development of an effective democratic peacebuilding process: reductionism (for example, the propensity to reduce democracy to the organization of elections); the political appeal of "instant democracy"; overlooking vital components of democracy-building (for example, credible power-sharing arrangements); the creation of "expectation gaps" by overselling democratization; a soft contextual approach; elitism; the use of double standards; taboos (for example, the sanctity of boundaries for the treatment of international democracy as a non-starter) or the silence about the colonial past; the so called cultural or economic prerequisites of democracy; and the conceptual separation of the political and economic domain.

A formidable mental wall is the belief in Asian democracy. Is there such a thing as Asian democracy? Many Asian leaders reject aspects of the Western democratic tradition. They claim to be building a distinctive form of Asian democracy. They favour an approach in which more weight is given to Asian values, including respect for authority, avoiding public conflict and accepting the primacy of the group. Democracy is defined in almost familial terms, with the elected leader adopting a parental style. The State leads the society, and democracy therefore depends less on the independent groups and associations that provide checks and balances in the Western State. The consequences of the Asian style include a subservient media and judiciary. In addition, the police and the security forces become more aggressive in their approach to criminals and dissidents. Among the arguments for non-Western models one finds (1) historically developed cynicism vis-à-vis Western hypocrisy, and (2) the fact that the Asian model has delivered economic growth by allowing leaders to focus on long-term modernization, free from electoral pressures. Critics allege (1) that "non-western democracy" is simply a justification for authoritarian rule or an excuse for failing to move beyond semi-democracy, (2) that economic and political progress will require a reduction in the State's role in the economy and a more liberal form of democracy; and (3) that rape is not something that is done in a "non-western way". Rape is rape, torture is torture and human rights are human rights (Hague et al. 1998: 27).

8.5   The Political Economics of Democratic Transition

The democratization transition cannot be understood without taking into account the role of economic factors (discussed above in 3.3.2). The standard of living of the people, the economic development, the economic system, the existence of an economic society and the role of business, the resource distribution, the nature of the privatization process, the "simultaneity dilemma", the costs of democratization, and the effectiveness of the 'conditionality' policy are important factors. Empirical evidence shows that economic development and increases in resource distribution regularly support popular efforts to democratize political systems. Therefore, one could argue that democratization should be linked to economic development efforts and to attempts to distribute economic, intellectual and other relevant power resources more widely among various groups in society. External actors should have a better grasp of the limits and possibilities of their democratization efforts. The carrots and sticks associated with the 'conditionality' policy are frequently applied without a thorough impact analysis. In many cases, they are not successful and cause significant collateral damage. If sanctions are considered, they need to be refined to influence the targets more effectively. There is a need for a democratization policy which goes "beyond conditionality".[5]

8.5.1 Leadership

Successful democratic peacebuilding requires leadership at different levels, including the elite, middle and local levels. The identification, training, and support of people who play a leadership role in the democratic peacebuilding process are of crucial importance. The development of field diplomacy at the local and middle levels of a country should be considered an investment in democratization (Reychler 1997a). Equally important are the identification of spoilers and the selection of appropriate strategies to discourage undemocratic behaviour. Spoilers are leaders who believe that democracy threatens their power, world-view and interests, and who use non-democratic ways and means to undermine attempts to achieve it.

8.5.2 Strengthening Transitional Democracies Through Conflict Resolution

In communist and authoritarian countries, the State did not accept dissent or conflict (Stonholtz/Shapiro 1997). By refusing to acknowledge the value, utility, and healing power of conflict, non-democratic regimes did not create the cultural, institutional or psychological foundations for constructive transformation of conflicts. Democratization challenges an authoritarian approach to conflicts and makes the expression of conflict acceptable. To cope with new and old conflicts in a constructive way, conflict resolution education and training are needed to lay the

---

[5]Overall, the influence that economic assessments have on regime support falls far short of endorsing a theory of economic determinism. Although economic attitudes matter, other influences matter as well. The economy is important but only as part of the explanation of support for current and alternative regimes.

intellectual foundations of democratic peace-building. Acquiring conflict resolution skills is very important during the fragile democratic transition period. Conflict resolution efforts are needed to overcome the limits of modernization, focusing predominantly on modernizing political and economic structures. Projects should aim at training people in group decision-making processes, public participation, conflict resolution, and mediation. The general purpose is to entrust citizens with decision-making power. Conflict resolution offers the possibility of assisting divided societies to better understand their needs, create new alternatives for social, political, and economic organization and development, and find ways to constructively manage social change and conflict. Non-governmental organizations play a significant role in the process of challenging old values and introducing new ones with open communication (Bianchi 1997). See Box 8.1.

Box 8.1 Two democratic transition approaches: modernization and conflict resolution. *Source* The author

Modernization	Conflict resolution
Outcome orientated: Focuses on prescribed end states of liberal democratic and free market institutions and policies Positivist objective truth Emphasis on shared universals Nation-state as primary form of political organization Representation in decision-making processes Free market economics as a necessary ingredient of technological progress Professional (outsiders) as experts Concerned with system maintenance, stability and order Adjustment through conflict Fairness: principled, objectively defined Adversarial, competitive Ethics: rights and justice Values: rights, laws, interests, independence Micro-processes: debate, vote Causes of conflict: pursuit of interest and innate human aggression Primary metaphor: organism (biological system)	Process orientated: Focuses on means that are inclusive, participatory, empowering, analytic and dialogic Constructivist multiple truths Emphasis on context-related particulars Flexible forms of political organization Direct participation in decision-making processes Flexible forms of economic organization for improving material standards of living Participants (insiders) as experts Concerned with system change/innovation and activity Transformation through conflict Fairness: pragmatic, defined by parties' needs Cooperative, collaborative Ethics: responsibility and relationship Values: responsibility, relationship, needs, interdependence Micro-processes: dialogue, consensus Causes of conflict: unsatisfied needs and learned aggression Primary metaphor: community (interdependent relationships)

8.5.3 Civil Society

The degree, type, and distribution of civil society have a significant impact on chances of installing and consolidating a democracy. The enthusiasm for supporting the development of civil society needs to be calibrated by identifying the civil and less civil society. External funding, for example, has stimulated the growth of a great number of pseudo-NGOs. Attention must be paid to the secular non-governmental organizations, and also to religious organizations. In Cameroon, for example, an ecumenical group called Service Humanus (1998), consisting of Christian and Muslim denominations, is playing a major role in getting the country out of its political impasse and moving the democratic peacebuilding process forward. Reinforcing the following is vital: (a) the economic decentralization and consequent expansion of the private sector, laying the material basis for civil associations that are fully independent of the State; (b) new information and communication technologies; and (c) the knowledge of one another within the country and across the borders, and deepening collective awareness of the pivotal role people must play in fostering democratic governance.

8.5.4 Strengthening Institutional Learning

The inadequate institutional learning of national and international institutions remains a serious problem. This is the result of (a) a decision-making process in which poor communication between functional and geographic units and between the central administration and the field inhibits synergetic policy-making, and (b) the lack of significant conflict prevention and democratic peacebuilding expertise. More effective networking among politicians, the administration, people in the field, and academics needs to be developed. Experts in conflict prevention and democratic peacebuilding should be engaged more actively the decision-making of governmental and non-governmental organizations. They could help to analyse and evaluate conflict prevention and democratic peacebuilding initiatives and make suggestions on quality. One tends to rely solely on country experts for analysis. The expertise of both country experts and democratic peacebuilding experts is needed. In addition, training in conflict prevention and democratic peacebuilding should be made available to policy-makers in the international community and the European Union. Finally, organizations need to learn more effectively from field experiences. Field workers should be debriefed in a way that helps the organization learn cumulatively from their experiences. Impact analyses are not only necessary at the policy, but also at the project level. The aim is to learn from the experience and use the lessons learned to design better projects in the future.[6]

---

[6]A good example is the evaluation plan suggested by Paula Garb of the Global Peace and Conflict Studies programme at the University of California (UCI) for a project on "Cross conflict NGO

8.5.5   *Electoral Assistance*

The last few years we have seen a boom in different kinds of democratic assistance (IDEA and Field Diplomacy Initiative 1999). The most visible of these is the sending of election monitors to transition countries. On the whole, these efforts are assessed as positive. There are, however, some problems which need to be overcome to enhance the impact of these efforts on the democratic peacebuilding process. These problems can be clustered into eight groups.

The first problem relates to the promotion of democracy. Democracy still seems to have an image problem. Indicative of this is the debate about the universal validity of democracy and the limited transferability of Western political systems to other parts of the world. This endless discussion could become more productive if it focused on the instrumental value of democracy. The empirical evidence tells us that democracy is more strongly correlated with respect for human rights, well-being and security than any other political system. The growing awareness of the usefulness of democracy will make it a universal political commodity.

The second problem concerns the overall approach of the democratic transformation process. A distinction could be made between (1) the promotion or the prescription of democracy by outside actors, and (2) a more facilitative approach which tends to be more elicitive. The key to success is not the imposition of democracy, but the appeal of consolidated democracies and efforts to facilitate the transition process.

Thirdly, there are a series of organizational problems concerning (a) the professionalism of election monitors and the presence of election cowboys whose credentials are questionable; (b) the question of the requirements for functioning effectively as an election monitor: is it enough to know the electoral laws or does one also need political experience or savvy?; (c) coordination inadequacies; (d) the credibility or the perception of the monitors by the voters (when election monitors come across as election tourists or the monitoring is seen as a rich man's pastime, there is a problem); and (e) the flat institutional learning curve. The debriefing could be used more effectively in order to learn lessons and feed them back into the monitoring activities.

Fourthly, there are some problems of reporting. More attention should be paid to the impact of reports on the democratic transition. Providing a good report is like issuing a political driving licence. It attributes internal and external legitimacy to the

---

capacity-building in Tbilisi and Sukhumi" in Georgia. Her evaluation consists of: (a) a content analysis of diaries kept by the project director and the coordinators, all correspondence among the partners and with others, and media reports on the project activities, and on all other NGO activities. These data will be used to document changes and the degree to which this project and related NGO activities have impacted the NGO sector's development and facilitation of the democratic peace process; (b) surveys of the participants in the joint workshops and parallel seminars. These are intended to help determine what the participants have learned from the planned activities and whether they have incorporated the newly gained experiences in their organizational activities to the benefit of NGO education in human rights and humanitarian law.

elected politicians. Milosevic once received such a licence. To prevent misuse of elections, reports should be more precise and indicate strengths and weaknesses.

The fifth issue concerns the scope of monitoring. A distinction can be made between shallow and in-depth monitoring of the democratization process. In shallow monitoring, attention is restricted to the monitoring of election behaviour during elections. In-depth monitoring also pays attention to the pre-election and post-election phases and to the other nine building blocks of democracy.

The sixth problem concerns the decision over when to accept an invitation to send election monitors. Should we respond positively when there is an invitation from the government and when security is assured? These are minimal criteria. Or should we insist that other requirements be satisfied, such as respect for human rights or pre-electoral power-sharing arrangements? A minimal requirement invitation could be risky, because the government could exploit this to present itself as democratically minded and use it as part of its international impression management.

The seventh problem concerns the tendency to overlook the non-political aspects of a democratization process, such as the privatization of the economy.

Finally, there is no clear correlation between the costs of the electoral assistance and the outcomes. The development of a more sophisticated impact assessment could help to make the efforts more cost effective.

8.6 Conclusion

If recent developments are any guide, the period of quick and easy gains for democracy in the world is over. We seem to have entered a long period of considerable regime instability. The new democracies are now facing the more difficult challenge of building solid democratic institutions and developing a democratic culture. Most countries will probably make incremental progress toward consolidation, others will continue to regress or perform very poorly, and some will break down. While it is good to provide further international assistance to foster and consolidate democracy, it should be clear that the installation of an internal democracy does not necessarily guarantee a high quality society (Linz/Stepan 1997: 30–31). More and other efforts at different levels are needed. No democracy can assure successful entrepreneurs, skilful bankers, creative scientists, good physicians, line artists, honest judges or a free and fair international trading system or a secure international environment. Despite the limits of democratization, consolidating third wave democracies remains one of the major challenges for building a stable and secure international environment. The greatest challenge of the twenty-first century will be to integrate whole regions democratically and to initiate the democratic transition of the world order.

References

Barkan, J. (1997). Can Established Democracies Nurture Democracy Abroad? In A. Hadenius (Ed.), *Democracy's Victory and Crisis*. Cambridge: Cambridge University Press.

Bianchi, G. (1997). Training in Skills for Coping with Democracy. *The Annals of the AAPSS, 552*.

Conference IDEA and Field Diplomacy Initiative. (1999). Belgian Parliament Brussels.

Di Palma, G. (1990). *To Craft Democracies: An Essay on Democratic Transitions*. Berkeley: University of California Press.

Hague, R., Harrop, M., & Breslin, S. (1998). *Comparative Government and Politics* (4th edn.). London: Macmillan Press Limited.

Hakim, P., & Shifter, M. (1977). New Beginnings: The Promise of Democracy and Prosperity. *Harvard International Review*, 8–11.

Harris, P., & Reilly, B. (Eds.). (1998). *Democracy and Deep-Rooted Conflict: Options for the Negotiators*. Stockholm: International Institute for Democracy and Electoral Assistance.

Linz, J. (1997). Some Thoughts on the Victory and Future of Democracy. In A. Hadenius (Ed.), *Democracy's Victory and Crisis*. Cambridge: Cambridge University Press.

Linz, J., & Stepan, A. (1997). Toward Consolidated Democracies. In L. Diamond, M. Plattner, Y.-H. Chu, & H.-M. Tien (Eds.), *Consolidating the Third Wave Democracies: Themes and Perspectives*. Baltimore, MD: Johns Hopkins University Press.

March, J., & Olsen, J. (1995). *Democratic Governance*. New York: The Free Press.

Reychler, L. (1997a). Field Diplomacy: A New Conflict Prevention Paradigm. *Peace and Conflict Studies*.

Reychler, L. (1997b). *Field Diplomacy Initiative and Democratic Peacebuilding in Cameroon*. Leuven: Centre for Peace Research and Strategic Studies.

Reychler, L. (1999). Lessons Learned from Recent Democratization Efforts. In L. Reychler (Ed.), *Democratic Peace-Building and Conflict Prevention: The Devil is in the Transition* (pp. 133–146). Leuven: Leuven University Press.

Service Humanus. (1998). *Sondage d'opinion sur la portée de la démocratie et des elections au Cameroun, Yaoundé*.

Sisk, T. (1998). Power-Sharing Democracy. In P. Harris & B. Reilly (Eds.), *Democracy and Deep-Rooted Conflict: Options for the Negotiators*. Stockholm: International Institute for Democracy and Electoral Assistance.

Stonholtz, R., & Shapiro, I. (1997). Strengthening Transitional Democracies through Resolution. *The Annals of the AAPSS, 552*.

Vanhanen, T. (1997). *Prospects of Democracy: A Study of 172 Countries*. London: Routledge.

Chapter 9
Peacebuilding: Conceptual Framework

9.1 How to Prevent Violent Conflict

9.1.1 Conflict Is a Driving Force in Human History

Conflict has led to destruction, but it is also a strong motivating force for peace-building.[1] In the first half of the twentieth century, Europe was one of the most violent places in the world; in the Guinness book of violence, it scored all the records. It caused two world wars, set up totalitarian regimes, built concentration camps, and had civil wars. In the second half of the twentieth century, a European community was created. It became one of the most free, secure, and well-off places on the globe.

Conflicts signal problems that need to be taken care of. If serious conflicts are not resolved effectively, they can become destructive and cause a great deal of suffering. Today one can visit nearly 200 places in the world where people kill people. A great number of seemingly peaceful countries suffer from serious tensions and latent conflict.

9.1.2 Violence Is Costly

The high costs of violence have led to an increase in the efforts to prevent conflicts from crossing the threshold of violence. This is called *proactive violence prevention*. This is more cost-effective than *proactive violence prevention*. The aim of reactive conflict prevention is to prevent a further escalation of the conflict by controlling the intensity of the violence, reducing the duration of the conflict, and containing or

---

[1]This text was first published as: Reychler, L. (2001). Peace building: conceptual framework. In L. Reychler, & T. Paffenholz (Eds.), *Peace Building: A Field Guide* (pp. 3–15). Boulder, CO: Lynne Rienner Publishers. Permission to republish this text was granted by Rienner.

© The Editor(s) (if applicable) and The Author(s), under exclusive license to Springer Nature Switzerland AG 2020
L. Reychler and A. Langer (eds.), *Luc Reychler: A Pioneer in Sustainable Peacebuilding Architecture*, Pioneers in Arts, Humanities, Science, Engineering, Practice 24, https://doi.org/10.1007/978-3-030-40208-2_9

preventing geographic spillover. Also in the post-conflict phase, violence-prevention efforts could be needed to avert a new flare-up of the conflict.

When a conflict crosses the threshold of violence, the costs and the difficulty of managing them increase significantly, Violence becomes the cause of more violence. A comprehensive analysis of the costs of the violence gives an idea of the reconstruction and peacebuilding efforts that will have to be made when the violence ends. Not only are there human and economic costs but also social, political, ecological, cultural, psychological, and spiritual destruction (see Fig. 9.1).

9.1.3 What Is Violence?

To understand violence prevention, it is essential to analyse the violence that exists and the instruments for committing violence. The term *violence* refers to a situation in which the quantitative and qualitative life expectancy of individuals or communities is intentionally reduced. Indicators of this are the average life expectancy of the members of a group, infant mortality, daily calorie intake, access to schooling, and so on.

There are different ways to hurt people. The most visible is armed violence. Armed violence is intended to deter, coerce, wound, and even kill people. Psychological violence aims at the minds and hearts and tries to incapacitate the emotional strength of people. Psychological violence is often intended to produce mental suffering or spread fear and hate. Structural violence differs from the two preceding types in that it is an indirect type of violence. Here, violence is built into

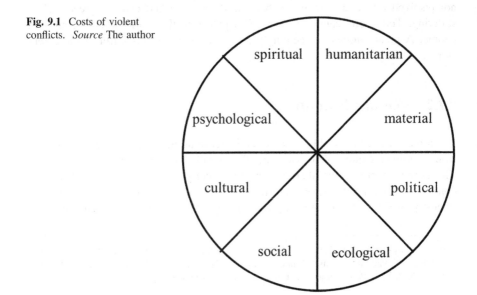

Fig. 9.1 Costs of violent conflicts. *Source* The author

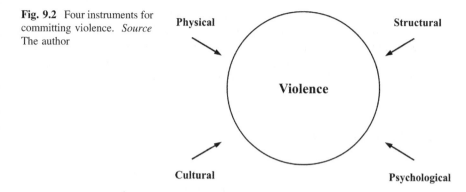

Fig. 9.2 Four instruments for committing violence. *Source* The author

the social structure and is less visible than physical or psychological violence. The previous apartheid regime in South Africa and the remaining caste system in India are textbook examples. A fourth instrument, cultural violence, refers to those aspects of the culture that legitimize the abuse of the instruments of violence cited above. Violence is approved of in the name of revolution, in the name of religious fanaticism, and in the name of political ideologies, such as nationalism and communism (see Fig. 9.2).

To achieve sustainable peace, we must first study how and to what extent each of the four types of violence are used. Peace requires more than the absence of armed violence. In fact, the absence of armed violence may mask all sorts of frustrations and even a potential for armed violence. An analysis limited to the study of armed violence can lead to surprises.

9.2 Diagnosis of a Conflict

By *conflict* we mean the pursuit of incompatible goals by different groups. For a diagnosis of a conflict, one can make use of the 4 + 1 model. This model indicates four necessary preconditions for a violent conflict: (1) interdependent parties, (2) who experience the interdependence as negative, (3) who have the opportunity to use armed violence, and (4) who consider the use of violence as the most cost-effective policy option. Once a conflict becomes violent, violence begets violence.

9.2.1 Actors

The first thing one needs to identify are the parties involved and their constituents: the owners and the stakeholders. This is not always an easy task. There are internal and external parties. Among the internal parties, one can distinguish the internal core parties who are directly involved in the conflict, the internal concerned individuals and groups who wish to play a role in facilitating conflict resolution, and the internal uninvolved. Among the external parties, one can single out the external involved parties, the external concerned parties who wish to and can play an active role in resolving the conflict, and the external uninvolved.

9.2.2 Interests and Issues

After identifying the parties, we need to learn what issues or interests are at stake for the parties. Competing interests are at the centre of any conflict, yet there can also be common interests. Conflicts of interest can be expressed in different ways. For example, they can be presented as scarce goods (territory, power, resources); as a definition of various situations, means, ends, values, and collective identity; or even as irrational issues.

For the sake of problem-solving, it is useful to define the terms *problem, interpretation, interest, position,* and *issue* in this context. The problem is the immediate source of the conflict (barking dog); the interpretation explains the other party's behaviour (my neighbour is unfriendly, inconsiderate); the interest refers to why is x a problem (I am not sleeping well); the position refers to demands, threats, fixed solutions, proposals, or points of view (buy a muzzle); and the issue is the topic the parties need to discuss and decide (how to control the barking at night) (Beer/Stief 1997).

9.2.3 Opportunity Structure

What does the opportunity structure look like for the parties involved in the conflict? Here we examine the power relations between the parties and the objective conditions in the conflict environment that enhance or inhibit the use of violence. The political, strategic, legal, economic, geographic, and cultural environment in which the conflict is embedded can raise or decrease the chances of war and peace. The chances of violence are higher in a non-democratic than a democratic environment. In a security community, conflicts tend to be handled constructively. In a crisis climate, good decision-making becomes more difficult.

9.2.4 Strategic Thinking

How do the actors define the present situation and how do they see the future? How do they assess different ways and means of handling the conflict? What do they consider to be their best alternatives to a negotiated agreement (leave the situation as it is, refer to a higher authority, or use power)? What are possible solutions for the parties in question? Solutions can be defined as actions that will resolve a given piece of the problem or the whole problem.

Most analysts identify five major styles (and variants) of handling conflicts (Folger et al. 1997). They can be distinguished in terms of *assertiveness,* defined as behaviours to satisfy one's own concerns, and *cooperativeness,* defined as behaviours intended to satisfy the other party's concerns. The *forcing/competing style* is high on assertiveness and low on cooperativeness: the party places great emphasis on his or her concerns and ignores those of the other. The basic desire is to defeat the other. The *accommodating style* is unassertive and cooperative: the party gives into the other at the costs of his or her concerns. This orientation has also been called appeasement or smoothing. An *avoiding style* is unassertive and uncooperative: the party simply withdraws and refuses to deal with the conflict. This style has also been called flight. A *compromising style* is intermediate in both assertiveness and cooperativeness: both parties give up some and split the difference. It has been referred to as sharing or horse trading. A *collaborating style* is high on both assertiveness and cooperation: the party works to attain a solution that will meet the needs of the parties involved. This style has been called the problem-solving or the integrative style.

All the styles are considered useful under different circumstances. The selection depends on the answer given to five questions: How important are the issues for the party? How important are the issues to the other party? How important is maintaining a positive relationship? How much time pressure is there? To what extent does one party trust the other? (Folger et al. 1997: 199–203).

9.2.5 Conflict Dynamics

Finally, the outcome of a conflict is determined by the dynamics of the interaction between the parties. Some conflicts transform constructively; others end up in a violent confrontation. Some are transformed in a mutually satisfactory way (win-win); others end up frustrating one or all parties involved (win-lose, lose-lose). One should not only make a static conflict analysis but also a diagnosis of the conflict's dynamics. Once a conflict has acquired a certain dynamic (for example, escalatory, de-escalatory, violent, or peaceful), the dynamic tends to feed on itself and become prolonged.[2] Escalation tends to further escalation, and war is the cause of further war. Peacebuilders ought to monitor indicators of positive or negative developments (Anstey 1993). Changes in each of the 4 + 1 variables can be classified as forces for or against a constructive transformation of the conflict

---

[2]This phenomenon could be called conflict inertia. Without external intervention some conflicts seem to stay on the same track.

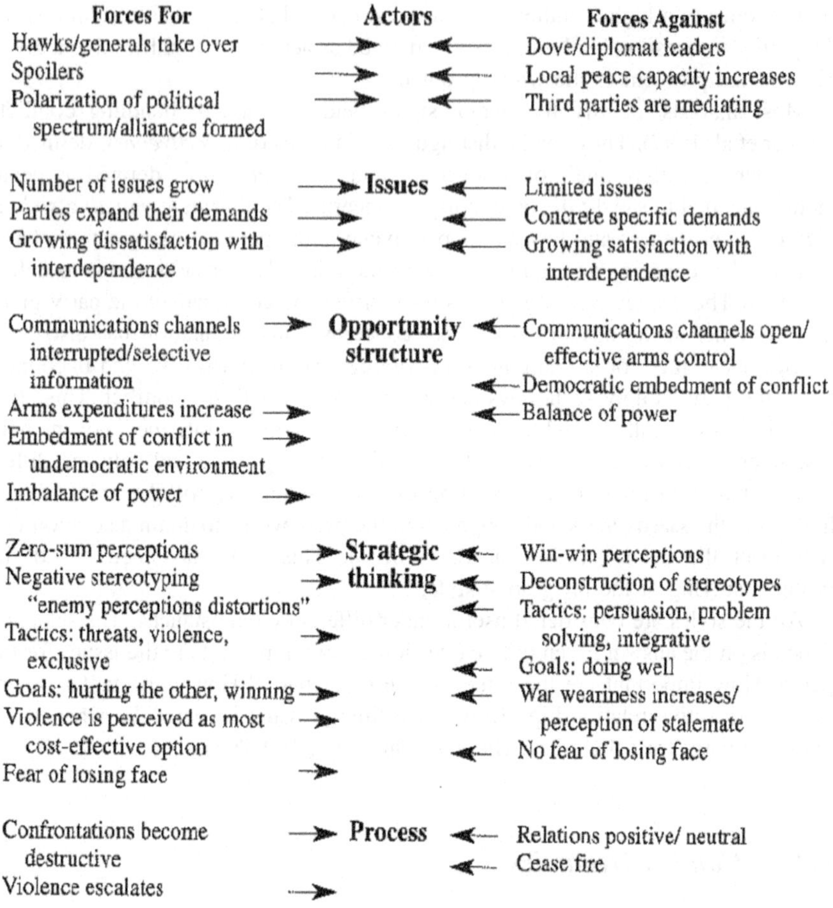

Forces For	Actors	Forces Against
Hawks/generals take over	⟶ ⟵	Dove/diplomat leaders
Spoilers	⟶ ⟵	Local peace capacity increases
Polarization of political spectrum/alliances formed	⟶ ⟵	Third parties are mediating

Number of issues grow	⟶ **Issues** ⟵	Limited issues
Parties expand their demands	⟶ ⟵	Concrete specific demands
Growing dissatisfaction with interdependence	⟶ ⟵	Growing satisfaction with interdependence

Communications channels interrupted/selective information	⟶ **Opportunity** ⟵ **structure**	Communications channels open/ effective arms control
	⟵	Democratic embedment of conflict
Arms expenditures increase ⟶		⟵ Balance of power
Embedment of conflict in undemocratic environment ⟶		
Imbalance of power ⟶		

Zero-sum perceptions	⟶ **Strategic** ⟵	Win-win perceptions
Negative stereotyping	⟶ **thinking** ⟵	Deconstruction of stereotypes
"enemy perceptions distortions"	⟵	Tactics: persuasion, problem
Tactics: threats, violence, exclusive ⟶		solving, integrative
	⟵	Goals: doing well
Goals: hurting the other, winning ⟶		⟵ War weariness increases/
Violence is perceived as most cost-effective option ⟶		perception of stalemate
Fear of losing face ⟶		⟵ No fear of losing face

Confrontations become destructive	⟶ **Process** ⟵	Relations positive/ neutral
	⟵	Cease fire
Violence escalates ⟶		

Fig. 9.3 Forces influencing escalation and de-escalation. *Source* The author

(see Fig. 9.3). The length of the arrow could depict the strength of these forces. The longer it is, the greater the influence it exerts.

9.3 Conflict Impact Assessment

9.3.1 What Is Conflict Impact Assessment All About?

The aims of a conflict impact assessment system (CIAS) are (1) to assess in time the positive and/or negative impact of different kinds of intervention or the lack thereof on the dynamics of the conflict; (2) to contribute to the development of a more coherent conflict prevention and peacebuilding policy; (3) to serve as a sensitizing tool for policy shapers and policy-makers, helping them to identify weaknesses in their approach (such as blind spots, incoherence, bad timing, and inadequate priority setting); and (4) to further the economy of development and peacebuilding efforts (Reychler 1996). A CIAS is a reminder for people involved in matters of development and conflict prevention.

9.3.2   Concerns and Critics

Despite the self-evident usefulness of a CIAS, one still senses resistance and reservations with respect to the development and the implementation of such a tool. Some of the reticence is caused by the preference of governmental and non-governmental organizations for leaving popular and appealing terms such as *peace* and *conflict prevention* undefined. Other reservations are based on the conviction that conflict dynamics are much too complex to be tracked by a CIAS. Still others fear that an adequate assessment will be a time-consuming exercise and/or inhibit flexibility. There is also the apprehension by some, stated less explicitly, that a CIAS would expose the limited or negative impact of one's policies on the peace process and thereby reduce the chances of funding. All these comments suggest that the road to the development and implementation of an assessment system will not be smooth. Anyway, the overall aim of a CIAS is to alert decision-makers and opinion leaders to the complexity of peacemaking, peacebuilding and peacekeeping and to help them formulate or design a more coherent, efficient and effective peace policy.

9.3.3   Levels of Assessment

Conflict assessment could be done at three levels. At the policy level, one studies the overall impact of the different peacebuilding efforts on the realization of a sustainable peace in a country or region. At the sector level, one looks at the impact of different peacebuilding efforts on one of the sectors of a sustainable peace in a country or region. At the project level, one focuses on the impact of a particular project on the realization of a sustainable peace. In the following section, some guidelines are suggested for performing a conflict impact analysis at the policy and project levels.

9.3.3.1   Conflict Impact Assessment at the Overall Policy Level

A CIAS at the policy level uses six criteria to evaluate policies with respect to their peacebuilding potential. Ideally, a good policy should have the following:

- A clear and compelling definition of the peace one wants to achieve and a valid conceptual framework indicating the conditions enhancing or inhibiting the realization of the aim.
- A comprehensive assessment of the needs or of the presence or absence of the above-mentioned conditions in the conflict region.
- A coherent action plan.
- An effective implementation of the action plan.
- Recognition and inclusion of the owners and stakeholders in the conflict-transformation process.
  - An awareness and dismantling of the sentimental walls inhibiting the satisfaction of all of the above.[3]

9.3.3.2   Conflict Impact Assessment at the Project Level

- *Step 1*: How and to what extent does the project contribute to the goals of the peace and development policy and satisfy the needs? An answer to this question necessitates adequate conflict analysis of the country and region where the project will be implemented, a clear idea of the peace one intends to realize, and a comprehensive needs analysis.
- *Step 2*: What is the impact of the conflict on the project? In this phase of the CIAS at the project level, the attention is focused on the concrete environment in which the project will be implemented. Here we are looking at a series of variables in the situation, which could negatively and/or positively influence the implementation of the project.
- *Step 3:* What is the impact of the project on the conflict dynamics? In this phase of the conflict impact analysis, we research how and to what extent the intervention reinforces or weakens the preconditions of sustainable peace

---

[3]To analyse and transform conflicts, more attention needs to be paid to political-psychological variables. More efforts in particular should be undertaken to identify and dismantle sentimental walls. The term *sentimental wall* refers to concepts, theories, dogmas, attitudes, habits, emotions, and inclinations that inhibit democratic transition and constructive transformation of conflicts. The existence of sentimental walls increases the chances of misperceiving the situation, misevaluating the interests at stake, lowering the motivation to act on the opportunity to do something about it, and developing the necessary skills and know-how. From the words *sentiment* and *mental,* 'sentimental' makes people aware of the emotional roots of many cognitions and attitudes. Making people aware of the existence of sentimental walls and efforts to dismantle them can provoke lots of resistance. Anyway, conflict prevention requires not only learning but also a lot of unlearning.

(consultation and negotiation system, political and economic structures, security and moral-political climate); the effectiveness of targeting the people; and the coherence and synergetic quality of the project.

- *Step 4*: Generating alternative options and decision-making. In this phase, a decision is made with respect to the project proposal. The project can be approved and given a green light, amended and linked to supporting other measures, postponed, turned down, or stopped. Efforts are also made to redesign the project to prevent or reduce the negative impact of projects and to raise the "peace added value" of the interventions. Development workers could become field diplomats. They could help to set up an adequate early warning system, sensitize the donors to the benefits of conflict prevention, and help people to build a durable peace. They could, using Anderson's (1999) terms, support existing capacities for peace and promote incipient ones. Projects can be redesigned to convey the message that cooperation is mutually beneficial; they can bring people together to discuss the problems and reduce misunderstandings; they can help to create a safe space where people can air and resolve their problems; they can help to build peace-enhancing political and economic structures; they can help to heal the past and raise confidence and hope in the future. For generating alternative options and redesigning projects, active listening and creativity are essential.

9.3.4 For Whom Is the CIAS Intended?

Ideally, all peacebuilders in conflict-prone regions should use the CIAS. Donors might use it to guide project selection, funding decisions, and monitoring; operational organizations could use it to design projects and guide implementation; and the recipient communities in conflict-prone areas could make use of the CIAS to assess the utility and relevance of outside-sponsored development projects.

9.3.5 Who Should Undertake Conflict Impact Assessment?

Taking into account the sensitivity and complexity of the CIAS, the assessment should be performed by a limited team, consisting of people with adequate conflict expertise and knowledge of the project and of the country and region where the project will be implemented. An adequate analysis should also make use of the input of the recipient parties, the stakeholders (who will have to live with the consequences and/or whose interests are at stake), and other donors who operate in the area.

9.3.6   When and Where Should the CIAS at the Project Level Be Applied?

Undertaking a CIAS should not be determined by the types of project but by the nature of the environment in which the intervention is planned. CIAS should be considered in environments characterized by incipient, latent, and manifest violence. CIAS should be performed in the pre-conflict, conflict, and post-conflict phase.

9.3.7   Where Does the CIAS Fit in the Project Process?

Conflict impact assessment should be done during all phases of the project cycle. CIAS should be done before the project is implemented, during its implementation in the field, when the project has ended, and even much later (ten to twenty years) to monitor long-term effects. In the pre-project phase a CIAS allows us to decide whether or not to go into the conflict-prone region. It also helps us to redesign project proposals to minimize negative side-effects and maximize peace added value. The aim of conflict impact assessment in the implementation phase is to monitor progress and identify and resolve unexpected problems more effectively. Equally important is conflict impact assessment after the project has finished. The purpose is to learn from the experience and to use the lessons to design better projects.

9.3.8   Methods?

An effective CIAS makes use of several sources and methods: existing country studies, active listening, informal in-depth interviews and discussions, question-naires, and problem-solving workshops. To assess the impact of the project in the implementation phase, one can invite the project leader and other coordinators to keep a diary.

9.4   Sustainable Peacebuilding

9.4.1   Transforming Conflicts

The overall aim of peacebuilding is to transform conflicts constructively and to create a sustainable peace environment. Transforming a conflict goes beyond problem-solving or managing a conflict.[4] It addresses all the major components of the conflict: fixing the problems which threatened the core interests of the parties; changing the strategic thinking; and changing the opportunity structure and the ways of interacting. Through peacebuilding, the conflict is not merely resolved – the whole situation shifts. It tries to make the world safe for conflicts. The term *peacebuilding* refers to all the efforts required on the way to the creation of a sustainable peace zone: imagining a peaceful future, conducting an overall needs assessment, developing a coherent peace plan, and designing an effective implementation of the plan. A useful overview of peacebuilding tools is presented by Lund (2001).

9.4.2 Sustainable Peace

The term *sustainable peace* refers to a situation characterized by the absence of physical violence[5]; the elimination of unacceptable political, economic, and cultural forms of discrimination; a high level of internal and external legitimacy or support; self-sustainability[6]; and a propensity to enhance the constructive transformation of conflicts.

9.4.3 Preconditions for a Sustainable Peace

The essential requirements for the creation of a sustainable peace can be clustered in four groups.

9.4.3.1 Effective Communication, Consultation, and Negotiation at Different Levels

The first precondition for establishing a sustainable peace is the presence of an effective communication, consultation, and negotiation system at different levels and between the major stakeholders. Effective negotiations can be distinguished from less effective negotiations by looking at the process, outcome, and implementation: there is no destruction or wasting of time and means; the relations

---

[4]The term *conflict management* refers to efforts to limit, mitigate, and contain a particular conflict. Conflict resolution goes further and implies that the root causes of a conflict are addressed and resolved.

[5]This could also be positively defined as a situation characterized by objective and subjective security (people are and feel secure).

[6]Although external support may be necessary in a particular phase of the peacebuilding process, in the end people should be able to support it themselves.

between the parties do not deteriorate but improve; the outcome is perceived mutually satisfactorily, and – with the exception of initial external support – the agreement is self-sustaining (Reychler 1999).

9.4.3.2   Peace-Enhancing Structures

Essential for the establishment of a sustainable peace is the establishment of series of peace-enhancing structures.[7] The first is of a political nature: it is the establishment of a consolidated democracy. Such a democracy consists of ten building blocks and internal and external support systems. The second structure necessary for the establishment of a sustainable peace is an effective, legitimate, and restorative justice system. The creation of a restorative justice system has been strongly promoted by Zehr (1990). The third structure that needs to be built is a social, free market system. The chances of establishing a sustainable peace are greater in a social, free market system than in a centralized or pure free market economic environment. Of great importance are the privatization process and the creation of a vibrant economic society. The fourth structure that needs attention is the education, information, and communication system. Here we look at the degree of schooling, the level of discrimination, the relevance of the subjects and the attitudes held, the control of the media, the professional level of the journalist, the extent to which the media play a positive role in the transformation of the conflicts, and the control of destructive rumours. Finally, structures are needed to cope with refugee problems in a satisfactory way.

9.4.3.3   An Integrative Moral-Political Climate

The term *integrative moral-political climate* refers to a political-psychological environment characterized by a 'we-ness' feeling or the existence of multiple loyalties, expectations of mutual benefits as a consequence of cooperation, reconciliation of the past and the future, a dismantling of sentimental walls, a reconciliation of the values that will guide the future, and a commitment to cooperate. An integrative moral-political climate enables a community to solve its problems constructively.

Is there hope for a better future? Does one expect mutual benefits as a consequence of cooperation? What hope-raising measures have been taken?

How strong is the nationalism? Is the political commitment inclusive or exclusive? Is there a we-ness feeling? Is the political engagement of the people characterized by multiple loyalties? How strong is and how far does the we-ness feeling reach? How much social capital is available? The term *social capital* refers to a

---

[7]For a more elaborate description of the peace-enhancing structures, including a series of checklists, see Reychler (1999).

culture of trust and cooperation that makes collective action possible and effective. As Putnam (1993) says, it is the ability of a community to develop the 'I' into the 'we.' A political culture with a fund of social capital enables a community to build political institutions with a real capacity to solve collective problems. Where social capital is scarce, even an elected government will be viewed as a threat to individual interests (Putnam 1993).

How much progress has been made with respect to the healing of past wounds and reconciliation? Is the reconciliation process orientated toward the future, here and now, and/or the past?

Which sentimental walls obstruct the way to a sustainable peace? Victims with despair, pluralistic ignorance, and political inefficacy? Offenders with historical falsification, stereotyping, dehumanization, distrust, indifference, wrong assessment of the consequences of decisions (optimism/pessimism), preference falsification and pluralistic ignorance, obstacles of a religious nature? Third parties with neutralism/passivity/non-intervention, cultural arrogance, a moral-legal approach, a wait until the conflict is ripe? Analysts with a one-dimensional approach who use invalid theories, the influence of 'scientific' doctrines/myths/taboos, an elitist analysis, a wrong assessment of future developments?

How compatible are the values of the parties? What are the basic values (peace, economic welfare, democracy, justice, truth, forgiveness, and dignity) that will guide the future of the community when reconciled?

9.4.3.4 Objective and Subjective Security

Here we look at the ingredients necessary for enhancing real and perceived security. Both could be enhanced by several measures, such as cease-fires, arms control and disarmament, confidence- and security-building measures, the creation of regional security arrangements, and the like.

Very important in the creation of a sustainable peace environment are efforts to also create these conditions at the regional level. Internal and external peace is indivisible. One of the secrets of the European Union is that a great deal of its political, economic and security problems are managed at a multilateral level. Another secret is the role played by a new type of leadership. The creation of a sustainable peace environment requires a new type of internal leader, such as Jean Monet, George Marshall, or Nelson Mandela. Such leaders are able to project a clear and compelling image of the future and the ways and means to get there. Equally important are courageous followers standing up to and for the leaders. Irresponsible followership remains an obstacle to the successful transformation of conflicts. Evil is stoked and fanned by followers who abandon their empathy with the suffering of others, either through fear or through seduction by a leader. Ira Chaleff suggests four principles on the basis of which a courageous follower operates within the group and a fifth dimension outside the group: the courage to assume responsibility, the courage to serve, the courage to challenge, the courage to participate in transformation, and the courage to leave (Chaleff 1995).

References

Anderson, M. (1999). *Do No Harm: How Aid Can Support Peace or War.* Boulder: Lynne Rienner.

Anstey, M. (1993). *Practical Peace Making: A Mediator's Handbook.* Cape Town: Juta and Co.

Beer, J., & Stief, E. (1997). *The Mediator's Handbook.* Gabriola Island, Canada: New Society Publishers.

Chaleff, I. (1995). *The Courageous Follower: Standing Up To and For Our Leaders.* San Francisco: Berrett-Koehler Publishers.

Folger, J., Poole, M. S., & Stutman, R. K. (1997). *Working Through Conflict: Strategies for Relationships, Groups, and Organizations* (3rd edn.). New York: Longman.

Lund, M. (2001). A Toolbox for Responding to Conflicts and Building Peace. In L. Reychler, & T. Paffenholz (Eds.), *Peace Building: A Field Guide* (pp. 16–20). Boulder, CO: Lynne Rienner Publishers.

Putnam, R. (1993). *Making Democracy Work: Civic Traditions in Modern Italy.* Princeton, NJ: Princeton University Press.

Reychler, L. (1996). Conflict Impact Assessment. Paper presented at the *IPRA conference, Durban.*

Reychler, L. (1999). *Democratic Peace-building and Conflict Prevention: The Devil is in the Transition.* Leuven: Leuven University Press.

Reychler, L. (2001). Peace Building: Conceptual Framework. In L. Reychler, & T. Paffenholz (Eds.), *Peace Building: A Field Guide* (pp. 3–15). Boulder, CO: Lynne Rienner Publishers.

Zehr, H. (1990). *Changing Lenses: A New Focus for Crime and Justice.* Waterloo, Canada: Herald Press.

Chapter 10
Peacebuilding Software

10.1 An Overview of the Peacebuilding Climate

In the last ten years several peace agreements have been signed.[1] At first glance this development seems positive. However, if we analyse the successfulness of the peace-building processes, the picture becomes less rosy. Some peace processes, like those in Northern Ireland and South Africa, seem to have travelled the difficult road to sustainable peace. Others, like those in Bosnia-Herzegovina, Rwanda, the Middle East and Angola, seem, at best, to have stopped the fighting for a time. How can these differences be explained? We could research the problems with respect to the installation of the necessary peacebuilding blocks: the quality of the negotiations, the consolidation of democracy, the functioning of the socio-economic system, the security arrangements, the support of the international community and leadership, and the integrative nature of the political-psychological climate (Reychler 1999). The latter building block focuses on the emotional, psychological and socio-psychological issues involved in these peace-building processes.

The 'subjective' issues like (1) future expectations, (2) reconciliation, (3) political commitment, (4) trust/confidence-building, (5) perceptions and (6) feelings tend to be widely underrated in the current peace-building practice. An important reason for this neglect stems from the implicit conceptual framework of the international community's peacebuilding community in which negotiations and political, economic and security structures predominate. The underlying assumption is that when there is a peace agreement and the necessary structures are installed, conflicts will be solved in a peaceful and constructive manner. Another explanation is that conflict management is a predominantly male-dominated field in which *soft* variables tend to be overlooked. Finally, there is the fact that the "software of peace" is more difficult to research and quantify than structures.

---

[1]This text was first published as: Reychler, L., & Langer, A. (2003). Peacebuilding Software. *Peace Research: The Canadian Journal of Peace Studies, 35*(2), 53–73. Reprinted with permission.

L. Reychler and A. Langer (eds.), *Luc Reychler: A Pioneer in Sustainable Peacebuilding Architecture*, Pioneers in Arts, Humanities, Science, Engineering, Practice 24, https://doi.org/10.1007/978-3-030-40208-2_10

The creation of an integrative climate is one of the necessary building blocks for a sustainable peacebuilding process. The underlying assumption is that all six peacebuilding blocks are mutually reinforcing and therefore need to be installed simultaneously. Although the climate may be less tangible and observable than the other building blocks, it is observable through its consequences. An integrative or disintegrative climate can express itself in the form of attitudes, behaviours and institutions. A high level of trust, for example, can be assessed by looking at attitudes towards other members of society (confidence/trust), behaviour (security controls), or institutional expressions of trust and distrust (security departments) (Fig. 10.1).

10.1.1 What Is an Integrative Climate?

The term 'climate' refers to the quality of a relatively enduring environment that is perceived by the occupants, influences their behaviour, and can be described in terms of the attributes of the environment (Reychler 1979). We use the term to accentuate the subjective character of the phenomenon and also to indicate that we are focusing not on sharply defined images of the environment but on the

Fig. 10.1 Necessary conditions for sustainable peacebuilding. *Source* The authors

perception of the highly generalized content of the context. The concept of climate, for example, has been used to analyse the moral and political qualities of the international climate. Reychler, for example, distinguishes in order of preference six types of international climate:

A *Hobbesian climate* could be characterized as a very rudimentary power interaction between countries pursuing their self-interests and indifferent to the fate of others. It is an anarchic place.

A *Marxian climate* refers to an environment with a high level of perceived structural violence; it is a highly structured system with top and bottom dogs.

An *instrumental exchange climate* refers to a situation in which right action consists of that which instrumentally satisfies one's needs and occasionally the needs of others.

A *we-ness climate* is marked by a growing feeling of we-ness at global level, as can now be observed between subgroups of countries, such as the rich man's club or the European Union.

In an *international law climate,* international national law is perceived as the attribute of the international community.

A *shared principle agreement climate* refers to an international environment in which decisions are made by mutual consent, without sacrifice of one or another country. The mutual consent is based on reciprocity between the partners and equality between them prior to the agreement.

10.1.2   Integrative Versus Disintegrative Climate

The term 'climate', as used here, refers to the total experience of the social-psychological environment in which the conflict transformation and the peacebuilding processes take place. The existing climate can enhance or inhibit the peacebuilding process. When the climate enhances or reinforces the peacebuilding process, it is called integrative; when it inhibits peacebuilding or regresses the process, it is called disintegrative. The climate is composed of six mutually influencing elements:

- the expectation that cooperation will provide a mutually beneficial future (hope/trust in the future versus cynicism/hopelessness/despair)
- human security or the feeling of having a satisfactory level of physical, political, economic, health and cultural security (human security versus human insecurity)
- the feeling that all groups are being included (inclusion versus exclusion)
- the feeling that one can depend on or trust others (trust versus distrust)
- the willingness to cooperate (cooperativeness versus uncooperativeness)
- the perception that the conflicting parties have been reconciled or are undertaking serious efforts (reconciliation versus no reconciliation)

- the absence of major "sentimental walls" that inhibit the peacebuilding process (open-mindedness versus close-mindedness/feelings or mindsets which inhibit progress).

The term "sentimental wall" refers, in this context, to concepts, theories, dogmas, attitudes, habits, emotions and inclinations that inhibit the democratic transition and constructive transformation of conflicts.

Some assumptions:

- there are several degrees of integrativeness
- a country, as a whole, may have a high-level integrative climate but be confronted with extremely violent local riots (Bradford, Burnley and Oldham, three communities with large Asian populations in the UK)
- a country can have communities with high integrative climates but lack an overall integrative climate (Northern Ireland)
- a climate is manageable and changeable
- changes in the climate could be used for assessing the likelihood of a violent conflict or the sustainability of a peace process
- the components of the climate are mutually reinforcing.

10.2 Constituent Elements of an Integrative Climate

10.2.1 Expectation of an Attractive Common Future

When people or their leaders don't believe in an attractive common future, sustainable peacebuilding will not be easy. Cynicism, despair and defeatism inhibit the mobilization of hearts and minds to build a new future. Hope-raising measures are needed to transform despair into hope. David Cooperrider, the inventor of appreciative inquiry (AI) (Cooperrider/Whitney 1999), writes:

> There are three basic conclusions about the affirmative basis of organizing: (1) organizations are products of the affirmative mind; (2) when beset with repetitive difficulties or problems, organizations need less fixing, less problem solving, and more reaffirmation – or precisely more appreciation; (3) the primary executive vocation is to nourish the appreciative soil from which new and better guiding images grow on a collective and dynamic basis (Global Excellence Initiative [GEM] 1999) (Fig. 10.2).

Essential for moving forward is the identification of the existing or potential strengths, which still exist in the conflict region, and the anticipation of a compelling future. In many conflicts too much attention is being paid to the negative and the past (Fig. 10.3).

> The anticipatory principle of AI says that the most important resource we have for generating constructive organizational change or improvement is our collective imagination and our discourse about the future. It is the image of the future that in fact guides the current

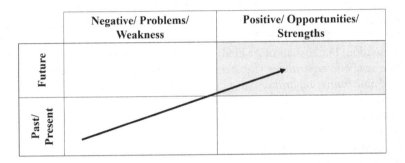

Fig. 10.2 Integrative and disintegrative climates. *Source* The authors

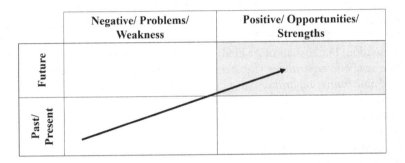

Fig. 10.3 Forward and backward conflict resolution. *Source* The authors

behaviour of any person or organization. Much like a movie projector on a screen, human systems are forever projecting ahead of themselves a horizon of expectation that brings the future powerfully into the present as a mobilizing agent (GEM 1999: 27).

10.2.2 Reconciliation: The Observation and Feeling that the Conflicting Parties Have Been Reconciled

"Reconciliation is a joint process of releasing the past with its pain, restructuring the present with reciprocal respect and acceptance, and reopening the future to new risks and spontaneity" (Augsburger 1992). Reconciliation is of vital importance for the success of sustainable peacebuilding. Reconciliation releases the necessary energy to build a new future. It requires the cooperation of the conflicting parties and involves a series of distinct but interdependent elements. Although there is no standard operational formula for reconciliation, it normally involves actions which reconcile the competing needs and values for security, justice, truth, amnesty, economic development, freedom and forgiveness.[2]

Justice or the punishment of perpetrators of violence (retributive justice, lustration): Depending on the interaction between the parties involved, the justice opted for can be retributive or restorative. Retributive justice is adversarial, focuses on guilt and blame, delivers pain and suffering and is past-orientated. Restorative justice is participatory, focuses on needs and obligations, tries to heal and resolve problems, and is future-orientated (Zehr 2001). Restorative justice assumes that justice can and should promote healing.

Truth: Confronting and reckoning with the past is vital for the transition of a society from violence to sustainable peacebuilding. "When communities have been victimized by the government or by another group during a conflict, underlying feelings of resentment and the desire for revenge cannot be alleviated unless the group is allowed to mourn the tragedy and senses that the wrongs have been acknowledged, if not entirely vindicated" (Lund 1998).

Restoration or compensation for damage: Restoration or compensation refers to all measures in the form of reparation, *ex gratia* payment, restitution, rehabilitation or recognition. Thus, restoration and compensation include not only monetary compensation, but also all other kinds of reparation strategies or measures. These include *symbolic reparations* (e.g., reburials, the building of memorials, etc.), *legal and administrative reparations* (e.g., expunging criminal records, the issuing of declarations of deaths, etc.), as well as community *reparations* (e.g., programmes for better access to health services, etc.). Reparation can be helpful or even necessary for both collective and individual reconciliation processes.

Forgiveness is the mutual recognition that repentance is genuine and right relations have been restored or achieved. Repentance is not a precondition for forgiveness, and it is not a desired consequence (Augsburger 1992: 284).

Reassurance that it will not happen anymore in the future. This can be assured through measures which prevent physical, structural, psychological and cultural violence.

Commitment to future cooperation: Reconciliation about the future does not replace the necessity to reconcile the past, but it could enhance the latter considerably.

Reconciliation requires the conflicting parties to assume responsibility: the offender, for example, by recognizing responsibility, expressing regret, and asking for forgiveness; the victim, by switching from external blaming to internal inquiry and reflection on the nature of the conflict. This is necessary to move the reconciliation process forward, because blaming is essentially a form of accepting the status quo of a victim and striking out at the victimizer (Rothman 1997).

---

[2]The competing character of four values (truth, peace, mercy and justice) has been discussed by Lederach (1997).

Reconciliation is about healing distress, at the individual and communal level, and about the restoration of harmony.

10.2.3   Multiple Loyalties or the Feeling of Being Excluded

A country in which the conflicting groups have developed exclusive loyalties is not fit to develop a successful sustainable peace process. A minimum requirement is the development of an overall we-ness feeling or multiple loyalties. The intensity and the nature of the commitment or attachment (instrumental or sentimental) may differ at different levels. The political attachment of many European citizens is spread over different levels (local, regional, national and European). Moreover, it is the idea of "unity in diversity". For example, in the South African context this would mean that it should be possible to be simultaneously Zulu, black, African and South African. These different identities should be mutually compatible and reinforcing, and should foster the development of an overarching common identity.

10.2.4   Human Security

Another characteristic of an integrative climate is the absence of fear and insecurity. In this context the concept of human security is widely used (Naidu 2001; Tadjbakhsh 2002). Naidu (2001: 1) defines human security analogically with national security as "a situation in which the life, body and the well-being of the human person have been protected through the use of physical violence". In our approach the concept of human security will be broadened. More particularly, it refers not only to the objective threats to life, body and well-being, but also to individual *feelings* of security and insecurity in different domains, such as physical, health, political, economic, social, and cultural, that could be the result of per-ceived, imagined or unreal threats.

In a country with a high level of human insecurity, such as Nigeria, people will attach themselves to the group they know best (community, ethnicity, state, reli-gion), and some politicians try to exploit these fears to mobilize people against other people and consequently keep the vicious circle of communitarian, ethnic or religious conflicts going. An expression of human insecurity is fear. Bar-Tal (2001) defines fear as an automatic emotion grounded in the perceived present and often based in the memorized past that leads to the freezing of beliefs, conservatism, and sometimes pre-emptive aggression. Hope, in contrast, involves mostly cognitive activity, which requires the search for new ideas and is based on creativity and flexibility. Because hope is based on thinking, it can be seriously impeded by the spontaneous and unconscious interference of fear. It is assumed that societies involved in intractable conflict are dominated by a collective orientation of fear, which is functional in their coping with the stressful and demanding situation. But such an orientation serves as a psychological obstacle to a peace process once it starts (Bar-Tal 2001). A hopeful, peaceful future involves uncertainty, ambiguity and risk-taking. Societies involved in an intractable conflict, like the Palestinian and Israeli societies are today, know how to cope with violent conflict: some have very successfully adapted to this situation in the past. The hope for peace, however, demands new approaches to the new situation of peacebuilding, which the collective orientation of fear inhibits from developing.

10.2.5   Social Capital

This component of the integrative climate, social capital, refers to trust and reciprocity between people that enable them to collaborate (Herriot et al. 1998). Gouldner (1960) identifies as the most fundamental manifestation of trust:

- the confidence that another will fulfil their obligations to you. These may have been explicitly undertaken and promised, or they may have been implicitly assumed.

Other manifestations of trust are:

- the confidence that the other will not try to deceive you;
- the expectation that people can do things which their position, qualifications, experience and achievements suggest they can do;
- the expectation that people will not harm us and maybe even care for our welfare (Herriot et al. 1998).

Trust itself is often reciprocal: if others trust me, I may be more likely to trust them; if someone has helped me without the expectation of a subsequent favour, I may be more inclined to behave in the same way myself.

10.2.6   The Absence of Other Sentimental Walls Which Inhibit the Peacebuilding Process

Sentimental walls are attitudes and feelings, perceptions and expectations, causal analyses and attributions of responsibility, strategic analyses, values, preferences, taboos, and social-psychological pressures (such as conformity pressures, groupthink and political correctness) which stand in the way of sustainable peacebuilding (Herriot et al. 1998: 43). The existence of sentimental walls increases the chances of misperceiving the situation, misevaluating the interests at stake, and lowering the motivation to act on an opportunity and develop the necessary skills and know-how. The hyphenation of 'sentiment' and 'mental' in 'sentimental' is intended to show that mental walls tend to be reinforced by emotions, and that efforts to dismantle them tend to be confronted with different kinds of emotional resistance (Reychler 1999: 54). They can be found in the victims, offenders, third parties and analysts.

- victims: despair, pluralistic ignorance and political inefficacy.
- offenders: historical falsification, stereotyping, dehumanization, distrust, indifference, wrong assessment of the consequences of decisions (optimism/pessimism), preference falsification and pluralistic ignorance, and obstacles of a religious nature.
- third parties: neutralism/passivity/non-intervention, cultural arrogance, moral-legal approach and waiting until the conflict is ripe.
- analysis: one-dimensional approach, using invalid theories, influence of 'scientific' doctrines/myths/taboos, elitist analysis and incorrect assessment of future developments.

10.3 Researching the Role of Integrative Climate in Peacebuilding Efforts

Studying the impact of the climate on the peacebuilding process requires (a) the development of measures or indicators for the components of the climate, and (b) an assessment of the impact on the success or failure of different peacebuilding efforts. To assess the level of integrativeness of the climate, we could make use of two types of measures: (1) a questionnaire to be filled in by experts who are well acquainted with the conflict regions being studied, and (2) a checklist of objective indicators.

In this section we shall present a questionnaire for experts and an accompanying evaluation sheet for assessing the integrativeness of a certain society or region. The aim of this questionnaire is obviously limited to providing a general view of the state of the climate. Further, the evaluation sheet should enable the researcher or practitioner to quickly determine the main problem areas or biggest obstacles. The evaluation sheet is constructed in such a way that, simply put, lower scores indicate more problems or obstacles. The second way to assess the integrativeness of a certain society or region is based on objective indicators. In the case study on Bosnia and Herzegovina in Sect. 10.4, we will present possible objective indicators that could be used to infer something about the integrativeness of the climate (Figs. 10.4 and 10.5).

10.4 A Preliminary Assessment of the Integrative Climate in Bosnia and Herzegovina

Hereafter we will look at the integrative climate of Bosnia and Herzegovina. The underlying idea is to illustrate the empirical workability, comparability and value of the concept of integrative climate. Moreover, we will tentatively try to evaluate the

Questionnaire for Experts	Strongly disagree (0)	Disagree (1)	Somewhat disagree (2)	Somewhat agree (3)	Agree (4)	Strongly agree (5)	General tendency
1. People are prepared to work towards a better future.							
2. The antagonists have developed good relations.							
3. There is sufficient tolerance and acceptance towards the cultivation of identity groups.							
4. Most people feel economically secure.							
5. Society has become more open-minded.							
6. Most people feel there is sufficient health security.							
7. People have overlapping we-ness-feelings and loyalties.							
8. A commonly accepted truth has come out.							
9. Most people want to leave the past behind.							
10. People or groups relate through a dense web of civic engagements and networks.							
11. Most people feel that justice has been done.							
12. Voluntary interaction and cooperation is a widespread feature of society.							
13. People feel they are empowered to build an attractive future.							
14. The negative stereotyping has largely disappeared.							
15. People feel they can rely on other people.							
16. Most people believe in a better common future.							
17. Most people feel politically secure.							
18. People believe in a better future for their children.							
19. People feel they have sufficient cultural security.							
20. There exists a culture of tolerance towards different views and opinions.							
21. Hate and animosity towards each other have disappeared.							
22. There is an overarching national or common identity.							
23. There are no attitudes or perceptions that significantly inhibit dialogue and cooperation.							
24. Different groups or communities have compatible visions.							
25. The society exhibits a great deal of trust and mutual helpfulness.							
26. The victims have been rehabilitated and compensated in satisfactory way.							
27. The society is characterized by inter-communal or inter-ethnic trust relations.							
28. People appreciate the multicultural character of their society.							
29. Most people feel physically secure.							
30. People feel they are empowered to build an attractive future.							
31. People believe they have leaders who know how to build a better future.							
32. There has been a great deal of understanding and mercy/forgiveness.							
33. The loyalty vis-à-vis the respective identity group is not exclusive.							
34. People feel environmentally secure.							
35. There are a great number of trans-communal NGOs or networks.							
36. There is a high level of conflict resolution know-how.							

Fig. 10.4 Questionnaire for experts. *Source* The authors

1. HUMAN SECURITY		2. RECONCILIA-TION		3. SOCIAL CAPITAL		4. FUTURE PERSPECTIVE		5. MULTIPLE LOYALTIES		6. SENTIMENTAL WALLS	
Q.4		Q.2		Q.10		Q.1		Q.3		Q.5	
Q.6		Q.8		Q.12		Q.9		Q.7		Q.13	
Q.17		Q.11		Q.15		Q.16		Q.20		Q.14	
Q.19		Q.21		Q.25		Q.18		Q.22		Q.23	
Q.29		Q.26		Q.27		Q.24		Q.28		Q.30	
Q.34		Q.32		Q.35		Q.31		Q.33		Q.36	
TO.		TO.		TO.		TO.		TO.		TO.	

TOTAL INTEGRATIVE CLIMATE: *(1+2+3+4+5+6) =*   [...]

DESINTEGRATIVE   0   180   INTEGRATIVE
CLIMATE   CLIMATE

Fig. 10.5 Table of software scores. *Source* The authors

different aspects that constitute an integrative climate. Obviously, the assessment of the integrative climate only intends to give us a broad indication of certain positive and/or negative social-psychological developments within a specific country.

10.4.1   Bosnia and Herzegovina

The Republic of Bosnia and Herzegovina has been part of the Federal Republic of Yugoslavia for over forty years. During this period the three major ethnic groups – Bosnians, Serbs and Croats – lived together in a peaceful and relatively harmonious way. Inter-ethnic trust was sufficiently high for a vibrant, ethnically heterogeneous society to exist. Further, inter-ethnic trust relations were sufficiently stable and developed to emanate in a high number of inter-ethnic marriages. However, all this changed when, at the end of the 1980s – in an environment of economic downturn and 'democratization' – the Federal Republic of Yugoslavia started to dissolve as a consequence of the upsurge of nationalist ideas and parties. The trigger for the Bosnian war was the republic's referendum concerning the question of independence. Following the results of this referendum, which was largely boycotted by the Bosnian Serbs, the president of the Republic of Bosnia and Herzegovina, Izetbegovic, declared independence on 2 March 1992. Shortly thereafter large-scale fighting broke out. The war officially ended in December 1995 with the signing of the Dayton Peace Agreement in Paris. The sad balance of the Bosnian war can be summarized as follows: over 200,000 casualties, more than half of the Bosnian population was made into refugees, large parts of the Bosnian territory were actively or passively ethnically cleansed, and inter-ethnic relations and networks were completely destroyed. During the post-Dayton years the international community has invested vast amounts of money, time and energy in rebuilding the physical infrastructure of Bosnia and Herzegovina. The question we pursue in this context deals with the state of the rebuilding process of the social-psychological infrastructure of Bosnia and Herzegovina.

10.4.2   Human Security

As a consequence of the continued presence of the Stabilization Force (SFOR) and the International Police Task Force (IPTF), the objective security situation has been generally good. However, inter-ethnic violence and attacks were still being reported regularly after the signing of the peace agreement (Human Rights Watch 2002: 296–301), and returnees-related violence remained commonplace (International Crisis Group [ICG] 2002b: 27). Obviously, this increased the subjective insecurity feelings of most refugees and displaced persons. The small numbers of refugees and displaced persons returning to their pre-war homes confirm this fact. Although in recent years the number of returnees has increased, the total number of refugees and displaced persons remains over a million people (Alfaro 2000). An UNHCR survey concludes that refugees and displaced persons list 'security' as the most important issue in deciding whether or not to return (Alfaro 2000: 13). Another important issue relates to the health care situation. Since the end of the war, health care has improved considerably and most people seem to feel sufficiently secure about this. In a recent opinion poll only 5% of the respondents stated Health/Social issues as the "Most important issues in deciding how to vote" (National Democratic Institute for International Affairs [NDI] 2002). More problematic are the issues concerning people's political and economic security. Clear illustrations of the problems in this context are the widespread corruption, especially in the Bosnian-Serb Republic, and the extremely high unemployment rates (estimates vary from 40–60%).

10.4.3   Reconciliation

At the time of writing, there has not been an official and/or public truth-recovering process. Truth-finding and clearing up what happened in the past mainly takes place within the judicial framework of the International Yugoslavia Tribunal (ICTY) in The Hague or as a result of academic research and other inquiries. For many people the truth about what happened in the past remains unclear and obscure. This situation is exacerbated by the nationalist parties, such as the Bosnian-Serb SDS, that still proclaim their nationalist rhetoric and frequently invoke the past to stoke up fear of other ethnic groups. In other words, a commonly accepted truth about the past is lacking. Further, acknowledgement of mutual suffering, pain and loss is extremely rare. Justice has been done in a limited number of cases only. The ICTY has to focus on the major cases due to limited resources and time. It is estimated that by 2007, the ICTY in The Hague will have heard between 120 and 200 cases (ICG 2002a: 31). In addition, more than six years after Dayton the two major architects of the ethnic cleansing in Bosnia and Herzegovina, Karadzic and Mladic, are still at large.[3] Only very few trials of lesser war crimes have taken place in

---

[3]Karadzic was eventually tried at The Hague in 2008 and Mladic in 2011. Both are now serving life sentences for their war crimes.

Bosnia. Such trials initially took place exclusively in the Federation, with approximately 35 verdicts against some 70 accused (ICG 2002a, p. 31). The Bosnian Serb Republic (RS) was extremely reluctant to cooperate constructively with the ICTY in The Hague. Further, most people were insufficiently compensated and rehabilitated. The fact that, in the decade after the war, there were still more than a million people registered as internally displaced or refugees, unable to return home because of security reasons and without receiving compensation for lost or destroyed property, underscores this. A process of forgiveness has been largely absent. Extreme nationalist parties still enjoy large popularity. In the RS and in the cantons of the Federation where the Bosnian-Croats constitute a majority, the extreme nationalist parties, such as SDS and HDZ, continue to be the largest parties by far.

10.4.4   Future Perspectives

Nationalist parties and movements still play the most prominent role within the Bosnian-Serb and Bosnian-Croat constituency. The nationalist parties, in particular the SDS, SRS, HDZ and SDA, assert that the only way towards a better future is through more independence, self-reliance and self-rule. Obviously, these future visions are anything but compatible. The Bosnian-Serb SDS and SRS and the Bosnian-Croat HDZ both spread visions of fear and oppression at the prospect of a unified Bosnia. Their visions of the future emphasize that integration would be dangerous for their ethnic group. Especially during election time these different future visions clash. In 1999 the High Representative even had to ban the Bosnian-Serb SRS party from participating in the November 2000 elections, because of the continued dismissal of the Dayton Peace Agreement and its hate-promoting discourse. During the election campaign the SDS "successfully galvanized RS voters around fears of a unified Bosnian army and a unified Bosnian education system, in which Serb children and youth would be harmed by exposure to other ethnic groups" (ICG 2000: 7). The HDZ, similarly, spread the view that the Croat people in Bosnia were threatened by extinction at the hands of other, larger groups and a hostile international community (ICG 2000: 7). This view was clearly captured in their election slogan "Determination or Extermination". In contrast, the moderate SDP is the only multi-ethnic party that has measurable support among all three of Bosnia-Herzegovina's ethnic groups (NDI 2002). The extensive mandate of the High Representative makes it possible to ban or remove persons or parties from participating in elections or executing certain public functions. This is what happened in April 2001 when Radisic was removed from the federal presidency. Positive developments at the national level are the accession of Bosnia and Herzegovina to the Council of Europe (April 2002) and the initiative to further the integration of Bosnia and Herzegovina within the European Union.

10.4.5   Multiple Loyalties

Tolerance towards the views of other ethnic groups is low. The violence directed at the minority returnees is an indication of this lack of tolerance. A further illustration is an incident that took place on 7 May 2001. On this day "several thousand rioters prevented the laying of the foundation stone for the rebuilding of the Ferhadija Mosque in Banja Luka" (ICG 2001: 33). There was strong suspicion that the SDS and SDS-related wartime veterans played a significant role in setting up this riot (ICG 2001: 33). Quite alarming was the fact that a large number of the secondary-school pupils and other teenagers participated in these riots (ICG 2001: 33). Moreover, the beating and stoning of the elderly Bosnians was in particular committed by these young people. Further, as a consequence of the ethnic cleansing most cantons are relatively ethnically homogeneous. Thus, inter-ethnic cooperation and interaction is limited. This lack of inter-ethnic interaction, and the stereotyping of other ethnic groups by nationalist parties and leaders, inhibits the development of more overlapping loyalties. Loyalty and we-ness feelings are exclusively based on ethnicity. An overarching national or common identify is largely invisible or absent. People's identities are based on being Serb, Croat or Muslim, not on being 'Bosnian'. The lack of a national identity was clearly demonstrated by the reluctance and opposition towards creating national identity-promoting features such as a uniform number plate, a national flag or a national anthem. Any progress in this context was made only under tremendous pressure from the international community.

10.4.6   Social Capital

Between the different ethnic groups exists a great deal of mistrust, exacerbated and stimulated by the nationalist parties. These parties need this atmosphere of distrust in order to hold on to their power and influence. Mutual helpfulness is also lacking in many places, especially urban areas. The fierce competition for jobs and resources leads to severe discrimination against refugees and displaced persons in particular. The Bosnian war has led to the fragmentation of individual and group identity (Aplon/Tanner 2000: 8). In pre-war Bosnia, primarily in the villages, "the principal sources of social cohesion were Tito, the schools and bureaucracy, religious and cultural practices, and personalized family norms identity" (Aplon/Tanner 2000: 9). The war changed everything; compatriots and neighbours turned into enemies, cities and villages were destroyed, and people were made refugees in their own country. Civil society is still underdeveloped. In 1997, several major international NGOs started a project to improve the legal environment in which the local NGO community had to function (the LEA7LINK project). However, a local NGO collaborative effort never materialized (Belloni 2000: 29). Further, "the extremely interventionist role of the international community severely complicates the role civil society should supposedly play in advocacy and democratization" (Belloni 2000: 30). More importantly, civil society is largely fragmented along ethnic lines. Inter-ethnic trust relations are largely absent. Evidence for this is the uncooperative and ineffective way the different ethnic groups deal with each other. In most cases ethnic groups or parties are unable to co-operate and achieve agreement at all. Today's Bosnians cannot escape the political baggage of their ethnic identities. There is "no ethnically neutral identity, such as 'citizen': civil society development in Bosnia must start with some of the more basic foundations of democratic social organization" (Aplon/Tanner 2000: 9).

10.4.7   Sentimental Walls

Attitudes of hate and mutual distrust are widespread and continuously stimulated by nationalist parties. Negative stereotyping is being used by nationalist parties to rally their ethnic group against other ethnic groups. The nationalist parties have become "the arbiters, filling the void and helping people make sense of a chaos that, ironically, they created" (Aplon/Tanner 2000: 9). The nationalist discourse is characterized by intolerance, exclusiveness and narrow-mindedness. The international community, in particular, is trying to break down these sentimental walls in several ways. First, the international community inhibits the most radical disseminators and propagators of these sentimental walls from taking part in elections and, thus gaining legitimate access to political power (e.g., the banning of the SRS in the November 2000 elections). Second, the international community tries to stimulate and promote more moderate initiatives, parties, leaders or groups (e.g., through providing media support). Third, the international community is trying to break down these sentimental walls through public campaigns (e.g., "Vote for Change"). Many people are caught in the nationalist discourse of fear and hate. Even if some people are willing to engage in an intercommunal dialogue, it is often hard to get a constructive dialogue started, not to mention to achieve actual results and agreements.

10.5   Conclusions

Without the development of an integrative climate, the chances of building a sustainable peace process are slim. The integrative character of the social-psychological environment is influenced by changes of six variables relating to the perception of the future, the handling of the past, the nature and levels political commitment, subjective security, trust, and the existence of sentimental walls. The analytic tools presented in this paper are intended to help describe the current climate and assess the chances of the peace-building efforts. They need to be further tested and developed to produce lean and reliable measures that could be used to compare successful and unsuccessful peacebuilding processes. Equally important are further research about the roots of a disintegrative climate, and tools and measures that could be used to strengthen the different components of the integrative climate. A disintegrative climate limits the control of making intelligent choices in conflicts and increases the chances of unintended cruelty.

Human security	Significant progress, but serious problems persist. Amongst others, refugees, returnee-related violence and unemployment
Reconciliation	Extreme nationalist parties deliberately inhibit and obstruct the start of a sincere and durable reconciliation process
Social capital	The rebuilding of trust relations and voluntary cooperation between the different ethnic groups has been negligible as a consequence of the segregated lives, the nationalist parties and the accumulated distrust
Multiple loyalties	Loyalties are exclusively directed at ethnic groups. An overarching 'Bosnian' identity is absent
Future perspectives	Both politically and economically the future is unclear and uncertain. People have negative future perspectives. Extreme nationalist parties cause and actively stimulate this uncertainty
Sentimental walls	Widespread sentimental walls are present and actively stimulated by extreme nationalist parties
Integrative climate	Bosnia has made minor progress regarding the integrative climate. Unfortunately, there are too many parties actively and deliberately obstructing the peace process

References

Alfaro, M. (2000). *Returnee Monitoring Study: Minority Returnees to the Republika Srpska-Bosnia-Herzegovina.* Sarajevo, Bosnia and Herzegovina: UNHCR.

Aplon, J., & Tanner, V. (2000). *Civil Society in Bosnia: Obstacles and Opportunities for Building Peace.* Washington, DC: Winston Foundation for World Peace.

Augsburger, D. (1992). *Conflict Mediation across Cultures.* Louisville, KY: Westminster.

Bar-Tal, D. (2001). Why does fear override hope in societies engulfed by intractable conflict as it does in the Israeli society? *Political Psychology, 22*(3).

Belloni, R. (2000). *Building Civil Society in Bosnia-Herzegovina* (Human Rights Working Paper No. 2).

Global Excellence Initiative (GEM). (1999). *Appreciative Inquiry: Practitioner's Handbook.* Washington, DC: Case Western Reserve University.

Gouldner, A. W. (1960). The Norm of Reciprocity: A Preliminary Statement. *American Sociological Review, 25*(2), 161–179.

Herriot, P., Bevan, S., Hirsh, W., & Reilly, P. (1998). *Trust and Transition.* New York: Wiley.

Human Rights Watch. (2002). *Human Rights Watch World Report 2002: Bosnia-Herzegovina.* Retrieved from http://www.hrw.org/wr2k2/contents.html.

International Crisis Group. (2000). *Bosnia's November Elections: Dayton Stumbles* (ICG Balkans Report No. 104).

International Crisis Group. (2001). *The Wages of Sin: Confronting Bosnia's Republika Srpska* (ICG Balkans Report No. 118).

International Crisis Group. (2002a). *Courting Disaster: The Misrule of Law in Bosnia and Herzegovina* (ICG Balkans Report No. 127).

International Crisis Group. (2002b). *Policing the Police in Bosnia: A Further Reform Agenda* (ICG Balkans Report No. 130).

Lederach, J. P. (1997). *Building Peace: Sustainable Reconciliation in Divided Societies.* Washington, DC: United States Institute of Peace Press.

Lund, M. (1998). Reckoning for Past Wrongs: Truth Commissions and War Crimes Tribunals. In P. Hams, & B. Reilly (Eds.), *Democracy and Deep-Rooted Conflict: Options for Negotiators.* Stockholm, Sweden: International Institute for Democracy and Electoral Assistance.

Naidu, M. V. (Ed.). (2001). *Perspectives on Human Security.* Brandon, Canada: Canadian Peace Research and Education Association.

National Democratic Institute for International Affairs. (2002). *A Survey of Voter Attitudes in Bosnia-Herzegovina* (Summary Report).

Reychler, L. (1979). *Patterns of Diplomatic Thinking: A Cross-National Study of Structural and Social-Psychological Determinants.* New York: Praeger.

Reychler, L. (1999). *Democratic Peace-building and Conflict Prevention: The Devil is in the Transition.* Leuven: Leuven University Press.

Reychler, L., & Langer, A. (2003). Peacebuilding Software. *Peace Research. The Canadian Journal of Peace Studies, 35*(2), 53–73.

Rothman, J. (1997). *Resolving Identity-Based Conflict in Nations, Organizations and Communities.* San Francisco, CA: Jossey-Bass.

Tadjbakhsh, S. (2002, April). Transition in Central Asia and Human Security. Paper presented at the United Nations Development Program Conference.

Zehr, H. (2001). Restorative Justice. In L. Reychler, & T. Paffenhoz (Eds.), *Peace Building: A Field Guide.* Boulder, CO: Lynne Rienn.

Chapter 11
Challenges of Peace Research

Abstract The Secretary General of the International Peace Research Association (IPRA) responds to one of the most frequently asked questions in the field of peace studies: "What are the challenges facing peace researchers in the twenty-first century?" In the first section he notes that, in some ways, the world is more peaceful now than at in any time in the past century, but then adds three sobering observations about the very high levels of manifest and potential violence, the predominantly reactive nature of most conflict prevention efforts, and the strong feelings of relative deprivation in the era of globalization. In the second section he states that if peace researchers want to make a greater difference, they must challenge the ways and means of the current practice of peacemaking, peacekeeping and peacebuilding. The first challenge is not to lose sight of the big picture. The macro-perspective gives an overview of the necessary peacebuilding efforts and allows the peacebuilders to oversee and coordinate what they are doing. The second challenge is to gain a better understanding of the sustainable peacebuilding architecture. Winning a war can sometimes be relatively easy – or at least rapid – but winning the peace can be a far more complex and time-consuming enterprise. The third challenge concerns the slow learning process. There is a need to build structures that support a better exchange of knowledge between the decision-makers, the practitioners in the field, and the research community. The fourth challenge is to deal more effectively with the peacebuilding context, which is characterized by uncertainty, unpredictability, competing values and interests, and the struggle for power. The article ends with a plea for reflecting on the meaning of professionalism in peacebuilding.

Waging peace is the greatest issue facing the international community – a question of life and death, of survival or extinction.[1] Such an issue demands thorough

---

[1]This text was first published as: Reychler, L. (2006). Challenges of Peace Research. *International Journal of Peace Studies, 11*(1), 1–16. Republished with permission.

reflection and analysis (Sawyer 1994).[2] Today, peacebuilding is spurred by the awareness that there are limits to violence. We can forget sustainable development if no serious efforts are undertaken to prevent violence and build sustainable peace.

Since the end of the Second World War, many scholars have been engaged in the study of war and peace. In the post-Cold War era there has been an increasing demand for peace research, and its findings are now being used by decision-makers and practitioners. The European Union (EU), one of the leading international bodies to affirm the importance of peacebuilding and conflict prevention, is building up its capacity for these activities (Moyroud et al. 1999). The same is true of the United Nations (UN). The Report of the High-Level Panel on Threats, Challenges and Change offers new ideas to improve the UN's performance, including the creation of a new inter-governmental body, the Peacebuilding Commission, whose task would be to assist states that are under stress or recovering from conflict to develop their capacity to perform their sovereign functions effectively and responsibly (United Nations 2004). Various ministries or departments in many countries have established conflict prevention units. Intervention projects, programmes and policies are subjected to proactive peace and conflict impact assessments (PCIAS). PCIAS is a process of determining the relevance of ongoing or proposed interventions and predicting their future effects on the conflict dynamics and peacebuilding process. This assessment is system-orientated and proactive. It is intended to increase both the conflict sensitivity and the peace benefits of the intervention. Governmental and non-governmental organizations dealing with a broad range of peacebuilding activities are mushrooming.

In their global survey of armed conflicts, Robert Gun et al. (2001) depict the world as more peaceful than at any time in the past century. The number and magnitude of armed conflicts within and among states have lessened since the 1990s by nearly half. Conflicts over self-determination are being settled with ever greater frequency, usually when ethnic groups gain greater autonomy and power-sharing within existing states (Gurr et al. 2003). The progress is attributed to the increase in conflict prevention efforts and the greater number of democratic efforts. This is the good news. But now let us take a look at three other, more sobering observations.

[2]I would like to thank all the people in the IPRA, the Centre for Peace Research and Strategic Studies (CPRS) and the Master of Conflict and Sustainable Peace (MaCSP) programme, along with the students and faculty of the University of Leuven, for stimulating and nurturing these reflections on peace research. I also would like to draw attention to the reflections on peace research made by Alger (1996) and Brock-Utne (1996).

11.1 Three Sobering Observations

11.1.1 High Levels of Manifest and Potential Violence

In contrast to these statistics, the overall level of manifest and latent violence is still high. Most databanks on conflicts monitor the most visible types of violence. A different picture appears when a broader definition of violence is used, namely: (a) when violence is defined as a situation in which the *quantitative and qualitative life expectancies* (measured, for instance, through the Human Security Index) of a particular group or groups within a community, state, region or the world are significantly lower than other groups, and (b) when this can be attributed to one or more sources of violence: physical violence, structural violence, psychological violence, cultural violence, bad governance, organized crime or extra-legal activities (Reychler and Jacobs 2004). The difference between armed violence and the other types of violence is that armed violence is direct and visible, and it kills faster. The other types of violence are indirect and less visible, and they affect more people. Gandhi (n.d.) called poverty the worst form of violence: "Earth provides enough to satisfy every man's need, but not every man's greed". It affects a billion people who live on $1 a day, and 2.8 billion who live on less than $2 a day. In the West poverty means living a bad life; in the Third World it means surviving in a state close to death (Zakaria 2005).

Mapping the whole *fabric of violence,* including its less visible forms, gives a more realistic picture of violence in the contemporary world. Paradoxically, the media and researchers continue to focus on the sensational violence (terrorism, irregular and conventional warfare), which kills less than the other forms of violence. Terrorism causes approximately 5000 deaths a year, and anti-terrorism and conventional warfare cause a hundred times this number (500,000); but structural violence shortens the lifespan of hundreds of millions of people, and bad governance reduces the life expectations of approximately 1.5 billion people. Calling terrorism the greatest threat in the world masks most of the violence in today's world. Bad governance has many faces. It can express itself as: (a) greed and corruption: infant mortality increases by 75% when the level of corruption increases from medium to high; (b) indifference and neglect: think of the ongoing genocidal conflicts in Chechnya and Sudan; (c) ignorance and stupidity: remember Mao's Great Leap Forward in China, which caused the death of millions of Chinese, and the retreat of the Blue Helmets from Rwanda in 1994 when the genocidal violence started; or (d) the harmful and negative side-effects of well-intentioned policies (Reychler and Jacobs 2004). Bad governance kills. Another source of violence is the activities of transnational organized crime. It is estimated that criminal organizations earn $300–$500 billion annually from narcotics trafficking, their single largest source of income.

The last strand in the fabric of violence is the "shadows of war". Nordstrom (2004) describes this as the complex sets of cross-state economic and political linkages that move outside recognized state-based channels and in many cases have greater power

Physical means of violence: terrorism, guerilla and conventional welfare. Visible direct-intentional	Psychological means of violence	Cultural means of violence / epistemic violence
Structural means of violence: political, economic and cultural exclusion. Less visible indirect-Intentional	Violence is about shortening life or significantly lowering the the quantitative and qualitative life expectancies of particular group(s)	Bad governance Maladministration Corruption Indifference and neglect Greed and self-interest Religious and ideologically inspired misgovernance Unintended negative impacts of well-intentioned interventions.
Organized crime	Extra-legal economic activities	Environmental violence

Fig. 11.1 The fabric of violence. *Source* The author

than some of the world's states. This set of economic and personnel flows ranges from the mundane (the trade in cigarettes and pirated software), through the illicit (gems and timber), to the dangerous (weapons and illegal narcotics). Initial inquiries estimate the amount of money generated per year through extra-state activities to be in the trillions of Euros (Nordstrom 2004). These amounts dwarf the budgets of international organizations such as the EU and the UN. The EU's budget in 2004 was €99.52 billion; the budget of the UN for 2002–2003 was $2.6 billion. Part of this money could be used to support the Millennium Development goals (MDGs), which require $135 million in 2006 and $195 million in 2015. These are huge opportunity losses for conflict prevention and peacebuilding.

To prevent violence more effectively, one has to look at the whole fabric of violence. Armed violence is intertwined with the other strands of the fabric. Before the genocide erupted in Rwanda with volcanic force, the country was considered a relatively secure place. A broader analysis of the violence would have warned us better about the growing tensions in the country (Uvin 1998). The price of a narrow analysis of violence is a surprise (Fig. 11.1).

11.1.2   The Reactive Nature of Most Conflict Prevention Efforts

Efforts to prevent violence have increased. In the 1980s, for example, there were five peacekeeping operations in the world, whereas in the 1990s there were 35. But the sobering observation is that most of the conflict prevention has been reactive in nature, being initiated only after the conflict has crossed the threshold of violence. Its aim is to limit further escalation (intensity, geographic and duration). Here, too, the price is high. Once a conflict turns violent it becomes not only more difficult, but also more expensive to de-escalate it and to build peace (Brown and Rosecrance 1999).

11.1.3 The Growing Gap Between the Perceived and the Preferred World Order

The third sobering observation relates to the impact of globalization on the perception of violence. The worldwide capacity to compare one's political, economic and cultural position with others is increasing the global sense of relative deprivation. Relative deprivation occurs when individuals or groups subjectively perceive themselves as unfairly disadvantaged over others perceived as having similar attributes and deserving similar rewards. This has lowered the legitimacy status of the international system and has increased the pressures towards international democratization. Spokespersons from Third World countries decry the structural violence and unfair intervention in the international environment (Reychler 1979).

The resilience of this gap causes discomfort, dissatisfaction and frustration among peace researchers. Some researchers have become cynical: they find it difficult to live with the wide gap and assert that all the research is a waste of time; it produces theories, but no results. Some are burned out and have stopped contributing to the field. Others continue to look on the bright side of life and stick to the "Law of the hammer" (if my only tool is a hammer, everything becomes a nail; Maslow 1966). They are convinced that more of the same will lead to peace. Most peace researchers, however, have turned the tension into a creative search for more cost-effective ways to build sustainable peace. Peace research requires strongly motivated people because: (1) It is a demanding field of study. (2) Peace still has image problems; it can evoke either hope and strength or despair and weakness. (3) Compared to the traditional fields of study, peace research is academically less embedded; it transcends the barriers of faculties and departments. (4) It deals with "senti-mental walls". Senti-mental walls are the theories, assumptions, expectations, attitudes, feelings, taboos, beliefs, norms and principles that stand in the way of conflict prevention and peacebuilding. It is important to identify and dismantle them. Frequently, this goes hand in hand with emotional and other kinds of resistance. Cynicism or defeatism can inhibit the peacebuilding process seriously. (5) Transforming violent conflicts can be exhausting. Four challenges can be distinguished.

11.2   Challenges of Peace Research

If peace researchers want to make a greater difference, they have to challenge the ways and means of the current practice of peacemaking, peacekeeping and peacebuilding.

11.2.1   Challenge 1: Seeing the Big Picture

The first challenge is not to lose sight of the big picture. Peacebuilding is the result of the activities of many people in different sectors and at different levels. These skills are acquired through education and practice. Effective peacebuilding also implies that the peacebuilders see where and how their efforts fit into the peacebuilding process. Seeing the big picture is vital for coherent peacebuilding efforts. The big picture or macro-perspective gives an overview of the necessary peacebuilding blocks. It enables the peacebuilders to oversee and coordinate what they are doing.

The essential requirements or preconditions for creating sustainable peace – which have been derived from the peace research – can be clustered into five peacebuilding blocks: (1) an effective system of communication, consultation and negotiation; (2) peace-enhancing structures and institutions; (3) an integrative political-psychological climate; (4) a critical mass of peacebuilding leadership; and (5) a supportive international environment. These peacebuilding blocks are all necessary and mutually reinforcing. The deficiency or absence of one of these building blocks can seriously undermine the stability or effectiveness of the entire peacebuilding process. In addition to these five clusters, there are the necessary support systems (legal, educational, health, humanitarian aid, information and environmental systems) that play an important role in the peacebuilding process. The first building block focuses on the establishment of *an effective communication, consultation and negotiation system at different levels* between the conflicting parties or members. In contrast to the negotiation styles used in most international organizations, the negotiation style in the European Union is predominantly integrative. Ample time and creativity are invested in generating mutually beneficial agreements. Without win-win agreements, the Union would disintegrate.

The second building block consists of *peace-enhancing structures*. In order to achieve a sustainable peace, (conflict) countries have to install political, economic and security structures and institutions that sustain peace. The political reform process aims at the establishment of political structures with a high level of legitimacy. The legitimacy status is influenced by two factors: (a) the ability of a regime to deliver vital basic needs, such as security, health services, jobs, and so forth; and (b) the democratic procedure. Initially, an authoritarian regime with high quality leaders and technocrats can get a high legitimacy score, but in the end, consolidated democracies provide the best support for sustainable peacebuilding. It is crucial to note that the transition from one state (e.g., non-democratic structures) to another (e.g., consolidated democratic environment) is not without difficulties: the devil is in the transition (Reychler 1999). The economic reform process envisions the establishment of an economic environment which stimulates sustainable development and the reduction of gross vertical and horizontal inequalities. The security structures safeguard and/or increase the population's objective and subjective security by effectively dealing with both internal and external threats. This implies a cooperative security system which produces a high level of human security, collective defence and security, and proactive conflict prevention efforts. In addition to these three core peacebuilding structures, there is also a need to build effective judicial, educational, health and environmental systems to sustain the peacebuilding process. As long as these structures are not in place, humanitarian aid will be needed.

The third necessary building block for establishing a sustainable peace process is an *integrative climate* (Reychler and Langer 2003). This is the software of peacebuilding. This building block stresses the importance of a favourable social-psychological environment. Although the climate is less tangible and observable than the other building blocks, it can be assessed by looking at the consequences. An integrative or disintegrative climate can express itself in the form of attitudes, behaviour and institutions. The characteristics of an integrative climate include: expectations of an attractive future as a consequence of cooperation; the development of a sense of 'we-ness', multiple loyalties, reconciliation, trust and social capital; and the dismantlement of senti-mental walls (Fig. 11.2).

The fourth building block is *a supportive regional and international environment*. The stability of a peace process is often dependent on the behaviour and interests of neighbouring countries or regional powers. These actors can have a positive influence on the peace process by providing political legitimacy or support, assisting with the demobilization and demilitarization process, and facilitating and stimulating regional trade and economic integration. However, these same actors can also inhibit the progress towards stability, for example, by supporting certain groups that do not subscribe to the peace process. Likewise, the larger international community plays a crucial role in most post-conflict countries. Via UN agencies or other international (non-)governmental organizations, the international community can provide crucial resources and funding or even take direct responsibility for a wide variety of tasks, such as the (physical) rebuilding process, political transformation, humanitarian aid, development cooperation, third-party security guarantor, and so forth.

The fifth building block is the presence of *a critical mass of peacebuilding leadership* (Reychler and Stellamans 2005). There are leaders in different domains (politics, diplomacy, defence, economics, education, media, religion, health, and so forth) and at different levels: the elite, middle and grass-roots levels (Lederach 1997). High on the agenda of architectural research should be research to identify the characteristics of successful peacebuilding leaders, such as Nelson Mandela, F. W. De Klerk, Mohandas Gandhi, Mikhail Gorbachev, Vaclav Havel, Jean Monnet, Helmut Kohl, George Marshall, Martin Luther King, Jacques Delors, and many others. This research involves differentiating between successful and unsuccessful peacebuilders and identifying the similarities and differences between successful and

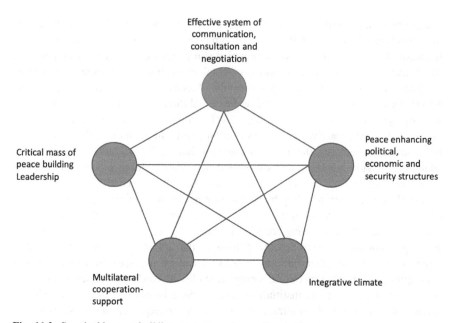

Fig. 11.2 Sustainable peacebuilding pentagon. *Source* The author

unsuccessful peacebuilders, and between peacebuilders and peace destroyers (Reychler and Stellamans 2002). Peacebuilding leaders distinguish themselves by the way they lead the conflict transformation process. They envision a shared, clear and mutually attractive peaceful future for all who want to cooperate; they do everything possible to identify and gain full understanding of the challenge with which they are confronted; they frame the conflict in a reflexive way (Rothman 1997); their change behaviour is adaptive, integrative and flexible; they are well acquainted with non-violent methods; they use a mix of intentional and consequential ethics and objectives; and they are courageous men and women with a high level of integrity.

11.2.2   Challenge 2: Mastering Sustainable Peacebuilding Architecture

The second challenge is to get a better understanding of the sustainable peace-building architecture. Despite a more supportive international environment, the costs and risks associated with peacebuilding remain high. The experiences from Germany to Iraq have shown successes and failures. Winning the war can some-times be relatively easy – or at least rapid – but, as RAND (2003; 2005) studies underscore, winning the peace that prevents a return to war can be a far more complex and time-consuming enterprise. The price of the regime change in Iraq is turning out to be higher than expected. The costs can be apprehended in terms of human losses (on the allied side approximately 2000 have died and 20,000 have been wounded; the number of Iraqis killed is estimated at 100,000), financial expenses (the Bush administration has spent $200 billion; this is more than the budget of the World Bank or the EU), material destruction and diplomatic damage. Iraq and other cases make it very clear that we need to learn to build sustainable peace in a more effective, cost-efficient and satisfactory way. This demands a better understanding of the architecture of peacebuilding. The term "peacebuilding architecture" refers to the art and science of successful peacebuilding (Reychler 2002). Peace architecture studies show how conflicts can be transformed in a more effective, efficient and satisfactory way. This requires coherent planning and a good understanding of the cross-impact of different peacebuilding efforts (diplomatic, political, economic, security, psychological, and others). The aim is to synergize these efforts and reduce possible negative side-effects. The key variables in architectural research are timing (simultaneous or sequential) and priority-setting (Reychler and Langer 2004). Time makes the difference between life and death. Comparative studies of successful peacebuilding efforts can further our insight into the peace architecture. One of the tools which needs further development is the peace and conflict impact assessment methodology (Reychler and Paffenholz 2004). The aim is to identify the relevance and future consequences of an ongoing or proposed intervention on the conflict dynamics and peacebuilding process. The assessment is system-orientated and proactive. It is meant to increase the conflict-sensitivity and the peace "added value" of the intervention. The peace and conflict impact assessment system could play a useful role at the different levels of intervention (project, programme, sector and broad policy levels) by increasing the awareness of the potential or actual impact on the conflict dynamics and peacebuilding process, and by helping to design more coherent interventions which do no harm and have a higher peace added value. This methodology can be used to improve the design and implementation of interventions ranging from development or humanitarian work to peace and reconciliation or democratization efforts taking place in situations of latent or manifest violent conflict – or in the aftermath of a violent conflict or war. The method consists of (1) preparatory steps to inform and commit the organization to the assessment process, (2) conflict analysis and prognosis, (3) a peacebuilding deficiency assessment, (4) an assessment of the peace relevance, (5) a conflict risks analysis, (6) an assessment of the impact of the intervention on the peace and conflict, and (7) concrete recommendations and follow-up.

11.2.3 Challenge 3: Synergizing the Know-How and Learning

The third challenge concerns the slow learning process. The learning of violence prevention and peacebuilding can be improved by (a) making use of different forms of scientific inquiry, (b) providing space for trans-disciplinary research, (c) creating structures which support a better exchange of knowledge between the decision-makers, the practitioners in the field and the research community, (d) inviting the actors involved in peacebuilding to reflect critically on their personal or organizational theories of violence prevention and peacebuilding, and (e) making more effective use of local expertise.

(a) Making use of different types of inquiry. Peace research should involve not only classical *empirical-analytical* research (searching for the causal explanations) and *interpretative research* (investigating how people perceive their experiences), but also participatory *peace action research.* The latter type of research is based on the principle that people – ordinary men and women – have a universal right to participate in the production of knowledge that affects their life. Smith (Smith et al. 1997) calls this "liberatory inquiry". People are not the passive subjects of research, but rather active subjects. Their needs should be the point of departure for knowledge production and justification of the research (Smith et al. 1997). The initiative of the mothers of soldiers from St. Petersburg to meet the Chechen rebels is a very good example of participatory peace action inquiry. They transcended their passive role as victims and assumed responsibility for transforming the conflict and making an end to that ugly war.

(b) Provide trans-disciplinary research capacity. If universities want to play a peacebuilding role, they will have to overcome their preference for uni-disciplinary studies or faculties. Peacebuilding involves changes in complex dynamic systems which can only be understood in a trans-disciplinary way. Progress in peace research means crossing boundaries. In each faculty you have scholars who want to change the world. For economists, development tends to be seen as the most important factor in the peacebuilding process; for lawyers, it's the rule of law; political scientists point out the importance of democracy; for strategists, security must come first; for psychologists and educators, peace is all about building peace in the minds and hearts of people; theologians stress the importance of mercy, forgiveness and reconciliation; medical doctors stress that a healthy mind resides in a healthy body; and artists believe in the aesthetics of peacebuilding. Most researchers are somewhat narcissistic. Narcissists work with a fervent passion. Their work is their life and their sense of destiny fuels their motivation. They have the ability to work through some of the toughest obstacles (Maccoby and Gschwandtner 2005). The challenge, therefore, is not to change, but rather to transform this narrow narcissism into productive narcissism by inviting all to play in the orchestra of peacebuilding.

(c) Build structures that support a better exchange of knowledge between the decision-makers, the practitioners in the field and the research community. One of the greatest causes of knowledge waste is the lack of dialogue and the weak connections between the decision-makers and the practitioners in the field (the *operari*) and the researchers (the *speculari*). To improve peacebuilding architecture, the development of dialogue and connections between these three sources of knowledge and know-how must be encouraged and rewarded. It is natural that theory and action should be complementary, that they should constitute harmonic aspects of one whole. In reality, however, there exists what has been called a "theory-practice gap" (Lepgold and Nincic 2002). This gap is caused by the disparities between the incentive systems of politicians, practitioners in the field and researchers (academics). The academic incentive system is characterized by a publish-or-perish mentality, by the recognition of originality, by the tendency for research methods to triumph over substance, by the preference for fundamental over applied research, by papers filled with jargon, and by the reinforcement of all of this by academic faddishness. The incentive system of policy-makers consists of their need to find timely solutions to concrete problems. Officials have less time to read and reflect. Joseph Nye (Lepgold and Nincic 2002), one of the few people to have acted as both a scholar and a policy-maker, was surprised at the "oral" nature of the communication and decision-making culture at the top levels of government service. One of the major challenges of peace research is to facilitate dialogue and connections between the decision-makers, the practitioners and the research community. Peace researchers have a bridging role: they should not only provide policy-relevant knowledge, but also effective dialogue with practitioners, especially in instances where the people on each side have interests in common. Peace research can provide instrumental knowledge (e.g., how to prevent groupthink in crisis situations), contextual knowledge (e.g., specifying the necessary conditions for sustainable peacebuilding) and consequential knowledge (e.g., anticipating the costs and consequences of policy options). The most productive exchanges might take place between researchers who have spent some time in government and in the field, and practitioners who have had some academic training in violence prevention and peacebuilding research. All of the above does not imply that the Ivory Tower should be dismantled. It exists for good reasons. It provides the academic freedom without which scientific research is impossible. It also allows intellectuals to reflect on the world from a certain distance, and not simply to do the work of policy commentators or journalists at a slower pace (Lepgold and Nincic 2002).

(d) Stimulate conscious and competent learning. The learning curve can be raised by stimulating *reflective peace research*. Policy-makers, practitioners in the field, researchers and educators involved in peacebuilding should reflect on the underlying normative, theoretical and epistemological assumptions of the personal and collective theories or mental models of peace. These mental models, frequently referred to as "common sense", influence the decision-making. Because mental models are usually tacit, existing below the level of awareness, they are often untested and unexamined. They are generally invisible for us unless we look for them (Senge 1994). A valuable contribution to reflective peace research has come from Jayne Docherty, who has been developing an interesting and creative approach to reflective peace research. She distinguishes between three types of personal theories: "Baby theories",

"Teenage theories", and "Big Grown-up theories" (Docherty 2005). The first are our gut level theories that guide behaviour, the second are theories based on practice, and the third are well-grounded theories based on systematic research.

(e) Make effective use of local expertise. Despite the existence of participatory action research rituals, donors and external interveners in conflicts rely predominantly on the advice of external consultants. The architecture of peace could significantly improve by eliciting local intelligence.

11.2.4 Challenge 4: Dealing with the Reality of the Peacebuilding Context

The fourth challenge is to deal effectively with the peacebuilding context, which is characterized by uncertainty, unpredictability, competing values and interests, and the struggle for power. Building peace requires not only analytic skills, but also imagination, creativity, reconciliation of competing values and interests, the implementation of arms and power control, the use of non-violent action, "in extremis" the use of arms, and a great deal of courage. Hope is not a strategy. Peace researchers can influence the context more favourably by: (a) developing a more accurate accounting system for the violence; (b) better assessing the costs of reactive and proactive violence prevention efforts (or of neglect); (c) generating concrete suggestions to increase the accountability of people who benefit directly or indirectly from the violence; (d) highlighting the negative role of epistemic violence and senti-mental walls; (e) improving the banding of peace and peace research; and (f) guaranteeing the academic freedom and the quality of peace research and education.

Let us now deal with each of these points in more detail:

(a) Developing a more accurate and comprehensive system for violence monitoring and accounting: Today's databanks have three needs: (1) They need more reliable information about the numbers of people killed. For some conflicts and groups of victims, the data on people killed are rough and unreliable. There are first-, second- and third-class victims. (2) They need information about the whole fabric of violence, including so-called sensitive or forbidden statistics, for example, about horizontal inequality in multi-ethnic countries. (3) They need data on the costs and benefits of violence.

(b) Anticipating and comparing the costs and benefits of neglect and/or reactive or proactive violence prevention: This is a complex methodological problem of counterfactuals. Brown and Rosecrance (1999) made an important contribution towards what will be a continuing critical debate in the years to come.

(c) Increasing the accountability of people who commit and/or profit from violence: If violence grieved everybody, it would have disappeared from the earth long ago. The problem is that for some people violence still pays off. Therefore it is of crucial importance to identify and name greedy people who profit from conflicts and make them more accountable for their deeds.

(d) Exposing the epistemic violence and the senti-mental walls which inhibit the peacebuilding process: This is very important because peacebuilding takes place within an environment of power and competing interests and values. In such a context, knowledge is a strategic asset. People in power try to manipulate this knowledge by controlling the media, research funding, school curricula and public opinion.

(e) Improving the branding of peace and peace research (Reychler 2000): One of the problems confronting the peace research community is the image of peace that they project. People choose peace on the basis of their mental image of a possible and desirable future state of their living environment. This image can be as vague as a dream or as precise as a goal or mission statement. If peace advocates articulate a view of a realistic, credible and attractive future for people, a condition that is better in some ways than what now exists, then this image will be a goal that beckons and motivates people to pursue it. Not all people are enticed by the idea of peace. For some, the term 'peace' has more negative than positive connotations. There are people in the European Community who take peace for granted. For them it is not a very important issue. Peace is banal and boring. It is not considered cool. Others do not like it because during the Cold War peace movements were misled and used by Communist regimes. Peace can also be distrusted when promoted by conservative regimes to prevent structural changes. Throughout history, people have fought to get rid of Pax Romana, Pax Germanica, Pax Sovietica, Pax Niponica and other imposed peaces. The negative image of peace can also be derived from the fact that some so-called peace gurus or experts are peace quacks (in Dutch the word is *paxzalver*). The association of peace with absolute pacifism tends to convey an image of passivity. For so-called 'realists' or 'cynics', peace is seen as an unrealistic and/or dangerous pursuit. Finally, people can become apprehensive when the 'peace' that someone is promoting is not defined clearly. Something certainly must be done to make the concept of peace more attractive. This could be accomplished by formulating clear and compelling definitions of peace; differentiating between different types of peace; convincing people that one can forget sustainable development when no efforts are made to create a sustainable peace; making people aware that there are "limits to violence"; and convincing them that peacebuilding is a serious enterprise that requires courage, professionalism and creativity. How can the idea or ideal of sustainable peacebuilding be disseminated more effectively? One could ask marketing specialists how to promote 'peace' as a product – how to give it "sex appeal". Perhaps *they* could give peace a chance. Even more important is the role of peacebuilding leadership. The leaders must capture the attention of the people through a clear and compelling vision of peace. Such a vision is necessary to bridge the present and the future. A third group that can promote peace more effectively is journalists. A distinction made by Galtung (n.d.) between "peace and conflict" journalism and "war and violence" journalism seems appropriate here. "Peace and conflict" journalism could help to enhance the image of peace and peacebuilding significantly. Peace researchers must give serious consideration to the issue of the branding of peace and peace research.

(f) Protecting the academic freedom and the quality of peace research and peace education: Academic freedom is of vital importance for peace researchers. Researching 'truth' in conflict zones is a risky business. In a climate of political correctness, academic taboos, spin doctors, groupthink, embedded journalism and epistemic violence, the pressure to conform can be very high. An essential ingredient of sustainable peacebuilding is professionalism and a critical mass of leaders who can raise hopes, generate ways and means to build peace, and commit people to the peacebuilding process. This necessitates not only peacebuilding skills, but also the will to take some risks in order to achieve one's ends. Peacebuilding is one of the fields of expertise in which there is no systematic control of the training and selection of people for conflict prevention and peacebuilding functions. As a consequence, one finds peace quacks in the peacebuilding sector. It's high time to have a discussion about the meaning of professionalism in peacebuilding. The plea for more professionalism does not imply that professionals should be in charge of the peacebuilding process. This would be unacceptable in a democratic peacebuilding process. Professionals, however, should help the people and the decision-makers to make better informed decisions; they should make initial conflict impact assessments of the policies under consideration; and they should help generate more effective peacebuilding alternatives. This has become the normal procedure in other fields, where health, employment, gender and environmental impact studies have become normal practice.

References

Alger, C. (1996). Introduction: Reflections on Peace Research Traditions. *International Journal of Peace Studies, 1*(1), 1–4.

Brock-Utne, B. (1996). The Challenges for Peace Educators at the End of the Millennium. *International Journal of Peace Studies, 1*(1), 37–55.

Brown, M., Rosecrance, R. (Eds.). (1999). *The Cost of Conflict: Prevention and Cure in a Global Arena.* Lanham: Rowman & Littlefield.

Docherty, J. (2005). *Growing Theories: A Guide for Reflective Practitioners of Conflict Resolution and Peace Building.* Harrisonburg: Eastern Mennonite University (EMU).

Gandhi, M. (n.d.) [Gandhi Quotes Compiled by Mark Shepard.]

Galtung, J. (n.d.) Available at https://www.transcend.org/galtung/#publications. Last accessed 20 March 2020.

Gurr, T., Marshall, M., & Khosla, D. (2003). *Peace and Conflict 2003: A Global Survey of Armed Conflicts, Self-Determination Movements, and Democracy.* University of Maryland: Center for International Development and Conflict Management.

Gurr, T., Marshall, M., & Khosla, D. (2001). *Peace and Conflict 2001: A Global Survey of Armed Conflicts, Self-Determination Movements, and Democracy.* University of Maryland: Center for International Development and Conflict Management.

Lederach, J. P. (1997). *Building Peace: Sustainable Reconciliation in Divided Societies.* Washington, D.C.: United States Institute of Peace Press.

Lepgold, J., & Nincic, M. (2002). *Beyond the Ivory Tower.* New York: Columbia University Press.

Maccoby, M., & Gschwandtner, G. (2005). Productive Sales Leaders. *Selling Power, 25*(1), 58–65.

Maslow, A. H. (1966). *The Psychology of Science: A Reconnaissance.* New York: Harper and Row.

Moyroud, C., Lund, M., Mehler, A. (1999). *Peace Building & Conflict Prevention in Developing Countries: A Practical Guide.* Ebenhausen: Stiftung Wissenschaft und Politik—Conflict Prevention Network (CPN).

Nordstrom, C. (2004). *Shadows of War.* Berkeley: University of California Press.

RAND. (2005). *The UN's Role in Nation-Building.* Santa Monica: RAND.

RAND. (2003). *From Germany to Iraq.* Santa Monica: RAND.

Reychler, L. (2002). Peace Architecture. *Peace and Conflict Studies, 9*(1), 26–35.

Reychler, L. (2000). The Promotion of Peace Building. In Y. S. Choue, *Proceedings of the International Conference on Global Governance in the 21st Century* (pp. 53–74). Seoul: Kyung Hee University.

Reychler, L. (1999). *Democratic Peace Building: The Devil is in the Transition.* Leuven: Leuven University Press.

Reychler, L. (1979). *Patterns of Diplomatic Thinking.* New York: Praeger Publishers.

Reychler, L., & Jacobs, M. (2004). *Limits to Violence: Towards a Comprehensive Violence Audit* (Cahiers of the Centre for Peace Research and Strategic Studies No. 68). Leuven: Centre for Peace Research and Strategic Studies.

Reychler, L., & Langer, A. (2004, July). Researching Peace Building Architecture. Paper presented at the *Conference of the International Peace Research Association,* Hungary.

Reychler, L., & Langer, A. (2003). The Software of Peace Building. *Canadian Journal of Peace Studies, 35*(2), 53–73.

Reychler, L., & Paffenholz, T. (2004). *Peace and Conflict Impact Assessment System (PCIAS)* (Document prepared within the framework of Field Diplomacy Initiative). Leuven: Centre for Peace Research and Strategic Studies.

Reychler, L., & Stellamans, A. (2005). *Researching Peace Building Leadership* (Cahiers of the Centre for Peace Research and Strategic Studies No. 71). Leuven: Centre for Peace Research and Strategic Studies.

Reychler, L., & Stellamans, A. (2002). Peace Building Leaders and Spoilers. Paper presented at the *Conference of the International Peace Research Association* (July), Seoul, South Korea.

Rothman, J. (1997). *Resolving Identity-based Conflict in Nations, Organizations and Communities.* San Francisco: Jossey-Bass.

Sawyer, R. (1994). *Sun Tzu: Art of War.* Boulder: Westview Press.

Senge, P. (1994). *The Fifth Discipline Fieldbook.* New York: Doubleday.

Smith, S., Willms, D., & Johnson, N. (Eds.). (1997). *Nurtured by Knowledge: Learning to do Participatory Action-Research.* Ottawa: International Development Research Centre & Apex Press.

United Nations High-Level Panel on Threats, Challenges and Change. (2004). *A More Secure World: Our Shared Responsibility.* New York: United Nations.

Uvin, P. (1998). *Aiding Violence.* West Hartford, CT: Kumarian Press.

Zakaria, F. (2005). The Education of Paul Wolfowitz. *Newsweek* (28 March).

Chapter 12
Intellectual Solidarity, Peace and Psychological Walls

A world without walls is unthinkable.[1] Some people or communities find them necessary and useful. In Robert Frost's poem *Mending Walls*, a neighbour even reiterates that "Good fences make good neighbours". Walls can give a feeling of safety and protect what one does not want to share. This conference, however, focuses on fences or walls which are generally considered bad and would be better dismantled. The fall of the Berlin Wall allowed the unification of a country, the widening of the European Community and détente within greater Europe.

Last month I visited another wall, the demilitarized zone (DMZ), between North and South Korea. North Korea is the last living museum of Stalinism and the DMZ of the Cold War. The same people, the Koreans, with a 5,000-year history, created two different worlds with extremely different political and economic systems: South Korea has a liberal free market system and North Korea has a dictatorship and a command economy. Both peoples long for the normalization of their relationship, which would hopefully end in the reunification of a divided people.

There are other even more newly erected walls today, which I will not discuss now. Instead, this presentation focuses on psychological walls (Reychler 1994). They are less visible and stand in the way of building sustainable security and peace in today's world; they can also be called mental or senti-mental walls. These walls can be perceptions, theories, doctrines, taboos, attitudes, values or emotions. Efforts to dismantle them tend to be accompanied by strong resistance and emotions.

Let me illustrate some senti-mental walls.

There is the illusion of being a superpower. After the Cold War, and especially during the presidency of George W. Bush, it became *bon ton* to depict the world as a unipolar system dominated by American power and neoconservative principles. Many neoconservatives argued in late 1990 that the United States and its

---

[1]This paper was first delivered at the international conference *A World Without Walls 2010: Peacebuilding, Reconciliation and Globalization in an Interdependent World* (6–10 November 2010), Berlin.

L. Reychler and A. Langer (eds.), *Luc Reychler: A Pioneer in Sustainable Peacebuilding Architecture*, Pioneers in Arts, Humanities, Science, Engineering, Practice 24, https://doi.org/10.1007/978-3-030-40208-2_12

like-minded allies should use its military preponderance to assert "benevolent hegemony" over strategically important parts of the world. The illusion of "superpower" is now being undermined by the high costs and lack of success of the military interventions in the Middle East (Radman 2010).

Another example of a mental wall is the indiscriminate and disrespectful depiction of conflicting parties or adversaries. Think of the "axis of evil" and the way the Islamic world has been stigmatized as backwards and the cradle of today's terrorism. Before 9/11, Kristol and Kagan called for a regime change not only in "rogue" states like Iraq, North Korea and Iran, but also in China. Before 9/11, China constituted their central opponent in the international system (Fukuyama 2006).

The demonization and humiliation of parts of the world community is counterproductive. It prevents the development of sophisticated realism in international relations. It does not pay attention to the geopolitics of emotions (Moïsi 2009) or the impact of humiliation on the development of non-violent and violent resistance, such as terrorism. Humiliation has been called the atom bomb of emotions.

None of this is new (Crooke 2009). On 17 September 1656, Oliver Cromwell, a Protestant Puritan who had fought a civil war in England and deposed and executed the King, asked the Parliament, "Who are our enemies and why do they hate us?" He claimed to know the answer. There was, he said, an "axis of evil" abroad in the world. He asserted that our enemies hate us because they hate God and all that is good; "they hate us from that very enmity that is in them against whatsoever should serve the glory of God and the interests of the people". This axis of evil has a leader, he told them, a great power – Catholic Spain – that refused simple liberties and freedom.

It would be useful to map today's psychological or mental walls in a more systematic way. A comprehensive mapping would have space for misperceptions, denial, reductionism (or the absence of the big picture), lack of creativity and innovation, lack of empathy, the absence of the ethical mind, and the professional responsibility for what we do and for what we do not do.

12.1   Sustainable Peacebuilding

The dismantling of the psychological walls is necessary to build sustainable peace in the world. Without sustainable peacebuilding, it will be nearly impossible to prevent and resolve today's and tomorrow's interlocking crises. The environmental climate, for example, will not improve if nations do not learn how to handle human conflicts more constructively, thereby improving the human climate.

Sustainable peace refers to a situation characterized by (a) the absence of armed violence and other sources of violence that reduce the quantitative and qualitative

life expectations of people,[2] (b) the support of the people (which endows it with a
high level of legitimacy), and (c) the establishment of changes in five sectors:
communication, consultation and negotiation; political, economic and security
structures (the hardware); the political-psychological climate (the software);
peace-support systems[3]; and the international neighbourhood. This is called the
architecture of peacebuilding or the peacebuilding pentagon (Reychler 2010). The
building of global sustainable peace will demand drastic changes in each of the
above sectors. In this paper, we are especially interested in changes to the
political-psychological climate or the software of peacebuilding. Sustainable
peacebuilding requires the development of an integrative political-psychological
climate, characterized by the perception of a mutually benefiting future, a "we-ness"
feeling, dealing with past, and the dismantling of psychological or mental walls.

12.1.1   Intellectual Solidarity

Building sustainable peace is the work of diplomats, educators, psychologists,
engineers, politicians, lawyers, medical doctors, the military, business entrepreneurs, psychologists, religious leaders and many others. It is also the responsibility
of intellectuals, scholars and researchers. Intellectual solidarity can strengthen the
impact of the latter. The UNESCO constitution still reminds us "That a peace based
exclusively upon the political and economic arrangements of governments would
not be a peace which could secure the unanimous, lasting and sincere support of the
peoples of the world, and that the peace must therefore be founded, if it is not to
fail, upon the intellectual and moral solidarity of mankind". Intellectual solidarity
could be defined as the commitment of scholars to use research and education to
realize the necessary preconditions for building sustainable peace: in the first place,
the identification and dismantling of the psychological or mental walls. This implies
(a) the promotion of independent thinking and enlightenment, (b) raising the
research capacity in poor and less research-friendly arenas, (c) exposing and
counteracting the misuse of information and the distortion of the truth, (d) enhancing dialogue and empathy, (e) developing a more comprehensive understanding of conflict, security and peacebuilding, (f) researching how to transform
conflicts and build peace more effectively, and (g) raising hope and tuning into our
highest future possibility (Scharmer 2007).

---

[2]Political, economic and social exclusion; psychological violence; cultural violence; epistemic
violence; bad governance; and environmental deterioration.

[3]Such as education, the judiciary, health-care, and environmental protection.

12.1.2 Epistemic Violence

Mental walls can be created intentionally and remain operational because of neglect or insufficient efforts to dismantle them. In that case, we are dealing with epistemic violence. In this paper, "epistemic violence" is defined as the active or passive inhibition of knowledge and know-how that could be used to further international cooperation and sustainable peacebuilding. Four types of conditions which make epistemic violence possible can be identified: (a) the existence of a rough and unlevelled playing field, (b) a reductionist research process, (c) significant gaps in the research, and (d) the curtailment of the impact of critical thinking and data on opinion- and decision-making.

12.1.2.1 A Rough and Unlevelled Playing Field

Bounded Academic Freedom

Academic freedom is a cornerstone of scientific work. It is especially important for the humanities and behavioural sciences dealing with conflict and peace. It is necessary for relevant and high-quality scientific research. Academic freedom has always been contested by forces outside and inside universities or research centres. Since the time of Socrates, researchers and teachers/professors have been persecuted by state, religious authorities or powerful interest groups and lobbies which do not like dissenting views or uncomfortable truths (Altbach 2009). Academic freedom is concerned with protecting the conditions that lead to the creation of sound scholarship and good teaching, not with maintaining political neutrality (Atkinson 2004). Science is not neutral; it is about values, interests and politics. Despite the importance of academic freedom, problems exist almost everywhere. In countries which are not free and partially free, restrictions exist on what can be researched, taught and published. They can be explicit, but in most cases the red lines are implicit and academics are sanctioned if they cross them. The sanctions can be: threats, incarceration, death, and loss of career options. However, academic freedom also remains a problem in the so-called "free world". It is curtailed by political conformity pressures (political correctness and taboos), prescriptions of acceptable language, industrial forms of organization, and the heat of an ever-increasing workload and time pressure. Many colleagues have become people without time; timeless people (*des hommes sans temps*). Scholars who do not respect the red lines can be ostracized as unpatriotic, communist, or one or another anti-ite; they get less access to the media and to research grants, and experience tenure track or promotion difficulties. Only a courageous minority of scholars dares to tackle politically sensitive issues. The majority evades sensitive topics; they focus on less sensitive issues, prefer abstract theorizing or become methodological fundamentalists. The history of bounded academic freedom in the free world still needs to be written. There are no reliable comparative statistics on academic freedom.

Unlevelled Playing Field

A second problem is the gross inequality of the research capacities of the Global North and South and the Western dominance in framing the cultural, security and development issues of the rest of the world. The international community relies predominantly on external expertise; it excludes internal expertise. In the foreword to Dambisa Moyo's book, *Dead Aid: Why Aid Is Not Working and How There Is Another Way for Africa*, Niall Ferguson expresses his frustration as follows: "It has long seemed to me a problem, and even a little embarrassing, that so much of the public debate about Africa's economic problems should be conducted by non-African white men. From the economists (Paul Collier or Jeffrey Sachs) to the rock stars (Bono and Bob Geldof), the African discussion has been colonized as surely as the African continent was a century ago" (Ferguson 2009: ix). Illustrative of the unequal framing is the use of adult-child metaphors for other nations. The child nations are developing or underdeveloped, weak and failing, rogue or evil states (Lakoff 2004).

12.1.2.2 Narrow-Minded Research Process

Reductionist approaches and small history. Sustainable peacebuilding implies work by many actors in different sectors on different levels, at different layers (behaviour, cognitive, emotional) and in different temporal settings. It demands a good understanding of changes in different sectors (political, economic, security, etc.) but also comprehension of the big picture. The latter requires not only interdisciplinary but also transdisciplinary research and understanding. Most universities are still divided into faculties or disciplines; transdisciplinary research and education, especially in the humanities, is exceptional. Vandana Shiva[4] labels most sciences reductionist because they (a) create a sharp divide between the expert and the non-expert – a divide which converts the vast majority of non-experts into non-knowers, (b) perform most of their research for powerful interest groups, such as the war industry, (c) do not address the full causal picture of the problem (for example, of political terrorism), and (d) seem to be blind to the negative side-effects of the prescribed policies or remedies. Hoffman (2000) called the historical knowledge of the social sciences skin-deep and stressed the importance of diplomatic history for understanding the business of war and peace. In Belgian universities, diplomatic history begins in 1815. This may be useful for understanding the creation of the country, but it does not help us to grasp the broader picture of changes in international relations over a longer time period. A welcome reaction against the reductionist approach is the creation of "Big History" as a field of study.

---

[4]"Eighty percent of all scientific research is devoted to the war industry and is frankly aimed at large-scale violence" (Shiva n.d.), at: http://www.unu.edu/unupress/unupbooks/uu05se/uu05se0i.htm.

This arose from a desire to go beyond the specialized and self-contained fields that emerged in the twentieth century and to grasp history as a whole, looking for common themes across the entire timescale of history (Spier 2010). In *The American Scholar*, Laughlin (2010) stresses the importance of looking at the Earth's past and future in terms of geological time, and of understanding climate change not just in the present and the immediate future. "Climate change over geological time is something the Earth has done on its own without asking anyone's permission or explaining itself. People can cause climate change, but major glacial periods have occurred at regular intervals of 1,000,000 years... [There is] always a slow steady cooling followed by abrupt warming back to conditions similar to today's" (Will 2010).

Low epistemic empathy. The low level of context sensitivity in research into the effects of human behaviour on security and peace is astonishing. Context sensitivity requires empathy to see and feel the situation from the points of view of the parties involved in the conflict or peacebuilding process. The failure of the war on terror is the failure of empathy. Amativa Kumar, a Hindu married to a Pakistani Muslim, is very critical of the global war on terror. He pleads for empathy to overcome terrorist behaviour: "If we are to bridge the perilous divide that separates us from those poor and unnamed people who resent us, we first need to see them, to look into their eyes. We need to acknowledge that they exist" (Kumar 2010, cited by Garner 2010).

Weak integration of knowledge and know-how. Despite the availability of different forms of research (empirical-analytic, interpretive and liberatory or participatory action-research), the prevailing form of enquiry is empirical analytical. This makes it difficult to contextualize the meaning of data and to create movements to redress injustices, support peace and form democratic spaces (Smith 1997). Equally contributing to the narrow-mindedness is the lack of integration of the know-how from other disciplines, frontline practitioners, decision-makers and the people involved in the conflict or peace.

12.1.2.3   Significant Research Gaps

Shortage of critical data. There are serious gaps in the research. Reliable data on the costs of violent conflicts are missing. First-class victims get most of the attention, second-class less, and third-class victims are relegated to rough and unreliable statistics. Second, it is hard to find good data on the profits of wars and conflict profiteers. Such information is crucial, because violent conflicts last as long as one or more parties benefit or expect to profit from them. Third, most research maps conflicts in a specific time and space (static analysis). To understand conflict and peacebuilding processes, dynamic analyses of changes over different periods and in multiple dimensions are needed. Fourth, there is a shortage of research that estimates and compares the costs of alternative conflict- or crisis-management options. The pioneering research of Brown and Rosecrance (1999) is exemplary. They developed a methodology (using counterfactuals) to compare the costs of prevention and conflict. Top priority should be given to research to anticipate and compare the cost-effectiveness of alternative approaches to conflict management and peace-building. The costs tend to be underestimated. In Afghanistan and Iraq the aggressors expected short and cost-effective interventions. Instead, the wars became protracted wars and are very expensive. Stiglitz (2008) expects the overall costs to amount to three trillion dollars. Fifth, generational accounting is not popular because of the economic complexities and the social and political problems. Generational accounting is a tool for identifying the long-term implications of current fiscal and social policy. Taking into account future demographic development, generational accounting shows which effects the prolongation of a given policy will have on the tax and transfer payments of living as well as future generations (European Commission 2000; Volkaert 2010). Sixth, equally under-researched are follies or the pursuit of policy contrary to the self-interest of the constituency or state involved. Tuchman (1984) and Diamond (2005) have done pioneering work. "Social systems can survive a great deal of folly when the circumstances are favourable, or when bungling is cushioned by large resources or absorbed by sheer size, as in the United States during its period of expansion. Today, when there are no cushions, folly is less affordable" (Tuchman 1984: 21).

Additional minds needed. Peace research is not only a descriptive enterprise; it also deals with values, futures and prescription. The world of the future demands capacities which are undervalued at universities or that until now have been mere options. Gardner (2007) has provided lists of aptitudes that need to be cultivated to deal with change and conflict. He identifies five minds which are crucial for the future:

(a) The disciplined mind, which masters at least one discipline, for example law, economics, politics, engineering. It takes up to ten years to master a discipline. Today's universities spend most of their time on teaching this mode of cognition. A disciplined mind helps the individual to think independently. Without mastering at least one discipline, the individual is doomed to March to someone else's drums.
(b) The synthesizing mind takes information from different sources and puts it together in ways that make sense to the synthesizer and to other people. A synthesizing mind helps the individual to see and understand the big picture.
(c) The creating mind breaks new ground, presents fresh ideas, raises unfamiliar questions, explores alternative solutions, frames the problem in new ways, gives unexpected answers and opens new perspectives.
(d) The respectful mind notes and welcomes differences between human individuals and between human groups, tries to understand the 'others' and seeks to work effectively with them.
(e) The ethical mind reflects on the nature of one's work and the needs and desires of the society in which one lives. It conceptualizes how people can serve purposes beyond self-interest and how citizens (and nations) can work unselfishly to improve the lot of all. It also reminds us that we are responsible not only for what we do but also for what we do not do.

12.1.2.4   Curtailing the Influence of Critical Analysis

Blocking the way to an open-minded and informed opinion is the spin-doctor business. Spin doctors are paid to frame events in order to fit the values and interests of their clients. You find them in public relations offices, think thanks, lobbies and the corporate media. Their toolkit consists of:

- Censorship or efforts to control communication and information and to deflect attention from material that is considered harmful to the particular interests of governments, corporations and other organizations. Project Censored and independent journalists around the world have been tracking censorship and freedom of the media in the US (Phillips/Roth 2008).
- Semantic assault or the use of language to misinform and mislead people. In his book *Unspeak* Poole (2006) warns us to be mindful of the consequences of words. Unspeak is the use of language to persuade us of something and to shut down arguments before they happen. "Operation Iraqi Freedom" tries to persuade us that the 2003 Iraq war had no other motive than the liberation of the Iraqi people. Initially, CENTCOM dubbed the US invasion of Afghanistan "Operation Infinite Justice". One of the latest semantic assaults was the introduction by Newt Gingrich of the phrase "stealth jihad", suggesting efforts to replace Western Civilization with the radical imposition of Sharia. The word 'stealth' effectively conflates all Muslims with terrorists. In a stealth campaign you never know who your friends are (Miller 2010).
- The tobacco strategy. Some scientists specialize in disputing scientifically proven conclusions on the unhealthy nature of tobacco, the inadequacy of the strategic defence initiative (Star Wars), the warming up of the Earth, etc. They are called the merchants of doubt. Although doubt is part and parcel of scientific research, they create pseudo debates, use findings selectively, focus on details, and tell the mass media that they should shed light on both sides of the story (Oreskes and Conway 2010).
- Inflation of fear and appeal to national self-defence. What conservatives have learned about winning elections is that they have to activate the strict father model either by fear or by other means (Lakoff 2004). "The pretext for Washington's terrorist wars was self-defense, the standard official justification for just about any monstrous act, even the Holocaust" (Chomsky 2004: 146). 9/ 11 gave Bush a perfect mechanism for winning elections: the declaration of an unending war on terror. Before September 2001, it was a "war on drugs"; before that a "war on communism". The chief costs of terrorism derive not from the damage inflicted by the terrorists, but from what those attacked do to themselves and others in response (Mueller 2006; Montbiot 2009). Anti-terrorism has become a huge industry and has been more lethal and destructive than terrorism. Elections in the US for the House of Representatives and the Senate have been characterized by a great number of fear-mongers, such as Glenn Beck.

- The making of scapegoats and stigmatization. The Middle East has been the place for scapegoats. The Palestinians have suffered this role for nearly half a century. Afghanistan, Iraq and Iran have been singled out for (un)merited threats, assault, economic sanctions and blame. At both the global and the personal level stigmatization is part and parcel of conflict dynamics. The stigmatized are ostracized, devalued, rejected, scorned, shunned, sanctioned, threatened and destroyed. They are evil, underdeveloped, unpatriotic people, anti-semitic, unlawful combatants, socialists, communists, etc. The stigmatizers describe themselves as patriotic, democratic, civilized, and fighting wars through necessity.
- Conformity pressure. The construction of reality is a major concern of the powerful.[5] Bennett (1990) found evidence of the tendency of the mainstream media to 'index' the range of voices and viewpoints in both news and editorials according to the range of views expressed in mainstream government debate on a given topic. The index paradigm suggests that the media and government are closer to each other than either is to public opinion. But the public tends to absorb both messages. A July 2003 poll found that 72% of the respondents agreed with the erroneous statement that Iraq was harbouring al-Qaeda terrorists and helping them to develop chemical weapons (Entman 2004).
- Double standard. Pilger (2003) warns his readers about lethal double standards: "international law" and the "international community" are often merely the preserves of great power, not the expression of the majority. These double standards relate to several issues, such as nuclear weapons, haves and have-nots, democratization at international and national level, democracy and colonization, free markets and protectionism, terrorism and anti-terrorism, occupation and self-defence. He also points out the common use of 'we' and its appropriation by great power.
- Following intellectual fashions. Ignatieff (2006: 14) recognizes that the greatest challenge for academics in public life is to avoid becoming a prisoner of intellectual fashions. "It's amazingly hard to think an honest and independent thought in modern academic life".
- Don't know, don't hear and don't see. Not enough attention is given to what we do not know, to what is not said, or to what is not shown. To prevent violent conflicts or follies it is important to see the big picture, to listen to other perspectives, and to broaden what's 'conceivable' in the future. Taleb (2007) draws our attention to the impact of the highly improbable (black swans) and Gladwell (2000) to the impact of small changes (tipping points).

---

[5]Gramsci (prison notebooks, n.d), and Gitlin (2003) define hegemony as the ruling class's domination through ideology and through the shaping, engineering or manufacturing of consent.

12.2   Strengthening Intellectual Solidarity

In order to strengthen intellectual solidarity, measures could be taken to build capacity, raise empathy, and monitor and confront truth falsification efforts. This paper highlights two remedies: (a) the development of objective indicators of intellectual solidarity for a comparative study of countries or other actors, and (b) the creation of space to discuss sensitive-controversial issues.

12.2.1   Remedy 1: Assessing, Comparing and Ranking the Level of Intellectual Solidarity

To assess the quality of intellectual solidarity, objective criteria and indicators need to be generated and approved on several dimensions:

- Context: Is there academic freedom? High quality staff? A good incentive system? An adequate research infrastructure? Do the mission statement and ethical code favour intellectual solidarity?
- Support: Are significant efforts made to raise the capacity of researchers in poor and unfree countries?
- Minds: What kind of minds are available or being developed: disciplined, synthesizing, empathetic, respectful, ethical, symphonic, future-orientated, strategic, broad historical, design etc.?
- Synergies: Are efforts made to facilitate interdisciplinary-disciplinary work? To involve different generations? To be gender-friendly? Is there cooperation between researchers, practitioners, decision-makers, stakeholders and the public? What is the place of action-research?
- Reflection: Are the researchers aware of the "You stand where you sit" impact? Are they invited or expected to make explicit their normative, theoretical and epistemological assumptions?
- Sustainable peacebuilding relevance: Is the research relevant to sustainable peacebuilding? Are there courses on the architecture of sustainable peacebuilding, integrative peace negotiations, political transformation, socio-economic development, cooperative and human security, etc.?
- Countervailing action: Are steps taken to monitor and counteract efforts to curtail intellectual solidarity? To help to expose disinformation? To create alternative options? To demand from the author "You stand where you sit" information?
- Raising hope: How does the institution contribute to a better common future? How does it help its scholars to be courageous? Are there courageous scholars?

12.2.2   Remedy 2: Creating Spaces for an Open Discussion on Sensitive and Controversial Issues and the Generation of Alternative Solutions

To resolve conflicts constructively, the parties involved should (a) make explicit their positions, (b) confront each other with them, and (c) sincerely respond to a series of probing 'why?' questions. These questions help to make explicit the normative, theoretical and epistemological assumptions underlying disagreements. These steps can turn a polarizing debate into a dialogue, facilitate a common and comprehensive understanding of the problem(s), and identify more sustainable win-win solutions. The following statements could be used in exercises to enhance more enlightened opinion-making and dialogue.

Positions	Opinions		
1	Terrorism is the greatest threat in the world	Agree	Disagree
2	The war on terrorism is killing more innocent victims and is more destructive than terrorism	Agree	Disagree
3	Aid to Africa has not helped. It has ruined the continent	Agree	Disagree
4	Democracies are less violent than non-democracies	Agree	Disagree
5	Free trade is the roadmap to wealth	Agree	Disagree
6	The international system is very undemocratic	Agree	Disagree
7	Occupation is the best predictor of suicide terrorism	Agree	Disagree
8	There is a weapon against drug gangs: legalizing drugs	Agree	Disagree
9	The war in Afghanistan is folly	Agree	Disagree
10	We should prepare for a war against Iran	Agree	Disagree
11	The danger from Iran is a dangerous myth	Agree	Disagree
12	Awareness of the dangers of Islamo-fascism, Judeo-fascism and/or Christo-fascism should be raised	Agree	Disagree
13	The US military policy is contributing to world security	Agree	Disagree
14	The Democratic Republic of the Congo will be a stable and affluent country within 25 years	Agree	Disagree
15	Enemies are made to justify the Military-Industrial complex	Agree	Disagree
16	South and North Koreans belong to the same civilization	Agree	Disagree
17	Poverty is the worst form of violence	Agree	Disagree
18	The European Union is an example of successful sustainable peacebuilding	Agree	Disagree
19	Greater equality makes societies stronger	Agree	Disagree
20	The threat of political terrorism will last as long as regions are directly or indirectly occupied or colonized by foreign actors	Agree	Disagree

12.3   Conclusion

To deal with the interlocking crises in the twenty-first century, drastic changes in human behaviour are needed. The building of sustainable peace is a *sine qua non*. Many people, however (especially in rich countries), still take peace for granted because they fail to understand what sustains it.

The research community has a key responsibility. It needs to address sensitive issues and strengthen intellectual solidarity. Intellectuals[6] should express themselves more on issues of public concern and be more active in the new *forum humanum*. Reflecting on controversies in the domestic or international environment, public intellectuals may express their dissatisfaction with the existing state of their country or region or with the international system and seek to steer society towards better alternative futures.[7]

Efforts are needed to stop the debilitating impact of the political and academic environment (academization, specialization, reductionism, methodological fundamentalism etc.) on the independence of intellectuals.

In all fields of study, role models of public intellectuals should be studied. In *The Economic Consequences of the Peace*, the prescient economist Keynes (1919) warned decision-makers of the negative consequences of a revengeful peace. He had the capacity for mischief (Clarke 2009). On the basis of case studies, Barbara Tuchman, a historian, highlighted the phenomenon of governmental follies or the pursuit of policy of contrary to the self-interest of the constituency or state involved (Tuchman 1984). Jared Diamond, a scholar of many disciplines, including physiology, ecology and anthropology, pointed out the reasons why, in history, societies chose to fail or succeed (Diamond 2005). Another example of a role model is Habermas (2010), who commented on the debate in Germany about Islam, *Leitkultur* and the claim that Judeo-Christian tradition distinguishes us from "the foreigners".

Intellectuals must join forces and strengthen intellectual solidarity to fight the actual gods of our time – the god of conformity, as well as the gods of apathy, hubris, fear, despair, unsustainable growth and exploitative power (May 1975). They should lead from the future and help people to develop a deeper awareness of the dynamics of change. This requires, according to Scharmer (2007), the intelligences of an open mind, an open heart and an open will. An open mind is needed to recognize our own taken-for-granted assumptions and to start to see things that were not evident before. Premature judgement inhibits open-mindedness. An open

---

[6]'Intellectual' is not a synonym for intelligence. Many intelligent people may not have an interest in ideas or controversies, or their interests and mindsets could be too narrow to engage in a broader discussion. Public intellectuals may or may not be affiliated with universities. They may be full-time or part-time academics; journalists or publishers; writers or artists; politicians or officials; or they may work for think thanks or hold ordinary jobs.

[7]Ignatieff (2006) stresses that they should stick to the knowledge and know-how that they genuinely claim to have mastered, because all-purpose experts quickly become clowns and entertainers. Smart asses are not necessarily the intellectuals that are needed.

heart allows a deeper level of attention, one that allows people to step outside their traditional experience and feel or sense beyond the mind. Cynicism stands in its way. An open will implies a readiness to let go of dysfunctional relations or systems, and to create a country or world that could survive in the future – and only together can this be achieved. Fear closes the way to an open will.

References

Altbach, P. (2009). Global Academic Freedom: A Realistic Appraisal. *University World News* (20 September).

Atkinson, R. (2004). Academic Freedom and the Research University. *Proceedings of the American Philosophical Society, 2* (June).

Bennett, W. L. (1990). Toward a Theory of Press-State Relations in the United States. *Journal of Communication, 40*(2): 103–127.

Brown, M. & Rosecrance, R. (1999). *The Cost of Conflict Prevention and Cure in the Global Arena*. Lanham, MD: Rowman & Littlefield.

Chomsky, N. (2004). *Chomsky on MisEducation*. Oxford: Lanham, MD: Rowman & Littlefield.

Clarke, P. (2009) *Keynes: The Twentieth Century's Most Influential Economist*. London: Bloomsbury.

Crooke, A. (2009). *Resistance: The Essence of the Islamist Revolution*. London: Pluto Press, 2009.

Diamond, J. (2005). *How Societies Choose to Fail or Succeed*. New York: Viking.

Entman, R. (2004). *Projections of Power: Framing News, Public Opinion, and the US Foreign Policy*. Chicago: University of Chicago Press.

European Commission. (2000). *Reports and Studies No. 6/1999: General Accounting in Europe*. London: The Stationery Office Books.

Ferguson, N. (2009). Foreword. In: Dambisa Moyo, *Dead Aid: Why Aid Is Not Working and How There Is Another Way for Africa* (pp. ix-xii). London: Allen Lane.

Fukuyama, F. (2006). *After the Neocons: America at the Crossroads*. London: Profile Books.

Gardner, H. (2007). *Five Minds for the Future*. Boston: Harvard Business School Press.

Garner, D. (2010). The Global War on Small People. *New York Times* (5 August).

Gitlin, T. (2003). *The Whole World is Watching: Mass Media in the Making and Unmaking of the New Left*. London: University of California Press.

Gladwell, M. (2000). *The Tipping Point: How Little Things Can Make a Big Difference*. New York: Little, Brown and Company.

Habermas, J. (2010). Leadership and Leitkultur. *The New York Times* (29 October).

Hoffmann, S. (2000). *World Disorders: Troubled Peace in the Post-Cold-War Era* (revised edn.). Lanham, MD: Rowman & Littlefield.

Ignatieff, M. (2006). Michael Ignatieff on Academics in Public Life. *Academic Matters: the Journal of Higher Education* (Spring): 14.

Kumar, A. (2010). *A Foreigner Carrying in the Crook of his Arm a Tiny Bomb*. Durham, NC: Duke University Press.

Lakoff, G. (2004). *Don't Think of an Elephant: Know Your Values and Frame the Debate*. Hartford, VT: Chelsea Green Publishing.

Laughlin, R. B. (2010). What the Earth Knows. *The American Scholar* (1 June).

May, R. (1975) *The Courage to Create*. New York: Bantam Books.

Miller, L. (2010). The Misinformants. *Newsweek* (6 September).

Moïsi, D. (2009). *De Geopitiek van Emotie*. Amsterdam: Uitgeverij Nieuw Amsterdam.

Montbiot, G. (2009). *Bring on the Apocalypse: Six Arguments for Global Justice*. London: Atlantic Books.

Mueller, J. (2006) *Overblown: How Politicians and the Terrorism Industry Inflate National Security Threats and Why We Believe Them*. New York: Free Press.

Oreskes, N., & Conway, E. M. (2010). *Merchants of Doubt: How a Handful of Scientists Obscured the Truth on Issues from Tobacco Smoke to Global Warming*. London: Bloomsbury.

Phillips, P., & Roth, Andrew. (2008). *Censored 2009*. New York: Seven Stories Press.

Pilger, J. (2003). *The New Rulers of the World*. London: Verso.

Poole, S. (2006). *Unspeak*. London: Little, Brown.

Radman, D. (2010). New Iraq Resembles the Old. *International Herald Tribune* (5 November).

Reychler, L. (1994). *Nieuwe Muren: Overleven in een Andere Wereld* [*New Walls: Living in Another World*]. Leuven: Leuven University Press.

Reychler L. (2010). Sustainable Peace Building Architecture. In N. Young, *The Oxford International Encyclopedia of Peace* (pp. 2027–2044). Oxford: Oxford University Press.

Scharmer, O. C. (2007). *Theory U: Leading the Future as it Emerges: The Social Technology of Presencing*. Cambridge: Society for Organizational Learning.

Shiva, V. (n.d.). *Reductionist Science as Epistemological Violence*. At: http://www.unu.edu/unupress/unupbooks/uu05se/uu05se0i.htm.

Smith, S. (1997) *Nurtured by Knowledge: Learning to Do Participatory Action Research*. New York: The Apex Press.

Spier, F. (2010). *Big History and the Future of Humanity*. Chichester: Wiley-Blackwell.

Stiglitz, J. (2008). *The Three Trillion Dollar War: The True Cost of the Iraq Conflict*. New York: W.W. Norton & Company, Inc.

Taleb, N. N. (2007). *The Black Swan: The Impact of the Highly Improbable*. London: Penguin Books.

Tuchman, B. (1984). *The March of Folly*. London: Abacus.

Volkaert, K. (2010). Maak eens duidelijk wie het kind van de rekeing wordt. *DE Tijd* (30 October).

Will, G. (2010). The Earth Doesn't Care about What's Done To or For It. *Newsweek* (20 September).

Chapter 13
Peacemaking, Peacekeeping, and Peacebuilding

13.1 Introduction

13.1.1 Mainstreaming Peace

Peacemaking, peacekeeping, and peacebuilding may not have the punch and the means of national security, but they are receiving an increasing amount of attention in education, research, and politics.[1] There are a growing number of Master and Ph. D. programmes, new publications, and more research at universities and think tanks.[2] The number of peer-reviewed journals covering different facets of peacebuilding has doubled since 1992. Peacebuilding has become embedded in the organizational theory and praxis of national governments, non-governmental organizations, and regional and global intergovernmental organizations. It became part of the official discourse when the UN Secretary-General Boutros Boutros-Ghali (1992) introduced the concept of post-conflict peacebuilding in the Agenda for Peace. The agenda specified four areas of action, which, taken together, were presented as a coherent contribution towards securing peace: i.e. (1) preventive diplomacy is action to prevent disputes from arising between the parties, to prevent existing disputes from escalating into conflicts and to limit the spread of the latter when they occur, (2) peacemaking is action to bring hostile parties to agreement, essentially through such peaceful means as those foreseen in Chap. VI of the Charter, (3) peacekeeping is the deployment of a United Nations military and

---

[1]This text was first published as: Reychler, L. (2010). Peacemaking, Peacekeeping, and Peacebuilding. In R. A. Denemark (Ed.), *The International Studies Encyclopedia* (pp. 5604–5626). Oxford: Wiley-Blackwell. Reprinted with permission.

[2]This contribution is based on many experiences, discussions, and research over the years. It is impossible to acknowledge and thank all those who have influenced my thinking on sustainable peacebuilding. Let me express my appreciation for the valuable and timely help from two doctoral students, Julianne Funk and Nikos Manaras.

L. Reychler and A. Langer (eds.), *Luc Reychler: A Pioneer in Sustainable Peacebuilding Architecture*, Pioneers in Arts, Humanities, Science, Engineering, Practice 24, https://doi.org/10.1007/978-3-030-40208-2_13

civilian presence in the field to expand the possibilities for both the prevention of the conflict and the making of peace, and (4) peacebuilding is action to identify and support structures which will tend to strengthen and solidify peace in order to avoid a relapse into conflict.

The 2000 *Brahimi Report,* a response to the failures of complex UN peace-keeping in the 1990s, was an attempt to improve the theory and praxis of peace-building. Another report entitled *In Larger Freedom: Towards Development, Security and Human Rights,* presented in 2005 to Kofi Annan by the High-Level Panel on Threats, Challenges and Change, led to the establishment of the Peacebuilding Commission. Its aim was to draft long-term strategies to guarantee reconstruction, institution-building, and sustainable development. Conflict preven-tion and peacebuilding have also been mainstreamed in the European Union and in most of the foreign offices of the member states. The European Union, itself a successful case of sustainable peacebuilding, affirmed the importance of peace-building in a series of EU documents, such as the *Communication from the Commission to the Council on the European Union and the Issue of Conflicts in Africa: Peacebuilding, Conflict Prevention and Beyond,* of 6 March 1996. Since then, European capacity to deal with peacebuilding has been considerably enhanced (2008). Most regional intergovernmental organizations now have departments for peacemaking, peacekeeping, and peacebuilding.

13.1.2   Political and Intellectual Drivers

As a result of several changes in the political landscape, attention began to be paid to peacemaking, peacekeeping, and peacebuilding. There was the unprecedented increase of intrastate conflicts after the Cold War, when several frozen conflicts turned violent. Globalization raised human insecurity in developing, transitional, and rich and powerful countries. Increases in expenditure on defence and antiterrorist opera-tions reflect the perception that we live in a more threatening world. Globalization has brought with it a large, unregulated arms bazaar, easier spillover of intrastate conflicts, and feelings of relative deprivation and fear. Some peacekeeping missions of the UN turned into failures, which led to a search for more effective and better-coordinated peacebuilding intervention strategies. Finally, there is growing recognition that there are limits to violence and that proactive violence prevention is more cost-effective than reactive conflict prevention. Brown/Rosecrance (1999) contributed to this awareness by calculating concrete costs and benefits.

The research community moved into this field of study because it is jointly responsible for the building of a more sustainable, secure, and peaceful world. There is also the appeal of critical theory which is emancipatory and has a strong distrust of the coopting and misuse of peace-related concepts and methods to further domination. This means interrogating the concept of peace during the peace process and challenging the hegemonic discourse of peacebuilding theory and practice (Lambourne 2004a; Mac Ginty 2006), and also acknowledging the violence of non-intervention and sometimes the irresponsibility of protecting (Chopra 1999). Finally, although the field is still dominated by researchers from the North West, the peace research community is becoming more democratic and has been enriched by the input of scholars from other parts of the world.

13.1.3 Widening and Deepening Peacebuilding

In the academic discourse, the meaning of the term 'peacebuilding' has become broader; it now tends to cover all activities undertaken before, during, or after a violent conflict to prevent, end, and/or transform violent conflicts and to create the necessary conditions for sustainable peace. Peacemaking and peacekeeping are part of the peacebuilding process.

The desire for a sustainable, stable, durable, viable, lasting, self-enforcing, and perpetual peace is universal. In the footsteps of Immanuel Kant, and following the preliminary and definitive preconditions for perpetual peace, a great number of peace researchers have focused on sustainable peace. For Boulding (1978), the pursuit of stable peace is the objective of peace policy.

Peacebuilding research has been nurtured in the extensive literature on conflict resolution by Johan Galtung, John Burton, Adam Curie, Karl Deutsch, Elise and Kenneth Boulding, Chadwick Alger, Louis Kriesberg, Chris Mitchell, Edward Azar, Herbert Kelman, and many others. The fusion of the two commissions on international conflict resolution and on peacebuilding during the global conference of the International Peace Research Association (IPRA) in Valetta, Malta in 1994 illustrated the synergies between research on conflict resolution and on peacebuilding. Researchers of peacebuilding focus on the bigger picture of peacebuilding and on the interconnections between peace negotiations, peacekeeping, the installation of peace-enhancing political, economic and security structures, and the transformation of the moral-political climate. They recognize the added value of an appreciative inquiry of conflict transformation, including the envisioning of a common future and the study of the root causes of success (Boulding 1991; Sampson et al. 2003). Criss-crossing the literature, one finds several hidden or explicitly stated theoretical assumptions, such as the following: (1) The *global* approach, which is seen as necessary to conceive of our global situation as part and parcel of our individual existence as human beings (Fischer et al. 1989; Booth 2007). (2) The *holistic, integrating,* or *transdisciplinary* approach, in which compartmental thinking is incompatible with an understanding of the complexity and the cross-impacts between the many activities taking place in the peacebuilding process. Peace cannot be reduced to diplomacy, politics, economy, or security, but is the result of the synergy of efforts in different sectors (Alger 2007). Systems thinking is back (Wils et al. 2006). (3) The *critical* approach, where researchers have gone beyond positivist/empirical approaches and made more space for normative, critical, and post-positivist theories. Reflection on the mental models, cosmologies, or deep ideologies which inform the research work is part of the work.

Galtung (1981), Fischer et al. (1989), and Senge et al. (1994) stress the importance of reflecting on the usually unquestioned assumptions about all kinds of things and how they relate to each other. Peace is not only an operational reality, but also a social construct. A major task of social constructionism is to uncover the ways in which individuals and groups participate in the creation of their perceived and preferred reality. (4) The *intellectual solidarity* approach, which holds the conviction that peace research would benefit from (a) a better exchange of knowledge and know-how between researchers, practitioners, decision-makers, and citizens, and (b) domination-free scientific discourse and analysis (Reychler/Carmans 2006; Reychler/Langer 2006; Verkoren 2008).

13.1.4 Preventive Diplomacy and Peacemaking

Both preventive diplomacy and peacemaking are key components of peacebuilding. The aim of preventive diplomacy is to prevent violence and escalation in time, space, and intensity. Peacemaking aims to end violence and to get a peace agreement. Ramcharan (2008) offers a comprehensive examination of the evolution of preventive diplomacy and its tools at the UN. Special attention is given to the practice of preventive diplomacy by the Security Council, the Secretary-General, and the representatives of the Secretary-General and the UN subregional offices. Steiner (2004) goes further back in history to the beginning of the nineteenth century and researches the potential for major states to work together in the practice of preventive diplomacy between small state antagonists. He describes two types of preventive diplomacy: collective intervention, which defuses the conflict between the primary antagonists in a conciliatory or coercive fashion, and collective insulation, which, unrelated in itself to the needs of the primary conflict parties, defuses the conflict as an irritant to great power relations and tries to head off unilateral intervention. At the beginning of the 1990s, before peacebuilding became mainstreamed in international politics, conflict and crisis prevention was a fashionable political and research topic. A great deal of time was invested in the development of early warning systems, the understanding of successful and less successful peace negotiation and mediation efforts, and the refinement of unofficial diplomacy.

13.1.5 Early Warning

The work done at the University of Maryland by Gurr/Harff (1994), and at the RAND Arroyo Center by Tellis et al. (1997), exemplified the development of early warning systems. They made use of correlation, sequential, response, and conjectural models. Gurr's anticipation of communal conflict was a correlation model, while Harff's anticipation of genocide and politicide was both a correlation and a sequential one. The variables include international and internal background conditions, intervening conditions, and accelerators. Response models were developed by Fein (1993) and Tellis et al. (1997). Conjectural models specify alternative sequences or scenarios of events.

All these efforts produced a variety of warning lights and alarm bells. The problem, however, was that early warning did not easily translate into early and effective action. In addition, there was a series of blind spots in the early warning research. Most of the variables used were hard rather than soft variables, such as private perceptions and emotions. Attention was focused on anticipating threats rather than anticipating opportunities to intervene. Finally, practically no attention was paid to anticipating the negative and positive impacts of well-intentioned interventions on the conflict dynamics and peacebuilding process. Anderson (1999) was one of the first analysts who warned the international community about the negative consequences of well-intentioned interventions. This started the development of methodologies for anticipating the impact of military and non-military interventions on the conflict-transformation and peacebuilding process.

13.1.6 Peace Negotiations and Peacemaking

Peace negotiation, peacemaking, and mediation are efforts to bring the conflicting parties to a peace agreement. Many researchers have addressed the question "Why do peace agreements fail or succeed?", emphasizing how the process, the accord itself, and the implementation process affect the possibilities of achieving a durable peace. Examples of this are the research of Hampson, Crocker, Aall, Walter, Stedman, Rothchild, Hoglund, Lederach, Darby and Mac Ginty, Bercovitch, and Zartman. On the basis of a comparative study of five cases, Hampson (1996) puts forward four possible answers to explain success: the international nurturance of the peace process, the ripeness of the conflict or the desire of the parties to make peace, systemic regional power balances that enhance peace, and the quality of the peace agreement itself, in particular the inclusion of appropriate power-sharing arrangements. Crocker et al. (2004) stressed the pros and cons of multiparty negotiations in large-scale intractable conflicts. For Walter (2002), the key variable to explaining successful implementation of a peace agreement is a third-party security guarantee defined by an implicit or explicit promise given by an outside power to protect adversaries during the treaty implementation period. Stedman et al. (2002) studied several recurrent problems constraining the successful implementation of peace agreements: vague and expedient peace agreements, the lack of coordination between mediators and implementers of agreements, the incomplete fulfilment of mandated tasks, the short-time perspective and limited commitment of implementers, and the presence of spoilers – actors who use violence to undermine implementation. Hoglund (2008) studied the impact of high-profile incidents (the assassination of a high-ranking person, a mass casualty attack, a symbolic attack on the identity of a party to the conflict, and a symbolic attack on the peace process) on the dynamics of the negotiation process. Lederach (1997) showed the limits of the traditional official peacemaking approaches and offered a new way of dealing with peacemaking that is more holistic, aims at restoring and rebuilding relationships, and stresses the importance of an elicitive process and of engaging multilevel leadership. Darby/Mac Ginty's (2008) contribution to peacemaking is their comprehensive or big picture approach to the many activities which take place in the name of peacemaking: preparing for peace, the negotiation process, the impact of violence, the peace accord, and the implementation and post-war reconstruction. Zartman (1995) argued that violent conflicts are ripe for a negotiated settlement when there is a hurting stalemate and Bercovitch (1996) studied the impact of different types of mediation on the success or failure of peace negotiations.

The second question occupying many researchers is: "What are the alternatives to traditional official negotiation?" Burton (1969), Fisher/Ury (1981), Azar (1990), Kelman (1992), Rothman (1992), Mitchell/Banks (1996), Cooperrider/Whitney (1999), and others developed different types of track two, or citizen, diplomacy. This is a specific kind of informal diplomacy, in which non-officials engage in dialogue, with the aim of resolving conflict and building confidence. Burton (1969) promoted problem-solving workshops. Fisher/Ury (1981) distilled a set of principles which led to more effective and integrative resolution of conflicts. Cooperrider/Whitney (1999) drew attention to the potential applications of appreciative inquiry in peacebuilding, a methodology that pays a great deal of attention to a forward-looking orientation (envisioning what might be) and uncovering the positive peacebuilding capacity. Diamond/McDonald (1996) expanded track two into ten separate tracks: government, professional conflict resolution, business, private citizen, research, training and education, activism, religion, funding, and media or public opinion. Another strand of researchers explored traditional and indigenous approaches to peacemaking. Augsburger (1992) concluded his research with the observation that "so-called primitive societies often have conflict solutions that are more effective in bonding adversaries and blending goals than those groups who designate themselves as advanced, developed, or possessing far more data on human relations."

These research findings contribute to a better understanding of successful and failed peacemaking efforts. They also remind us of the close interconnection between peacemaking and other ways to build peace. Many analysts focus on one or more pet variables at the expense of a more systematic and comprehensive study of the relation between peacemaking and peacebuilding. Finally, most of the research is still conceived from the perspective of strong and rich countries and has not been reviewed by colleagues from weak and poor states.

13.1.7   Peacekeeping and Support Operations

Peacekeeping developed in the 1950s as part of what Dag Hammarskjold called preventive diplomacy. Later it became an essential component of conflict prevention and peacebuilding. Since then, the number of peacekeepers and peacekeeping operations has increased, especially since the end of the Cold War. The new interventionism was characterized by the number and changing nature of peace operations. Analysts distinguish between two or more types or generations of peace operations. The two most prevalent are: (1) the traditional "peacekeeping operations" that are backed by the UN Security Council, have the consent of the parties in the conflict, operate within a limited mandate (self-defence and defence of the mandate), and act with impartiality; and (2) "peace support operations" that do not need the consent of the conflicting parties, are not necessarily neutral, are impartial to the mandate, and can make use of the full spectrum of force to fulfil tasks such as countering peace spoilers and applying pressure for the peace operation to succeed, assisting interim civil authorities, protecting humanitarian relief operations, and guaranteeing or denying movement, etc. (Ramsbotham et al. 2006). Some authors, confusingly, include in this last type of peace operation a variety of different kinds of mechanism, such as second- and third-generation peacekeeping, humanitarian intervention, complex peace operations, wider peacekeeping, peace enforcement, peace support operations, peace maintenance, etc. This is not primarily the result of intellectual laziness on the part of researchers or practitioners. Virtually anyone has a personal sense of what peace operations are, but they are usually perceived as activities with extremely flexible boundaries (MacQueen 2006).

The research of peace operations focuses on several dimensions: (1) the peacekeeping and support tasks, (2) the difficult strategic environment, (3) contributors and motivations, (4) factors influencing success or failure, and (5) interconnections with the other activities of the peacebuilding process.

13.1.7.1   Peacekeeping and Peace Support Tasks

Researchers such as Chopra (1999), Berdal/Economides (2007), MacQueen (2006) have analysed the experiences and lessons learned, which led to the development of new types of peace operations and efforts to improve the international and regional organization. Most of the findings are based on thorough analysis of successful and less successful case studies, such as Cambodia, the Former Yugoslavia, Somalia, Rwanda, Haiti, East Timor, Kosovo, and Sierra Leone. Jeong (2005) offers an overview of confidence and security-building measures that are needed to create an environment conducive to good governance and development: (1) confidence-building by means of effective international verification measures, (2) demobilization, disarmament and reintegration (DDR), (3) building local capacity to enforce peace, (4) building a local police force to bring their law enforcement up to international standards, and (5) demilitarization of the internal security system.

13.1.7.2 Hazardous Operational Theatres

Analysts also identified the characteristics of war zones which seriously complicate peace operations. Stedman (1997) observes that peacekeepers can fall prey to spoiler leaders or parties who believe that peace emerging from negotiations threatens their power, world-view and interests, and use violence to undermine attempts to achieve it. Zahar (2008) added to the spoiler debate by focusing on the spoilers' intent, opportunities, and capabilities. Others focused on warlords, militias, paramilititary, and armies seeking control of resources through plundering, terror, and force. Duffield (2001) identified war-zone economies where civilians are a resource base to be corralled, plundered and killed. Nordstrom (2004) described the process whereby dirty war becomes the means by which economies of violence fuse with what she calls "cultures of violence". In most conflict zones organized crime crosses borders and has severe effects on peace and law enforcement (Giraldo/Trinkunas 2006).

13.1.7.3 Contributors and Motivations

Another part of the research deals with the question "Why so few troops from among so many?" Despite the fact that there has been an increase in peace operations, one cannot deny that demand continues to exceed supply. Planners are faced with considerable difficulty in finding appropriate military personnel to man and sustain missions (Daniel 2008a). Among the top ten contributors to UN peacekeeping operations (Blue Helmets) between 2001 and 2005 were Pakistan, Bangladesh, India, Nigeria, Ghana, Jordan, Nepal, Uruguay, and Ukraine (Heldt 2008). Daniel (2008b: 228–229) lists possible national motivations for contributing to peace operations: a sense of international obligation, regional ethos, prestige, repayment of a favour from a major power, outlet for surplus military capacity, remunerations, the desire for training and equipment, burden-sharing, and better control of their own destiny. Kerr (2007) looks at moral and humanitarian motivations, such as concerns about human insecurity and the "responsibility to protect". Marten (2004) observes that the international community has been reluctant to commit the necessary resources to support and maintain peaceful rule. She compares the motives for colonialism a century ago with the motives for complex peace operations, and concludes that, despite their differences, both were pursued for a combination of national interest and humanitarianism. Politically, the responsibility to protect is not widely accepted in developing countries.

13.1.7.4 Successes and Failures

Reports inside and outside the UN have identified (f)actors which contributed to failed peace operations, as in Somalia (1992–5) and Rwanda (1993–4). The *Brahimi Report* of 2000 put forward a wide-ranging set of recommendations. In

2005, the UN High-Level Panel on Threats, Challenges and Change identified a series of weaknesses and offered recommendations, such as the five criteria of legitimacy that the Security Council (and anyone else involved in these decisions) should always address when considering whether to authorize or apply military force (seriousness of threat, proper purpose, last resort, proportional means, and balance of consequences), and when creating a peacebuilding commission. The main task of the latter is to draft long-term strategies and facilitate the coordination of conflict prevention, mediation, peacekeeping, and all the other efforts needed for peacebuilding. All the peace operations have been evaluated. Among the short-comings of the UN mission in Liberia in 1999 (UNAMSIL), for example, Wilkinson (2000a, b) mentions a peacekeeping and not peace enforcement man-date, poorly equipped and trained troops, the lack of a "lead nation" to coordinate command and control structures, and inadequate support from the UN headquarters in New York. In the discourse on the causes of success and failure, attention has also been focused on the integration of long-term security concerns from the start (Jeong 2005), the effects of body bags and CNN coverage, the four pathologies of peace operations (Farrell 2007), the negative impact of HIV (Elbe 2007), the pre-occupation with exit strategies and fear of operation creep, the dilemma of selec-tivity, the violence of non-intervention (Chopra 1999), and civil-military cooperation (CIMIC).

13.2 Peacebuilding Architecture

Peacebuilding is about complex change; it involves concurrent activities by many people in different sectors, at several levels, in different time-scapes, and in different layers. Depending on the conflict, peacebuilding deals with actors at the local, middle, top, and international levels; it looks for synergies between the multiple transformations in diplomatic, political, economic, security, social, psychological, legal, educational, and many other sectors; it involves short-, medium-, and long-term activities, and impacts on institutional, behavioural, perceptual, and emotional layers. A tremendous amount of research has been produced in different disciplines (international relations, political sciences, strategy and security, eco-nomics, law, anthropology, psychology, humanitarian assistance studies, ecology, etc.). The work has been carried out under a great variety of headings, such as peacebuilding, conflict prevention, conflict resolution and transformation, security building, and nation building (Dobbins et al. 2003, 2005, 2007). This is due to the different backgrounds of the contributors, the compartmentalization of the academic environment, and the favourable political resonance of different labels. The research relates to six major components of the theory and praxis of peacebuilding archi-tecture: (1) the end state, (2) the baseline, (3) the context, (4) the planning of the peacebuilding process, (5) the peacebuilding coordination, and (6) the monitoring and evaluation.

13.2.1 The End State

The end state can be defined as the set of required conditions that defines achievement of the peace one wants to build. Without a clear operational definition of peace, it is impossible to develop a good theory about how to achieve it. As Mark Twain observed, "If you don't know where you're going, any road will get you there."

13.2.1.1 Definition of Peace

Do the researchers provide clear operational definitions of the peace they are studying? What types of peace are distinguished? What necessary conditions for peacebuilding have they identified? How valid is their theory? The definition of peace should be clear and the theory used valid. In a great deal of research work, the end state is left vague and undefined. Analysts define peace as a preferred reality (end state) in terms of negative and positive peace indicators and/or make a distinction between sustainable and less sustainable types of peace. Sustainable peace, for example, has been framed in terms of (1) outcome characteristics, such as the absence of armed violence, the near absence of other types of violence (structural, psychological and cultural), the handling of conflicts in a constructive way, and a high level of internal and external legitimacy of the achieved peace; (2) the resolution of the root causes of the conflict; and (3) the successful installation of the necessary conditions or peacebuilding blocks for sustainable peace.

13.2.1.2 Peacebuilding Blocks

Most researchers see peacebuilding as the result of transformations in multiple sectors. Chesterman (2004) studies state building as one of many other activities necessary for peacebuilding. Paris (1997) focuses on two peace-enhancing conditions: marketization and democratization. Cousens et al. (2000) specify five objectives of peacebuilding: a self-enforcing ceasefire (the armed conflict, just settled, will not recur), a self-enforcing peace (new armed conflicts will not occur), democracy, justice, and equity. Caplan (2005: 256) divides the chief functions of the transition administrations into "the establishment and maintenance of internal order and security; repatriation and reintegration of internally displaced persons and refugees; performance of basic civil administrative functions; development of local institutions and the building of civil society; economic reconstruction and development". Pugh (2000: 129) defines peacebuilding as a sustainable process which has, as its main purpose, the prevention of threats to human security which cause protracted violent conflict. Human security implies the need for intervening in the domains of political security and governance, community security and societal stability, personal security and human rights, and, lastly, economic security. Mason/Meernik (2006) define peace as a combination of negative peace and positive peace.

The latter involves a transformation of the conflict by means of democratization efforts, the establishment of truth commissions, the establishment of security, and long-term economic and social development. Jeong (2005) highlights four peacebuilding pillars: security and demilitarization, political transition, development, and reconciliation and social rehabilitation. Orr (2004) identifies four interrelated sets of tasks to rebuild countries and win the peace: security, governance and participation, social and economic well-being, and justice and reconciliation. Darby/Mac Ginty (2008) pay attention to peacemaking, demobilization, disarmament and reconstruction, democratization and power-sharing, refugees, and negotiating how to deal with past human rights violations. Ryan (2007) stresses the importance of peacemaking, reconciliation, political transformation, development, and sentimental education. In their study of nation-building, Dobbins et al. (2007) list five challenges related to the security, humanitarian, civil administration, democratization, and economic reconstruction conditions. Reychler's (2004) sustainable peacebuilding pentagon conveys the importance of (1) an effective system of communication, consultation and negotiation at different levels; (2) peacebuilding political, economic and security structures – called the hardware; (3) an integrative moral-political climate – called the software; (4) other systems (legal, educational, media, humanitarian) supporting the peacebuilding process; and (5) a supportive international neighbourhood. For these conditions to be realized, peacebuilding leadership is paramount.

An urgent task for the research community is to distil from this vast, but scattered, reservoir of knowledge a valid and comprehensive theory of peacebuilding. This will not be easy because (1) several analysts do not use an operational definition of peace – the meaning of peace in 'peacebuilding' needs to be made explicit, and different types of peace should be differentiated in a systematic and theoretically sound way; (2) there are wide differences in the content and the weight attached to the peacebuilding conditions – a systematic analysis of these differences would benefit the academic discourse considerably; (3) despite the fact that analysts recognize the importance of an integrated analytic framework for understanding the complex dynamic process of peacebuilding, most end up by describing the peacebuilding activities made in different sectors separately. There is more multidisciplinary than transdisciplinary research. This could be attributed to methodological difficulties associated with studying dynamic interactions between multiple transitions, but also to reality, where peacebuilding tends to boil down to a compilation of peacebuilding measures and efforts designed and implemented by different departments or actors.

13.2.2 The Baseline

The baseline is the situation at the starting point of a peacebuilding intervention. Before planning the intervention, it is important to conduct an accurate analysis of the conflict and the peacebuilding deficiencies and potential.

13.2.2.1   Conflict Analysis and Prognosis

The literature is flooded with all kinds of models to analyse and anticipate conflicts. These models require information about the parties involved, the issues, the positions, the alternatives to a negotiated agreement, the conflict environment, the strategic thinking of the parties, the current interaction, the legacy, and the costs and benefits. Despite the availability of these analytic and anticipatory tools, problems continue to hamper accurate analysis. There is a lack of accurate information about non-armed violence and the complex dynamics of conflicts. Violence tends to be defined narrowly, and information about second- and third-class victims is difficult to find. Another problem relates to the exclusion of parties from the analysis of the conflict. Exclusion is an obstacle to a comprehensive understanding of the conflict, which should involve empathy and recognition of the distinctive cultural understandings of the conflict and its resolution, which have to be clarified, elucidated, and enhanced through reflection and dialogue. Lederach (1997) calls this the elicitive approach. Third, the discourse is loaded with confusing terms, such as "conflict prevention", 'terrorism', "post-conflict situation", 'peacekeeping', "regime change" and 'self-defence', which complicate good diagnosis. Fourth, it is difficult to find data on the profits and profiteers of violent conflicts. Finally, peacebuilding research is still lacking in adequate conflict differentiation. A positive contribution to the analysis of the baseline has been the measurement of the difficulty of the conflict. The difficulty of a conflict has been labelled 'deep-rooted' (Burton 1987), 'protracted' (Azar 1990), or 'intractable' (Burgess/Burgess 2009). Stedman's (2001) factors that are commonly associated with a difficult conflict are the presence of spoilers, neighbouring states that are hostile to the agreement, a large number of soldiers, valuable natural resources, and secession-orientated conflict. Caplan (2005) correlates difficulty with the clarity and appeal of operational objectives. Chesterman (2004) links clarity of purpose to success. For Doyle (2002), hostile or incoherent factions are obstacles to peacebuilding operations. Reychler/Langer (2006) monitor seven clusters of variables to assess the degree of difficulty: the parties involved, the issues, the conflict styles, the internal opportunity structure, the legacy of the conflict, the internal readiness for peace, and the external involvement with and support for the peace process.

13.2.2.2   Peacebuilding Deficiency Assessment

To evaluate the relevance of peacebuilding efforts, a comparison has to be made between the situation at the start of the intervention and the necessary conditions to realize the envisaged peace. The quality of peacebuilding deficiency assessment depends on the clarity of the definition of peace, the validity of the peacebuilding theory, and the availability of reliable information. There are checklists for assessing the quality of the peace negotiation process, the accord, and the implementation (Reychler et al. 2008), and the same is true of deficiencies related to political legitimacy, good governance, genuine democracy, freedoms, human rights, gender democracy, and consolidation (Freedom House[3]). The economy of peace focuses on human development, poverty, vertical and horizontal inequality, trust and economic expectation, greed and grievance (Collier/Hoeffler 2004), relative deprivation, and the politico-economic perspectives of young populations. An important contribution to the assessment of horizontal inequality has been made by the research team of Stewart (2008). Horizontal inequalities are inequalities in economic, social or political dimensions or cultural status between culturally defined groups. In many countries, researching horizontal inequality between identity groups is considered risky and politically incorrect and is therefore not done. In security assessment, there are indicators of internal and external security, human security, demobilization, disarmament, social and military integration, modernization of the military forces, modernization and demilitarization of the police, and multilateral or cooperative security. The moral-political environment is more difficult to assess because it requires data on (1) hope-raising measures, (2) the development of a we-ness feeling and multiple loyalties, (3) dealing with the past and reconciliation, (4) trust, (5) social capital, and (6) the dismantlement of senti-mental walls. Senti-mental walls are attitudes and feelings, perceptions and expectations, causal analyses and attributions of responsibility, strategic analyses, values, preferences, taboos, and social psychological pressures (such as conformity pressure, group-think, and political correctness) which stand in the way of sustainable peacebuilding. The yoking of 'sentiment' and 'mental' is intended to make people aware that mental walls tend to be reinforced by emotions and that efforts to dismantle them tend to be confronted with different kinds of emotional resistance. Moisi (2008) describes how the cultures of fear, humiliation, and hope shape today's world. Lindner (2006, 2009) deals with emotions, especially humiliation in conflict and peacebuilding. For Wallensteen (2002), the agenda of peace is formed by trauma and hopes. For the fourth cluster of peace conditions, there are checklists assessing transitional, retributive, and restorative justice, the role of the media, peace education, and humanitarian aid. The fifth cluster of multilateral support looks at positive and negative roles enacted by external governmental and non-governmental actors. Remarkably, a great number of analysts focus on the domestic scene and its close neighbourhood. The donor community has the propensity to overlook or underestimate the role of the global international political and economic environment, especially the impact of imbalances of institutionalized military, political, economic, and cultural power and the roles of lobbies, interest groups, diasporas, and extralegal arms, drugs, and people dealers. Falk (1999) and Mearsheimer/Walt (2006) deal with these. Some label such research as politically incorrect, others as critical and emancipatory. In any case, conflict analysis and peacebuilding deficiency assessment would benefit from domination-free discourse and analysis.

---

[3]Freedom House, International Institute for Electoral Democracy and Electoral Assistance (EDEA). See at: https://freedomhouse.org/report-types/freedom-world.

13.2.2.3 Peacebuilding Potential

Lederach (1997), Anderson (1999), and others stress the importance of identifying the available and potential peacebuilding socio-economic and socio-cultural resources.

13.2.3 The Context

The lack of universal formulae and the complexity of conflicts require the development of a high level of context sensitivity. This requires a deep appreciation of the impact of the context on the peacebuilding process and vice versa. Contextual judgement can be more important than knowledge of the ten best peacebuilding practices in other situations. The contextual features are: scope, time, preservation, diversity, capability, capacity, readiness for change, and power (Balogun/Hailey 1999). *Scope*: Does the change affect the whole country as well as all sectors and levels, or does it impact only on part of the country or a particular sector? Does peace imply a radical transformation, a reconstruction or a realignment of the situation? *Time*: How much time does the peacebuilder have to build peace? Are the stakeholders expecting short-term results from the intervention? Do they see their intervention as crisis management or as a long-term peacebuilding process? Efforts have been undertaken to integrate crisis management with peacebuilding. *Preservation*: To what extent is it essential to maintain continuity in certain practices or preserve specific assets? Do these practices and/or assets constitute invaluable resources, or do they contribute towards a valued stability or identity within a country? One of the mistakes made by Paul Bremer (Director of Reconstruction and Humanitarian Assistance for post-war Iraq) was to disband the Iraqi army and equate the Baath party with Saddam Hussein. *Diversity*: Is the group of actors involved in the peacebuilding process diverse or relatively homogeneous in terms of its values, norms, and attitudes? Are there many cultures or subcultures within the country? Are there conflicting and common interests? Ramsbotham et al. (2006) identify three types of responses by peace researchers to the cultural variation in conflict zones: not relevant, should be taken into account, and is fundamentally significant in peacebuilding. *Capability*: How capable or competent are the peacebuilders at managing the peacebuilding process? Are the necessary kinds of expertise (internal and/or external) available? Is there enough expertise at the policy, management, and individual levels? Ingelstam (2001), Perrigo/Pearce (2005), and Fitzduff (2006) have been researching the qualifications expected and required of those involved in peacebuilding activities. Related to the search for qualities is the study of successful mediators and peacebuilding leadership (Reychler/Stellamans 2005). *Capacity*: What peacebuilding tools are available? What financial and human resources are available for peacebuilding? Lund/Mehler

(1999) have mapped measures and tools to remedy peacebuilding deficiencies. *Readiness*: Are the external actors willing and motivated for peacebuilding? How much support (domestic and international) is there for change? Are the internal actors ready for change? *Power*: Who are the major stakeholders? How much power do they have? Who are the stakeholders whose support must be canvassed? Sustainable peacebuilding requires not only hard, soft, and smart power, but above all integrative power. Integrative power is the power that binds humans together. Though it is seldom studied or discussed, Boulding (1989) argues that it is the strongest form of power, especially because exchange and coercive power cannot operate without integrative power.

13.2.4 The Planning of the Peacebuilding Process

The fourth component is the planning of the peacebuilding process. This is one of the most fascinating and complex areas of study, with a long way to go in terms of further research. In this component of the architecture of peacebuilding, several choices need to be made about how to build peace. This relates to the framing of time, entry and exit, priority-setting, pacing the process, creating synergies, and anticipating and reducing negative side-effects.

13.2.4.1 Framing Time

This involves choices about differentiating phases in the process, and the framing of the building process as a linear, circular, or procedural activity (Murnighan/Mowen 2002) All the authors perceive peacebuilding to be a *multi-phased* process, each phase characterized by its own priorities. The *Deutsche Gesellschaft fur Technische Zusammenarbeit* (2005) makes a distinction between stabilization (1–3 years), reorganization (4–7 years), and consolidation (8–10 years) The Centre for Strategic and International Studies (2002) works with three phases: initial response (short-term), transformation (mid-term), and fostering sustainability (long-term); no exact timelines are given to each phase. The peacebuilding strategy of the US Department of State (2005) uses the same peacebuilding phases. The New Partnership for Africa's Development (2005) frames post-conflict reconstruction in three phases: emergency (90 days-1 year), transition (1–3 years), and development (4–10 years). Lederach (1997) uses a four-phased approach: the crisis and issues stage (2–6 months); the people and relationships stage (1–2 years); the institutions or subsystem phase (5–10 years); and finally the phase that needs work for generations – the vision of peace and the desired future that all hope for and move towards. Lederach opts for a non-linear, procedural perspective; NEPAD uses a more linear perspective with clearly defined timetables.

13.2.4.2   Entry and Exit

The entry-exit decision has facets, such as when to intervene, the expected exit, when and how to exit (instant vs. phased withdrawal), assessing the impacts of withdrawal, and the choice of follow-up arrangements. According to Caplan (2005), "A good exit strategy depends on good entrance and intermediate strategies. An exit strategy cannot compensate, easily or at all, for major deficiencies in the design or implementation of a territorial administration, but by the same token, a poorly conceived exit strategy can jeopardize the achievements of the international administration and imperil the viability of the new state or territory." Chesterman (2004) focuses on the timing of the elections, criticizes the timing of the Dayton peace agreement, which provided for elections to be held between six and nine months after the conclusion of the peace, and the perception that the troops would be home in a year. "After the elections, politics became the continuation of war by other means."

13.2.4.3   Pacing the Peacebuilding Process

Changes can be implemented, either in an all-at-once, big bang fashion, or in a more incremental, step-by-step, stage-by-stage fashion. The interventions in Bosnia and the latest war in Iraq were handled in a big bang fashion, but each turned into "operation creep". Most intervention tends to take time and be handled in incremental ways.

13.2.4.4   Setting Priorities

In the different phases of the conflict transformation, which tasks get priority or are allocated more resources and time than others? Although there is a general consensus on the need for complementarity, most authors tend to prioritize one or more areas of intervention (Llamazares 2005). First there are analysts who give security priority; they claim that peacebuilding doesn't go anywhere without basic security. Security is considered the key to successful post-war peacebuilding and vital for freedom of movement, the absence of personal or group threats, and safe access to resources in the post-war setting. Schnabel/Ehrhart (2005), for example, prioritize efforts to reduce the military/security deficit foremost so that internal security structures become an asset and not a liability in the long-term peacebuilding process. A second group of analysts sees economic development as the path to success and claims that economic vulnerability should be tackled from the beginning. For a third group, social welfare and civil society is of vital importance for the regeneration of societies and peacebuilding. Pugh (2000) points to imbalance between short-term, hard, visible reconstruction measures and soft, long-term social-civil programmes. A fourth group stresses the economic agendas of war as a key source of conflict. Collier/Sambanis (2005) state that good peacebuilding must include disincentives for those benefiting from war in order to reduce their influence over the process. A fifth group claims that priority should be given to the remediation of the political and institutional deficits. Cousens et al. (2000) consider the "fragility or collapse of political processes and institutions" to be the main catalyst for war. A sixth group highlights the importance of justice and reconciliation. Lambourne (2004b) argues that both justice and reconciliation are fundamentally significant goals that need to be addressed in the design of successful post-conflict peacebuilding processes and mechanisms, especially in the aftermath of genocide. Finally, there is the social-psychological approach. Rothstein (1999) points out the value of psychological and emotional components in the resolution of protracted conflicts.

13.2.4.5 Synchronicity and Sequencing

Are all the tasks implemented synchronically or is there a clear sequencing of the efforts? This is one of the least systematically researched aspects of the peacebuilding process. Several approaches can be distinguished: (1) the free-for-all approach: the underlying assumption is that more peacebuilding interventions will add up to more peace, (2) the ideology-driven approach, based on a belief in the primacy of security, development, democracy, or other types of interventions in peacebuilding; (3) the power-driven approach, which claims that power makes or breaks peace; (4) the theory-driven approach, based on the research of successful and unsuccessful sequencing of different activities within and between different sectors; (5) the reconciliation-driven approach, based on the belief that competing views and values need to be reconciled. Paris (2004: 289) claims that pushing war-shattered states into stable market democracies too quickly can have damaging and destabilizing effects. A sensible approach would be to establish a system of domestic institutions capable of managing the disruptive effects of democratization and marketization in a first phase, and only then to phase in political and economic reforms as conditions warrant. Mansfield/Snyder (1995) support Paris's vision that fast democratization is susceptible to instability. Furthermore, economic gains in the medium and long term can be created if, in the short term, macro-economic policies are socially sensitive (Collier et al. 2003).

13.2.4.6 Negative and Positive Cross-Impacts or Synergies

How much attention is paid to the positive and negative cross-impacts of efforts in different sectors and at different levels? Have the impacts been assessed proactively? The assessment of peace and conflict impacts is not new. In December 1919, J. Keynes's *Economic Consequences of Peace* appeared on the bookstalls. Keynes argued that the terms of the Versailles Treaty would be disastrous for both Germany and its allies. At the end of the book he presented an alternative policy – something like a Marshall Plan – providing Germany with resources that would enable it to pay a reasonable amount of restitution, but also to recover economically and socially.

13.2.5   Peacebuilding Coordination

All peacebuilders are interdependent, in that they cannot achieve peace by themselves (Lederach 1997).

13.2.5.1   Coherence Deficit and Dilemma

Despite growing demands for working with an integrated framework and coordinating peace efforts, there is still a coherence deficit. The Upstein study of peacebuilding, which analysed 336 peacebuilding projects in Germany, the Netherlands, the United Kingdom, and Norway in the 1990s, showed a lack of coherence at the national strategic level (Smith 2004). Similar findings were identified at the international level and in fragile states (United Nations 2006; Patrick/Brown 2007). Jones (2002) identifies three types of coordination problems in deadly conflicts: divergent and diffuse efforts in Bosnia, conflicting strategies in Rwanda, and a fragmented international presence in Burundi. Inadequate coordination increases the risks of duplication, inefficient spending, a lower quality of sendee, difficulty in meeting the goals, and a reduced capacity for delivery (De Coning 2008). On the other hand, greater coordination and coherence do not automatically mean better peacebuilding operations. This has been called the "coherence dilemma". De Coning (2008) highlights some potential negative side-effects: short-term political and security considerations may override long-term socio-economic rehabilitation; there may be undue pressure on internal actors; and the neutrality of humanitarian action may be negatively affected. Part of the research tries to identify the obstacles to fruitful coordination. Minear (2002: 22–32), for example, lists five factors which could inhibit effective and efficient coordination: a lead agency's lack of power (sticks and carrots), a lack of visibility and an inability to mobilize resources, high costs, the existence of ineffective structures (a multiplicity of actors who act autonomously), and the lack of leadership.

13.2.5.2   Dimensions of Coordination

An assessment of cooperation and coordination in peacebuilding implies an analysis of the (1) spaces of coordination, (2) participation, (3) elements of coordination, (4) degree of coordination, and (5) strategy formation.

Spaces of Coordination

Most analysts distinguish between four coordination spaces: (a) agency coordination or consistency between the politics and actions of an individual agency,

(b) whole-government coordination or consistency among the policies of different agencies in a country, (c) external-donor coherence or consistency between the policies pursued by external actors in a country, and (d) internal-external coordination, or consistency between the policies of internal and external actors in a conflict zone (Picciotto 2005; Owen/Travers 2007; De Coning 2008). Caplan (2005) emphasizes that coordination is needed at different levels: the strategic level, the tactical level, and the field level.

Participation

Who is involved in the peacebuilding process? What about local ownership? There is broad consensus on the need to involve inside and outside actors in most peace settlements. Hampson (1996) assessed the impact of several factors based on the success or failure of peace settlement negotiations and concluded that third-party intervention contributed greatly to successful post-settlement peacebuilding. Caplan (2005) observes that a minimum of local ownership is needed in a transition regime. Without local ownership, it is difficult to develop political responsibility; the wrong lessons would be remembered by the local population and the legitimacy of the transition process be called into question. This is illustrated by the events in Bosnia and Herzegovina, where the High Representative could dismiss elected or appointed officials. The HR could remove anyone from office who, in his view, was obstructing the implementation of the Dayton accord. This has been called despotic or transnational authoritarianism. Large (1998) believes that the unique resources brought by local actors to the process make it imperative that meaningful participation take place. Despite growing awareness of the links between gender-sensitive approaches and more sustainable and participatory responses in conflict-affected contexts, in current EU interventions, women continue to be marginalized in peacebuilding initiatives (Barnes/Lyytikainen 2008).

Elements of Coordination

Coordination can be focused on the articulation and implementation of the overall peacebuilding strategy and the operational/tactical level of field operations. To get a better understanding of the degree of coherence achieved in peacebuilding operations, it would be useful to study the nature of coordination and coherence in the six components of peacebuilding architecture: (1) the end state or definition of peace and theoretical assumptions about the preconditions to achieve that peace, (2) the baseline or analysis of the conflict and the peacebuilding deficiency, (3) the analysis of the context, (4) the planning of the peacebuilding process, (5) the nature of coordination, and (6) the monitoring and evaluation of the impact of the peacebuilding efforts (Reychler 2009).

The Degree of Coordination

Jones/Cherif (2003) distinguish between integrated, coordinated, parallel, and sequential peace operations. In integrated operations the full scope of operations is managed within a single chain of command. De Coning uses a scale going from coherence, cooperation and collaboration to coexistence. Coherence, the highest degree of coordination, refers to a coalition that acts upon a standard mandate, strategic vision, and objectives (Friis/Jarmyr 2008). Jordan/Schout (2006) measure coordination on a nine-level Metcalfe scale: independent policy-making, exchange of information, consultation, speaking with one voice, looking for consensus, conciliation, arbitration, setting margins, and working towards a specified objective.

Strategy Formation

Choices also need to be made about the management of the peacebuilding process. Hart (1992) identifies five modes of strategy-formation processes. This framework is built around who is involved in the strategy formulation and in what manner. In the *command mode,* a strong leader controls the process. The strategy is a conscious, controlled process that is centralized at the top. The end state, the baseline, and alternatives are considered, and an appropriate course of action is decided upon and implemented. This strategy formation mode can vary from being directive to coercive (using power to impose change) (Balogun/Hailey 1999). The *symbolic mode* involves the creation, by the actors who take the lead, of a clear and compelling vision and mission. The major task is to motivate and inspire and to provide the necessary focus to guide the creative actions of the actors involved. Education and communication are core activities. This mode requires a great deal of participation and commitment. The *rational mode* is a theory-driven strategy formation. Strategy is developed through formal analysis (and information processing) and strategic planning. The *transactive mode* is based on interaction and learning rather than on the execution of a predetermined plan. Strategy is crafted based upon an ongoing dialogue with the key stakeholders. Cross-sector and cross-level communication between the actors involved is very important in this mode. The last mode of strategy formation is the *generative mode.* This mode depends on the autonomous initiatives of the actors involved in the peacebuilding process. The donor community selects and nurtures initiatives with high peace potential.

These *ideal* types are not exclusive. In many cases, one notices a combination of several of these modes. The choice is influenced by several factors: the power relations between the actors, the level of complexity of the peacebuilding plan, the heterogeneity of the conflict environment, the phase the conflict is in, etc. Donini (1996) distinguishes between three types of strategy formulation and implementation: (1) coordination by command, (2) coordination by consensus, and (3) coordination by default. Some analysts, like Minear (2002), argue in favour of the coordination by command approach; others, like Stephenson/Kehler (2004), prefer coordination by consensus. Some researchers have focused on the unilateral versus multilateral organization of external interventions. Dobbins et al. (2003, 2005) observe that multiplicity tends to lead to more complex and time-consuming decision-making than the unilateral approach. The activities could be highly atomized and the administration unwieldy. Caplan (2005) stresses that in a postcolonial age it has become politically unacceptable (and too expensive) to entrust responsibility for the administration of a territory to a single state, even if elaborate accountability mechanisms would be created: "Although the US drew in other states to share the responsibility of administering Iraq, precisely in an effort to confer legitimacy on the interim regime, the dominant role played by a major western power is one reason why it encountered such fierce resistance." Mullenbach (2005) saw some evidence that the risk of military hostilities is at least somewhat lower when the UN or a regional IGO coordinates a multidimensional peacebuilding mission.

Coordination Mechanisms and Structures

Most researchers have analysed and evaluated existing coordination mechanisms and structures, and some have generated alternative models of coordination. There is, for example, a considerable amount of research into the United Nations peacebuilding and integrated missions. Jones (2002) lists among the successful cases the role of the Special Representatives of the Secretary General (SRSGs), the continuity of key actors, the role of friend groups, and coordination mechanisms. Jordan/Schout (2006) have produced an interesting and critical analysis of the coordination in the European Union. Ricigliano (2003) introduced the concept of a Network of Effective Action (NEA) as a set of practices for collaboration that is capable of facilitating integrated approaches to peacebuilding both on the ground and in terms of the theoretical development of the field.

13.2.6 Monitoring and Evaluation

Without an effective system for monitoring and evaluating the impact of interventions on the conflict dynamic and peacebuilding, it is difficult to adapt to new challenges and unpredictables effectively and to learn from experience. Evaluation of the conflict and the peace impact of peacebuilding interventions has become common practice. Evaluations have been done before, during and after intervention, and focus on different levels and sectors. Most evaluations look at part of the big picture. There are no macro evaluations of the peacebuilding activities in different sectors of all the major actors (internal and external) in a particular conflict setting. An essential part of the evaluation is the selection of objective criteria for evaluating the process and of benchmarks for progress and success. The OECD Development Assistance Committee (DAC) proposes nine criteria to assess conflict prevention and peacebuilding activities. (1) *Relevance/appropriateness*: Is the intervention based on an accurate analysis of the conflict and does it deal with the driving factors? (2) *Effectiveness*: Has the intervention achieved its stated purpose? (3) *Efficiency*: Were the resources used in an economic way? (4) *Impact*: What were the positive and negative impacts on the conflict and peace of specific interventions? (5) *Sustainability*: Will the hard-won results persist when the intervention stops? (6) *Connectedness*: Are there linkages between macro changes and individual/personal changes? (7) *Coherence*: Is there enough consistency or positive synergy between the interventions? (8) *Coverage*: Are there still hidden conflicts? (9) *Consistency*: Is the intervention consistent with conflict prevention and peacebuilding values? Has it succeeded in reconciling competing values (OECD/DAC).

More criteria could be added, such as the participation and ownership of national/local owners and stakeholders, the clarity of the definition of the preferred peace, and the validity of the underlying assumptions about how to realize that peace. The research community has contributed to evaluation by operationalizing the criteria used to assess good peacebuilding, developing methods for monitoring and evaluating interventions (Earl et al. 2001; Paffenholz/Reychler 2007) and studying the problems of researching in violently divided societies (Smyth/Robinson 2001).

13.3 Conclusion

The nexus between peacemaking, political change, development, peacekeeping, building, and reconciliation has become a central focus of the research, and peacebuilding the common framework within which the interactions between the activities are studied. Peacebuilding involves high-stake decisions that must be made when information is ambiguous, values conflict, and experts disagree. The research relates to six areas where decisions, choices, and judgements have to be made, regarding (1) the definition of the peace and the theoretical assumptions of peacebuilding (the end state), (2) the conflict and the peacebuilding deficiencies at the baseline, (3) the context, (4) the planning of the peacebuilding process, (5) the coordination of the process, and (6) the monitoring and evaluation of the intervention. Despite the progress made, there remain big gaps and challenges. Many analysts, for example, leave the end state vague and implicit and make no systematic differentiation between different types of peace. There is convergence in the identification of peacebuilding conditions, but the research community still needs to distil from the vast and scattered reservoir of knowledge a comprehensive and valid theory of peacebuilding. Considerable progress has been made in the analysis of the baseline, including conflict analysis, early warning, and assessment of the peacebuilding potential and the difficulty of conflict transformation. With respect to the context, two salient issues require more attention: the qualities of a peacebuilder and the role of integrative power. The widest research gap is found in the planning of the peacebuilding process. It is one of the most fascinating and complex areas of study, relating to the framing of time, entry and exit, priority-setting, pacing the process, synchronicity and sequencing, and positive and negative synergies. Higher-quality information and methodology for analysing complex dynamic behaviour are urgently needed. The fifth and sixth components, peacebuilding coordination and monitoring and evaluation, have recently experienced a boost in attention and produced new insights and methodologies.

More scientific research would help to shape and create more effective, sustainable peacebuilding policies. A better exchange between researchers, practitioners and decision-makers could raise the learning curve. This would involve overcoming several obstacles. First, diplomats and politicians often deride academics' lack of first-hand experience when it comes to the practice of managing conflicts and peacebuilding. They are perceived as being out of touch with the realities of a rapidly changing international landscape. This contains some truth, but distance can also be an advantage. The view from the academic balcony allows one to reflect dispassionately on perturbing foreign policy problems, to discern underlying patterns of behaviour, to anticipate future threats, and to forecast the consequences of different policy options. Second, there is the problem of *slow institutional learning*. Some countries learn, while others have a flat learning curve. A third obstacle that limits the impact of researchers is the *diminution of academic freedom* in democratic countries. In the academic environment, especially in the humanities, the incentives for transdisciplinary research remain very poor. The political environment, especially with respect to conflict, peace, and security issues, has had an extremely negative impact on academic freedom, in the form of political correctness, the influence of spin doctors, unspeakable truths, the use of euphemisms, and confusing language and taboos. Some scholars get around politically sensitive issues by engaging in pure theorizing and methodological correctness. The last obstacle is the foreign policy and security decision-making process, which is low on democratic checks and balances. All of this makes critical and sustainable peace theorizing essential.

References

Alger, C. (2007). Peace Studies as a Transdisciplinary Project. In C. Webel & J. Galtung (Eds.), *Handbook of Peace and, Conflict Studies* (pp. 299–318). London: Routledge.

Anderson, M. (1999). *Do No Harm: How Aid Can Support Peace or War.* Boulder: Lynne Rienner.

Augsburger, D. W. (1992). *Conflict Mediation across Cultures: Pathways and Patterns.* Louisville: John Knox.

Azar, E. (1990). *The Management of Protracted Social Conflict: Theory and Cases.* Aldershot: Dartmouth.

Balogun, J., & Hailey, V. H. (1999). *Exploring Strategic Change.* New York: Prentice Hall.

Barnes, K., & Lyytikainen, M. (2008). *Improving EU Responses to Gender and Peace Building: Priority Action Areas for the European Commission.* London: International Alert.

Bercovitch, J. (Ed.). (1996). *Resolving International Conflicts: The Theory and Practice of Mediation.* Boulder: Lynne Rienner.

Berdal, M., & Economides, S. (Eds.). (2007). *United Nations Interventionism 1991–2004*. Cambridge: Cambridge University Press.

Booth, K. (2007). *Theory of World Security*. Cambridge: Cambridge University Press.

Boulding, E. (1991). The Challenge of Imaging Peace in Wartime. *Futures 23*(5), 528–33.

Boulding, K. (1978). *Stable Peace*. Austin: University of Texas Press.

Boulding, K. (1989). *Three Faces of Power*. Newbury Park, CA: Sage.

Boutros-Ghali, B. (1992). *An Agenda for Peace*. New York: United Nations.

Brown, M., & Rosecrance, R. (1999). *The Costs of Conflict: Prevention and Cure in the Global Arena*. New York: Rowman and Littlefield.

Burgess, G., & Burgess, H. (n.d.). The Beyond Intractability Knowledge Base Project. Retrieved from https://www.beyondintractability.org/, accessed March 2009.

Burton, J. (1969). *Conflict and Communication: The Use of Controlled Communication in International Relations*. London: Macmillan.

Burton, J. W. (1987). *Resolving Deep Rooted Conflict: A Handbook*. Lanham: University Press of America.

Caplan, R. (2005). *International Governance of War-Torn Territories: Rule and Reconstruction*. Oxford: Oxford University Press.

Center for Strategic and International Studies. (2002). *Post-Conflict Reconstruction: Task Framework*.

Chesterman, S. (2004). *You, The People: The United Nations, Transitional Administration, and State-Building*. Oxford: Oxford University Press.

Chopra, J. (1999). *Peace-Maintenance: The Evolution of International Political Authority*. London: Routledge.

Collier, P., & Hoeffler, A. (2004). Greed and Grievance in Civil War. *Oxford Economic Papers, 56*, 563–595.

Collier, P., & Sambanis, N. (2005). *Understanding Civil War: Evidence and Analysis (Africa)*. Washington, DC: World Bank.

Collier, P., Elliott, V. L., Hegre, H., Hoeffler, A., Reynal-Querol, M., & Sambanis, N. (2003). *Breaking the Conflict Trap: Civil War and Development Policy*. Washington, DC: World Bank.

Cooperrider, D., & Whitney, D. (1999). *Appreciative Inquiry*. San Francisco: Berret-Koeller.

Cousens, E., Kumar, C., & Wermester, K. (Eds.). (2000). *Peacebuilding as Politics: Cultivating Peace in Fragile Societies*. Boulder: Lynne Rienner.

Crocker, C. A., Hampson, F. O., & Aall, P. (2004). *Taming Intractable Conflicts: Mediation in the Hardest Cases*. Washington, DC: United States Institute of Peace Press.

Curie, A. (1971). *Making Peace*. London: Tavistock.

Daniel, D. (2008a). Why So Few Troops from among So Many? In D. Daniel, P. Taft, & S. Wiharta (Eds.), *Peace Operations: Trends, Progress and Prospects* (pp. 47–62). Washington, DC: Georgetown University Press.

Daniel, D. (2008b). Conclusion. In D. Daniel, P. Taft, and S. Wiharta (Eds.), *Peace Operations: Trends, Progress and Prospects* (pp. 221–232). Washington, DC: Georgetown University Press.

Darby, J., and Mac Ginty, R. (Eds.). (2008). *Contemporary Peacemaking: Conflict, Peace Processes and Post-War Reconstruction*. Basingstoke: Palgrave Macmillan.

Debiel, T., & Terlinden, U. (2005). *Promoting Good Governance in Post-Conflict Societies* (Discussion paper). Eschborn: Deutsche Gesellschaft fur Technische Zusammenarbeit.

De Coning, C. (2008). *Coherence, and Coordination in United Nations Peacebuilding: A Norwegian Perspective*. Oslo: NUPI.

Deutsch, D. (1957). *Political Community and the North Atlantic Area*. Princeton: Princeton University Press.

Diamond, L., & McDonald, J. W. (1996). *Multi-Track Diplomacy: A System's Approach to Peace*. West Hartford, CT: Kumarian.

Dobbins, J., Jones, S. G., Crane, K., Rathmell, A., Steele, B., Teltschik, R., & Timilsina, A. (2005). *The UN's Role in Nation-Building: From the Congo to Iraq*. Santa Monica: RAND.

Dobbins, J., Jones, S. G., Crane, K, & DeGrasse, B. (2007). *The Beginner's Guide to Nation-Building*. Santa Monica: RAND.

Dobbins, J., McGinn, J. G., Crane, K, Jones, S. G., Lai, R., Rathmell, A., Swanger, R., & Timilsina, A. (2003) *America's Role in Nation-Building: From Germany to Iraq*. Santa Monica: RAND.

Donini, A. (1996). *The Policies of Mercy: UN Coordination in Afghanistan, Mozambique, and Rwanda* (Occasional Paper No. 22). Providence, RI: Thomas J. Watson Institute for International Studies.

Doyle, M. W. (2002). Strategy and Transitional Authority. In S.J. Stedman, D. Rothchild, & E.M. Cousens (Eds.), *Ending Civil Wars: The Implementation of Peace Agreements* (pp. 71–88). Boulder: Lynne Rienner.

Duffield, M. (2001). *Global Governance and the New Wars: The Merging of Development and Security*. London: Zed Books.

Earl, S., Carden, F., & Smutylo, T. (2001). *Outcome Mapping: Building Learning and Reflection into Development Programs*. Ottawa: IDRC.

Elbe, S. (2007). HIV/AIDS and Security. In A. Collins (Ed.), *Contemporary Security Studies* (pp. 331–345). Oxford: Oxford University Press.

European Commission External Relations Directorate-General. (2008). *From Early Warning to Early Action?* Luxembourg: Office for Official Publications of the European Communities.

EU Commission. (1996). The European Union and the Issue of Conflicts in Africa: Peace-building, Conflict Prevention and Beyond. *SEC (96)*332 (6 March).

Falk, R. (1999). *Predatory Globalization: A Critique*. Cambridge: Polity.

Farrell, T. (2007). Humanitarian Intervention and Peace Operations. In J. Baylis, J. Wirtz, C.S. Gray, & Cohen, E. (Eds.), *Strategy in the Contemporary World: Introduction to Strategic Studies* (pp. 286–308). New York: Oxford University Press.

Fein, H. (1993). *Genocide: A Sociological Perspective*. London: Sage.

Fischer, D., Nolte, W., & Oberg, J. (1989). *Winning Peace: Strategies and Ethics for a Nuclear-Free World*. London: Crane Russak.

Fisher, R., & Ury, W. (1981). *Getting to Yes*. Boston: Houghton Mifflin.

Fitzduff, M. (2006). *Core Competences for Graduate Programs in Coexistence and Conflict Work – Can We Agree?* (Woodrow Wilson Center Leadership Notes Series No. 1). Waltham: Brandeis University.

Friis, K., & Jarmyr, P. (2008). *Comprehensive Approach: Challenges and Opportunities in Complex Crisis Management*. Oslo: NUPI.

Galtung, J. (1981). Social Cosmology and the Concept of Peace, *Journal of Peace Research, 18* (2), 183–199.

Giraldo, J., & Trinkunas, H. (2006). Transnational Crime. In A. Collins (Ed.), *Contemporary Security Studies* (pp. 346–366). Oxford: Oxford University Press.

Gurr, T. R., & Harff, B. (1994). Early Warning of Communal Conflicts and Humanitarian Crises. *Journal of Ethno-Development* (Special Issue), *4*(1).

Hampson, F. O. (1996). *Nurturing Peace: Why Peace Settlements Succeed or Fail*. Washington, DC: United States Institute of Peace Press.

Hart, S. (1992). An Integrative Framework for Strategy Making Processes. *Academy of Management Review, 17*(2), 327–351.

Heldt, B. (2008). Trends from 1948 to 2005: How to View the Relation between the United Nations and Non-UN Entities. In D. Daniel, P. Taft, & S. Wiharta (Eds.), *Peace Operations: Trends, Progress and Prospects* (pp. 9–26). Washington, DC: Georgetown University Press.

Hoglund, K. (2008). *Peace Negotiations in the Shadow of Violence*. Leiden: Martinus Nijhoff.

Ingelstam, M. (2001). Motivation and Qualifications, interviewed by K. Kruhonja. In L. Reychler & T. Paffenholz (Eds.), *Peace Building: A Field Guide* (pp. 16–20). Boulder: Lynne Rienner.

Jeong, H. W. (2005). *Peace Building in Postconflict Societies*. Boulder: Lynne Rienner.

Jones, B. (2002). The Challenges of Strategic Coordination. In S.J. Stedman, D. Rothchild, & E. M. Cousens (Eds.), *Ending Civil Wars: The Implementation of Peace Agreements* (pp. 89–115). Boulder: Lynne Rienner.

Jones, B., & Cherif, F. (2003). *Evolving Models of Peacekeeping: Policy Implication and Responses* (External Study United Nations Best Practices Section).

Jordan, A., & Schout, A. (2006). *The Coordination of the European Union: Exploring Capacities of Networked Governance.* Oxford: Oxford University Press.

Kerr, P. (2007). Human Security. In A. Collins (Ed.), *Contemporary Security Studies* (pp. 91–108). Oxford: Oxford University Press.

Kriesberg, L., Northrup, A., & Thorson, S. J. (Eds.). (1989). *Intractable Conflicts and their Transformation.* Syracuse: Syracuse University Press.

Lambourne, W. (2004a, March). Challenging the Hegemonic Discourse of Peacebuilding Theory and Practice. Paper Presented at the *Annual Meeting of the International Studies Association,* Montreal, Canada.

Lambourne, W. (2004b). Post-Conflict Peacebuilding: Meeting Human Needs for Justice and Reconciliation. *Peace, Conflict and Development, 4,* 1–24.

Large, J. (1998). *The War Next Door: A Study of Second-Track Intervention during the War in Ex-Yugoslavia.* Stroud: Hawthorn.

Lederach, J. P. (1997). *Building Peace: Sustainable Reconciliation in Divided Societies.* Washington, DC: United States Institute of Peace.

Lindner, E. (2006). *Making Enemies and Humiliation in International Conflicts.* London: Greenwood.

Lindner, E. (2009). *Emotion and Conflict: How Human Rights Can Dignify Emotion and Help Us Wage Good.* London: Greenwood.

Llamazares, M. (2005). *Post-War Peacebuilding Reviewed: A Critical Exploration of Generic Approaches to Post-War Reconstruction* (Centre for Conflict Resolution Working Paper No. 14). Bradford: University of Bradford.

Lund, M., & Mehler, A. (1999). *Peace-Building and Conflict Prevention in Developing Countries: A Practical Guide.* Brussels: CPN Guidebook.

Mac Ginty, R. (2006). *No War, No Peace: The Rejuvenation of Stalled Peace Processes and Peace Accords.* Basingstoke: Palgrave Macmillan.

MacQueen, N. (2006). *Peacekeeping and the International System.* London: Routledge.

Mansfield, E. D., & Snyder, J. (1995). Democratization and the Danger of War. *International Security, 20*(1), 5–38.

Marten, K. Z. (2004). *Enforcing the Peace: Learning from the Imperial Past.* New York: Columbia University Press.

Mason, D. T., & Meernik, J.D. (Eds.). (2006). *Conflict Prevention and Peacebuilding in Post-War Societies: Sustaining the Peace.* London: Routledge.

Mearsheimer, J. J., & Walt, S. (2006). The Israel Lobby. *London Review of Books, 28*(6), 3–12.

Minear, L. (2002). *The Humanitarian Enterprise: Dilemmas and Discoveries.* Bloomfield, CT: Kumarian.

Mitchell, C., & Banks, M. (1996). *Handbook of Conflict Resolution: The Analytical Problem-Solving Approach.* London: Pinter.

Moisi, D. (2008). *La Geopolitique de Temotion.* Paris: Flammarion.

Mullenbach, M. (2005). Deciding to Keep Peace: An Analysis of International Influences of the Establishment of Third Party Peacekeeping Mission. *International Studies Quarterly, 49*(3), 529–555.

Murnighan, J. K., & Mowen, J. C. (2002). *The Art of High Stakes Decision Making: Tough Calls in a Speed Driven World.* New York: John Wiley.

New Partnership for Africa's Development (NEPAD). (2005). *African Post-Conflict Reconstruction Policy Framework.* Retrieved from http://www.nepad.org/2005/aprmforum/PCRPolicyFramework_en.pdf, accessed March 2009.

Nordstrom, C. (2004). *The Shadows of War: Violence, Power, and International Profiteering in the Twenty-First Century.* Berkeley: University of California Press.

OECD/DAC. (2007). *Encouraging Effective Evaluation of Conflict Prevention and Peace Building Activities: Toward DAC Guidance.* Paris: OECD.

Orr, R. (Ed.). (2004). *Winning the Peace: An American Strategy for Post-Conflict Reconstruction.* Washington, DC: CSIS Press.

Owen, T., & Travers, P. (2007). Peacebuilding while Peacemaking: The Merits of a 3D Approach. Paper presented at the *48th International Studies Association Convention,* Chicago.

Paffenholz, T., & Reychler, L. (2007). *Aid for Peace: A Guide to Planning and Evaluation for Conflict Zones.* Baden-Baden: Nomos.

Paris, R. (1997). Peace Building and the Limits of Liberal Internationalism. *International Security, 22*(2), 54–89.

Paris, R. (2004). *At War's End: Building Peace after Civil Conflict.* Cambridge: Cambridge University Press.

Patrick, S., & Brown, K. (2007). *Greater than the Sum of its Parts: Assessing "Whole of Government" Approaches to Fragile States.* New York: International Peace Academy.

Perrigo, S., & Pearce, J. (2005). *European Doctoral Enhancement in Peace and Conflict Studies Qualification Framework* (Internal Discussion Paper). Bilbao: University of Deusto.

Picciotto, R. (2005). *Fostering Development in a Global Economy: A Whole Government Perspective.* Paris: OECD.

Pugh, M. (2000). *Regeneration of War-Torn Societies.* London: Macmillan.

Ramcharan, B. G. (2008). *Preventive Diplomacy at the UN.* Bloomington: Indiana University Press.

Ramsbotham, O., Woodhouse, T., & Miall, H. (2006). *Contemporary Conflict Resolution.* Oxford: Polity Press.

Reiman, H. (1992). Informal Mediation by the Scholar/Practitioner. In J. Bercovitch & J. Rubin (Eds.), *Mediation in International Relations: Multiple Approaches to Conflict Management* (pp. 64–95). London: Macmillan.

Reychler, L. (2004). Peace Architecture: The Prevention of Violence. In A. H. Eagly, R. M. Baron, V. L. Hamilton (Eds.), *The Social Psychology of Group Identity and Social Conflict: Theory, Application, and Practice* (pp. 133–146). Washington, DC: American Psychological Association.

Reychler, L. (2009). Sustainable Peace Building. In N. Young (Ed.), *The Oxford International Encyclopedia of Peace.* Oxford: Oxford University Press.

Reychler, L., & Carmans, J. (2006). *Violence Prevention and Peace Building: A Research Agenda* (Cahiers of the Centre of Peace Research and Strategic Studies No. 74). Leuven: Centre of Peace Research and Strategic Studies.

Reychler, L., & Langer, A. (2006). *Researching Peace Building Architecture* (Cahiers of the Centre for Peace Research and Strategic Studies No. 75). Leuven: Centre for Peace Research and Strategic Studies.

Reychler, L., & Stellamans, A. (2005) *Researching Peace Building Leadership* (Cahiers of the Centre for Peace Research and Strategic Studies No. 71). Leuven: Centre for Peace Research and Strategic Studies.

Reychler, L., Renckens, S., Coppens, K., & Manaras, N. (2008). *A Codebook for Evaluating Peace Agreements* (Cahiers of the Centre for Peace Research and Strategic Studies No. 83). Leuven: Centre for Peace Research and Strategic Studies.

Ricigliano, R. (2003). Networks of Effective Action: Implementing an Integrated Approach to Peacebuilding, *Security Dialogue, 34*(4), 445–462.

Rothman, J. (1992). *From Confrontation to Cooperation: Resolving Ethnic and Regional Conflict.* Princeton: Princeton University Press.

Rothstein, R. L. (Ed.). (1999). *After the Peace: Resistance and Reconciliation.* Boulder: Lynne Rienner.

Ryan, S. (2007). *The Transformation of Violent Intercommunal Conflict.* Aldershot: Ashgate.

Sampson, C., Abu-Nimer, M., Liebler, C., & Whitney, D. (2003). *Positive Approaches to Peace Building.* Washington, DC: Pact.

Schnabel, A., & Ehrhart, H. G. (2005). *Security Sector Reform and Post-Conflict Peacebuilding.* Tokyo: United Nations University Press.

Senge, P., Kleiner, A., Roberts, C., Ross, R., & Smith, B. (1994). *The Fifth Discipline Fieldbook.* New York: Doubleday.

Smith, D. (2004). *Towards a Strategic Framework for Peacebuilding: Getting their Act Together* (Evaluation Report Royal Norwegian Ministry of Foreign Affairs). Oslo: PRIO.

Smyth, M., & Robinson, G. (Eds.). (2001). *Researching Violently Divided Societies.* Tokyo: United Nations University Press.

Stedman, S. (1997). Spoiler Problems in Peace Processes. *International Security, 22*(2), 5–53.

Stedman, S. J. (2001). International Implementation of Peace Agreements in Civil Wars: Findings from a Study of Sixteen Cases. In C. A. Crocker, F. O. Hampson, & P. R. Aall (Eds.), *Turbulent Peace: The Challenges of Managing International Conflict* (pp. 737–752). Washington, DC: United States Institute of Peace Press.

Stedman, S. J., Rothchild, D., & Cousens, E. M. (Eds.). (2002). *Ending Civil Wars: The Implementation of Peace Agreements.* Boulder: Lynne Rienner.

Steiner, B. H. (2004). *Collective Preventive Diplomacy: A Study in International Conflict Management.* New York: State University of New York Press.

Stephenson, M., & Kehler, N. (2004, May). Rethinking Humanitarian Assistance Coordination. Paper presented at the *International Society for Third Sector Research Sixth International Conference,* Toronto.

Stewart, F. (Ed.). (2008). *Horizontal Inequalities and Conflict: Understanding Group Violence in Multiethnic Societies.* Basingstoke: Palgrave Macmillan.

Tellis, A., Szayna, T., & Winnefeld, J. (1997). *Anticipating Ethnic Conflict.* Santa Monica: RAND.

United Nations (2006). *Delivering as One: Report of the UN Secretary General's High Level Panel on Systemwide Coherence.* New York: United Nations.

US Department of State. (2005). *Post-Conflict Reconstruction: Essential Tasks.* Office for the Coordinator of Reconstruction and Stabilization.

Verkoren, W. (2008). *The Owl and the Dove: Knowledge Strategies to Improve the Peace Building Practice of Local Non-Governmental Organizations* (Doctoral Dissertation). Amsterdam: University of Amsterdam.

Wallensteen, P. (2002). *Understanding Conflict Resolution: War, Peace and the Global System.* London: Sage.

Walter, B. F. (2002). *Committing to Peace: The Successful Settlement of Civil Wars.* Princeton: Princeton University Press.

Wilkinson, P. (2000a). *Peace Support under Fire: Lessons from Sierra Leone* (ISIS Briefing on Humanitarian Intervention No. 2). London: International Security Information Service.

Wilkinson, P. (2000b). Sharpening the Weapons of Peace: Peace Support Operations and Complex Emergencies. *International Peacekeeping, 7*(1), 63–79.

Wils, O., Hopp, U., Ropers, N., Vimalarajah, L., & Zunzer, W. (2006). *The Systemic Approach to Conflict Transformation.* Berlin: Berghof Foundation for Peace Support.

Zahar, M. J. (2008). Reframing the Spoiler Debate in Peace Process. In J. Darby and R. Mac Ginty (Eds.), *Contemporary Peacemaking: Conflict, Peace Processes and Post-War Reconstruction* (pp. 159–177). Basingstoke: Palgrave Macmillan.

Zartman, W. (1995). *Elusive Peace: Negotiating an End to Civil Wars.* Washington, DC: Brookings Institution.

Online Resources

Conciliation Resources (CR) is an international nongovernmental organization which publishes *Accord,* an international review of peace initiatives, and provides many links to other relevant organizations. At http://www.c-r.org/index.php, accessed July 2009.

European Peacebuilding Liaison Office (EPLO) is the platform of European NGOs, networks of NGOs, and think-tanks active in the field of peacebuilding, who share an interest in promoting sustainable peacebuilding policies among decision-makers in the European Union. At http://www.eplo.org/, accessed July 2009.

INCORE (International Conflict Research) is a joint project of the United Nations University and the University of Ulster. Its Peace Agreement Database lists over 640 documents in over 85 jurisdictions which can be termed "peace agreements." At http://www.peaceagreements.ulster.ac.uk/, accessed July 2009.

International Crisis Group (ICG) is recognized as the leading independent source of analysis and advice to governments and international organization. It publishes reports on conflicts and a crisis watch and provides databases and resources. Available at https://www.crisisgroup.org/, accessed on 20 March 2020.

OECD: DAC Network on Conflict, Peace and Development Co-operation (CPDC) is a unique decision-making forum which brings together governments and international organizations in order to support peacebuilding. It has produced "Monitoring principles for good international engagement in fragile states" and many publications about related key concepts, findings and lessons. At http://www.oecd.Org/department/0,3355,en_2649_33693550_1_1_1_1_L00.html, accessed July 2009.

The Stockholm Peace Research Institute (SIPRI) has the reputation of publishing objective data and analyses about everything related to arms, arms expenditures, arms control, and peace operations. Each year it produces a SIPRI yearbook on armaments, disarmament, and international security. Available at https://www.sipri.org/, accessed on 20 March 2020.

United Nations Development Program (UNDP). Its crisis prevention and crisis recovery unit provides a rich set of reports on issues such as DDR, economic recovery, rule of law, small arms, state building, and gender equality. UNDP also produces human development reports. Available at http://www.hdr.undp.org/, accessed July 2009.

UN Peacebuilding Commission provides UN documents and resolutions on peacemaking and peacebuilding. Available at https://www.un.org/peacebuilding/commission, accessed on 20 March 2020.

United States Institute for Peace (USIP) provides high quality books, issue papers, and practi-tioners' toolkits on the prevention and resolution of violent international conflicts and on peacebuilding. At http://www.usip.org, accessed July 2009.

World Bank: Conflict Prevention and Reconstruction. The aim of the Conflict Analysis Framework (CAF) is to enhance the conflict sensitivity and conflict prevention potential of the World Bank. CAF analyses key factors influencing conflict, focusing on six areas: social and ethnic relations; governance and political institutions; human rights and security; economic structure and performance; environment and natural resources; and external factors.

Chapter 14
Raison ou Déraison d'état: Coercive Diplomacy in the Middle East and North Africa

Abstract A first glance at today's world confirms the prevalence *of raison d'état* in international relations and shows a higher level of *raison* in the defence and promotion of national interests (This text was first published as: Reychler, L. (2012). Raison ou Déraison d'État. Coercive diplomacy in the Middle East and North Africa. In A. Fabian (Ed.), *A Peaceful World is Possible* (pp. 217–241). Sopron, Austria: University of West Hungary Press. Permission was granted.). We are living in the least violent period in history. Despite all of this, the "reason of the state" remains the focus of criticism. The bad press relates to:

- the challenges to the reason of the state by other interests at lower and higher system levels,
- the cost-ineffectiveness of ways and means used to defend or promote national interests and
- changes in the moral-political climate.

Most disturbing are the coercive diplomacy and military interventions of the West in the Middle East and North Africa (MENA). These interventions may benefit particular interest groups, but overall they are foreign policy failures. Within the decline of Western influence in the MENA lie seeds of hope for healthier relations in the future.

14.1 Reason of the State and the State of Reason

The concept *raison d'état* has many meanings. In essence, however, it refers to the primordial duty of a government to use its hard and/or soft power[1] to defend and promote the vital (existential) interests of the nation-state in a cost-effective way and in tune with the international moral-political climate (Nye 2008; Reychler 2010, 2011a, b). In today's world, practices like mercantilism, colonization or an unnecessary militarization of foreign policy have become unacceptable. The

---

[1]Luc Reychler has posted several articles that are related to this problem on the blog http://www.diplomaticthinking.com.

© The Editor(s) (if applicable) and The Author(s), under exclusive license to
Springer Nature Switzerland AG 2020
L. Reychler and A. Langer (eds.), *Luc Reychler: A Pioneer in Sustainable
Peacebuilding Architecture*, Pioneers in Arts, Humanities, Science, Engineering,
Practice 24, https://doi.org/10.1007/978-3-030-40208-2_14

primary concern of the *raison d'état* is the state's survival and security. While defending or promoting vital interests, other values may be temporarily devalued. Machiavelli, one of the fathers of the *raison d'état*, claims that "a prince ... cannot observe all the virtues for which men are reputed good, because it is often necessary to act against mercy, against faith, against humanity ... in order to preserve a state" (Machiavelli 1532).

A first glance at today's world confirms the prevalence of *raison d'état* in international relations and shows a higher level of *raison* in the defence and promotion of national interests. The interest of the state or nation is still a key for shaping, making and legitimizing foreign policy. The *raison d'état* prevails over competing values. The war on terror, for example, was justified in the name of the protection of the American people and implemented at the expense of the economic development and the civil rights of the American people. However, the *état de raison* (state of reason) has improved. Countries pursue their interests in a less violent way. Indicative are the statistics in the Canadian Human Security Report and recent book by Steven Pinker on the decline of violence in history and its causes. The 2005 Human Security Report reported a significant decline in the number of violent conflicts (Human Security Centre 2005). Steven Pinker, a Harvard professor of psychology, observes that the human race is losing its appetite for violence. He claims that the twenty-first century – the century of terrorism, genocide in Darfur, civil war in Somalia, etc. – is the least violent era in human history.

The advances are attributed to changes in human nature, especially enhanced self-control, empathy, morality and reason. The forces of civilization increasingly encourage the good in us to get the upper hand. Strong centralized governments, international trade and the empowerment of women all help to make us kinder, gentler human beings. Of great importance is what he calls "the escalator of reason" (Pinker 2012: 650), or the tendency of people to reframe conflict as a problem to be solved through the brain instead of brawn (Henig 2011).

14.1.1 Critiques of the "Reason of the State"

The use of *raison d'état* to justify questionable security or foreign policy decisions has prompted several critiques. The critiques relate to (a) the competition of the reason of the state with other interests at lower and higher system levels, (b) the cost-ineffectiveness of some of the ways and means of defending and promoting the national interest, and (c) changes in power relations and the moral-political climate.

14.1.2   Competing Interests (Reasons)

14.1.2.1   Particular Interests

The foreign policy-making process is the least democratic sector of decision-making. The lack of transparency in security and foreign policy-making processes in most countries of the West eases the disproportionate influence of particular groups and this at the expense of national interests (the interest of the citizens). The military industrial complex, the oil business, the financial intellectual complex,[2] religious organizations and foreign lobbies shape American policy towards the Middle East and North Africa (Mishra 2011). The lobbyists subvert the rule of law for the sake of the special interest they represent (Ferguson 2011). The resignation of Liam Fox, the UK's former minister of defence, resulted from the exposure of his links to his friend Wirrity (a lobbyist claiming to be an advisor) and a raft of businessmen, lobbyists and US Neoconservatives (The Guardian Weekly 2011c). The distinction between personal, particular interests and governmental activities became blurred (Preston 2011). Let us listen to some other critical voices.

- The electoral process is controlled by money. The 400 richest Americans control more wealth than half of the US population does (Edgecliffe-Johnson 2011; Krugman 2011b). Krugman (2011b) worries about the existence of an American-style oligarchy. "With American wealth increasingly concentrated in the hands of a few people, democracy is under threat."
- Conformity pressure and taboos: After 9/11, the conformity pressure with respect to policy regarding the Middle East increased significantly. Taboos and political correctness curtailed the freedom of opinion. The American House of Representatives voted 407 to 6 to call on the Obama administration to use its diplomatic capital to try to block the Palestinian initiative to seek statehood at the United Nations, while threatening to cut the Palestinians' funding if they proceeded to seek statehood. Similarly, when Israel stormed into Gaza in 2008 to halt rocket attacks, more than 1300 Palestinians were killed, along with 13 Israelis. As Palestinian blood flowed, the House, by a vote of 390 to 5, hailed the invasion as "Israel's right to defend itself" (Kristof 2011).
- Disproportionate impact of lobbies: In *A New Voice for Israel*, Jeremy Ben Ami calls for reasonableness in the Middle East. He is aghast at the way the United States facilitates hard-line Israeli politics that make peace less likely. He studied the impact of major Jewish organizations, like the American Israeli Public Affairs Committee (AIPAC), that embrace hawkish positions because the Jews that donate and vote are disproportionately conservative. The same is true, he writes, of Christians who are most passionate about Israeli issues. He concludes that the "the loudest eight per cent" have hijacked Jewish groups to press for

---

[2]The mantra of economic growth has justified government intervention on behalf of big business.

policies that represent neither the Jewish mainstream nor the best interests of Israel (Kristof 2011).
– Missing data: Data that are vital to evaluate foreign and security policies are very difficult to get (Mancassola 2011). This impedes the development of an informed and enlightened public opinion. Which interest groups shape the foreign and security policy in the Middle East? How do they influence the decision-making? How many people have been killed in the liberation of Libya? How many Iraqis or Afghanis? It is easier to find data of first-class victims than of second- and third-class victims. First-class victims receive detailed and personalized attention; the information about second-class victims is less accurate and personalized (Palestinian prisoners); third-class victims are faceless inaccurate statistics (civilians killed or seriously wounded in Libya or Afghanistan). What were the alternative policy options and their respective costs and benefits? How many member countries of the Arab League abstained, voted against, or were in favour of the Security Council "No fly zone" resolution of March 17?

14.1.2.2 Human-Centric Interests (Human Security)

The inclusion of human security came into currency in security studies in the mid-1990s and served useful purposes (Kerr 2007):

- it shows that the state-centric approach does not adequately address the security of the citizens within the state;
- it pays attention to the broad definition of violence[3] that looks at several factors which can reduce the life-expectancies of the people[4];
- it reminds policy-makers that the concept of security should encompass properly functioning states. States with low levels of human security may appear stable, but tend to be instable and vulnerable. For example, think of the implosion of the Warsaw Pact countries or the Arab Spring revolutions in the MENA.

14.1.2.3 The Interests of the United Nations

Although the United Nations is still a relatively powerless organization and dominated by the West, it continues to reflect the aspiration and the need for more

---

[3]The narrow definition of violence pays attention only to the number of people killed or wounded as a consequence of armed violence; the broad definition operationalizes violence as the reduction of the life expectancy of people resulting from the use of: armed violence, structural violence, psychological violence, cultural violence, ecological violence and bad governance.

[4]The life average expectancies of a people can be lowered by the use of physical violence, structural violence, psychological violence, ecological violence, cultural violence, bad governance and corruption.

effective global governance. Globalization and the increasing interdependency, sensitivity and vulnerability of countries require better governance at the global level. The demand for more effective and legitimate governance at the global level has been jolted by the negative impact of human behaviour on our natural environment (climate, natural resources etc.). It cannot be denied that even the most ethically acceptable notion of foreign policy, involving privileging the interests of the people of the state in question, may be regarded as illegitimate from a global perspective (Brown 2010).

14.1.2.4 *Raison Planetaire* or the Planetary Interest

In his book *The Planetary Interest*,[5] Kennedy (1999) suggests that contemporary political and institutional arrangements for problem-solving, based on sovereignty as a principle and the nation-state as a decision-making unit, are not equipped to handle planetary problems; he argues that, henceforth, the planet and humanity must be considered as a single-entity for decision-making purposes. He introduces and develops the idea of the planetary interest as a vital conceptual tool to assist citizens and policy-makers in understanding emerging global problems and approaching them with global solutions. The book includes contributions from current or former politicians, who discuss the particular global problems threatening their countries, looking at the ways in which the planetary interest can help them pursue a legitimate national interest.

14.1.2.5 The Demand for Cost-Effective Security, Failures and Limits of Military Power

- *The demand for cost-effective security policies*: The confluence of the many crises that confront the international community (financial, ecological, the marginalization of the majority, weaponization, the increasing number of weak and failing countries, competition for natural resources and energy etc.) demands more cost-effective and sustainable ways of resolving conflicts. War is inhuman; moreover, it is one of the least effective ways to handle these crises. They are very good at killing and destroying; and they predominantly benefit conflict entrepreneurs. The world will never be able to deal with the climate if it does not succeed in improving the human climate. This means resolving conflicts constructively and building sustainable peace. Strategic history describes many formulae for defending and promoting the interests of states or groups of states: the establishment of an empire, unipolarity, bilateral and multilateral balances of power, collective defence, collective security and security

---

[5]The term "raison de planete" was coined by Jacques Haers of the University of Leuven, Belgium.

community. After the Second World War, European countries transformed one of the most violent regions in the world into a security community, characterized by defence and economic cooperation and a high level of objective and subjective security. A security community is a cost-effective and durable system for assuring security within the European region (Reid 2004).[6]

- *Foreign policy failures, follies and collapses*: Several scholars are drawing our attention to foreign policy failures, follies and disastrous decisions. The research is associated with names such as Tuchman (1984), Diamond (2005), Jervis (1976), Stoessinger (2005) and Janis (1972). Recent examples of foreign policy failures are the armed interventions in Afghanistan and Iraq; the aims were not achieved and the operations turned out to be very expensive. Some foreign policy failures have been called follies, or policies pursued contrary to the self-interest of the constituency or state involved. To qualify as a folly, the policy adapted must have been perceived as counter-productive in its own time, given a feasible alternative course of action. Folly is universal and unrelated to the type of regime: democratic or undemocratic. The militarization of Western foreign policy in the Middle East, the deregulation of financial markets, and the inadequate policies vis-à-vis environmental crises are (or come close to being called) follies. Barbara Tuchman (1984: 21) warns us that "Social systems can survive a great deal of folly when the circumstances are historically favourable, or when bungling is cushioned by large resources … Today, when there are no more cushions, folly is less affordable".

- *The limits of armed violence and coercive diplomacy*: Coercive diplomacy seeks to resolve crises and armed conflicts short of full-scale war. "It is a reactive strategy relying on threats, limited force and inducements to influence an adversary to stop or undo the consequences of actions already undertaken" (Jakobsen 2007). Full-scale war or brute force (Shock and Awe) are influence strategies that deprive the adversary of any choice by forcing compliance upon it. It has become evident in the MENA that brute force and coercive diplomacy are not the most effective ways to resolve conflicts, change regimes or impose democracy. In addition, the assumptions that deficits do not matter and one does not have to choose between "guns and butter" have proven false. America and the most assertive European countries (the UK and France) lack the resources – economic, political and military – to support large scale, protracted conflict without, at the very least, inflicting severe economic and political damage upon themselves (Bacevich 2008: 11). The "guns versus butter" rule is back. Weaponized Keynesian economics must be rejected except in those cases where it is being used to defend lucrative contracts (Krugman 2011a). Finally, wars have become less useful to divert the attention of the people from political-economic problems and rally people around the flag. The current economic situation (growing pessimism and unemployment) in the West is creating uncertainty, dissatisfaction and tension (Pignal 2011). In October 2011,

---

[6]Karl Deutsch coined the term "security community".

a poll by the *National Journal* found that 59% of respondents either fully or strongly agreed with the aims of the Occupy Wall Street (OWS) movement and that 99% backed protests against social and economic inequality, corporate greed, corruption, and influence over the government – particularly from the financial sector (Luce 2011; Occupy Wall street [OWS] 2011). In America in 2011, 17.1% of those below the age of twenty-five were out of work. In Europe, youth unemployment averaged 20.9%. Britain's youth unemployment was over 20% (Senneth/Sassen 2011). In Spain, it was a staggering 46.2% (The Economist 2011a). The International Labour Organization (ILO) warned that a jobs crisis in the world caused by the global economy slowdown threatened a wave of widespread social unrest engulfing both rich and poor countries (Elliott 2011).

14.1.2.6   Changes in the Moral-Political Climate

International stability depends not only on balancing power, but also on the legitimacy of the internal and foreign policy behaviour of nation-states and of the international system.

- *Non-interference versus responsibility to protect*: In 1996 Francis Deng argued in *Sovereignty as Responsibility*: *Conflict Management in Africa,* that when nations do not conduct their internal affairs in ways that meet international standards, other nations not only have a passive right, but an active duty to do so (Deng et al. 1996). The most important normative principle of international law, established by the Treaty of Westphalia in 1648 – non-interference in another's internal affairs – was turned upside down. In effect, Deng replaced the long-standing definition of sovereignty as the "supreme authority within a territory" with a new definition of sovereignty as responsibility to protect people in a given territory. The adoption of the Responsibility to Protect (R2P) principle by world leaders assembled at the UN in 2005 is presented as one of the great normative advances in international politics since 1945 (Thakur 2011). The latest application of the R2P principle is the intervention in the Libyan civil war or "the escalation of an internal conflict to protect Libyan people". R2P is a doctrine born of good intentions, but there are some drawbacks:

  - it leads to fables of righteousness and evil,
  - war involves a descent into barbarism,
  - the aim can be distorted, as it was by NATO, and
  - R2P is only applicable to relatively powerless countries. We need to answer tough questions. Was the international community confronted with a large-scale, actual or expected loss of life or ethnic cleansing? Did the intervention respect the precautionary principles of:

    (a) right intention (exclusively humanitarian)
    (b) last resort,
    (c) proportional means,
    (d) reasonable prospects of success? (Rieff 2011).

It is too early to consider this a successful case of R2P. There are missing data, such as the "hypothetical number of lives saved", the number of people killed in the process and the destruction caused by the bombardments of NATO and the rebels.

- *Representative global international governmental organizations (GIGOs).* There are also growing demands for more representative global IGOs. These demands are furthered by

    (a) the changing power relations, as reflected by the BRIC countries, and
    (b) the cognitive dissonance between the promotion of democracy at the national level and the resistance of democratic changes at the international level by the West.

- *Violence prevention and sustainable peacebuilding*: After the Cold War, international governmental organizations, Departments of Foreign Affairs, and non-governmental organizations started to pay more attention to the prevention of violence and peacebuilding (Bauwens/Reychler 1994). Despite the aspirations and the commitment to make the world more peaceful, most efforts boiled down to crisis management and to reactive instead of proactive violence prevention.

14.1.2.7 Coercive Diplomacy and Armed Intervention in the MENA

The rosy picture of a more peaceful world suggested by the statistics of the Human Security Report of 2005 and Steven Pinker is contradicted by:

(a) a 25% increase of the number of armed conflicts from 2003 and 2008 after declining for more than ten years,
(b) an increase in military expenditures during the first ten years of the twenty-first century – democratic countries account for more 70% of the global military budget (Stockholm International Peace Research Institute [SIPRI] 2010) – and
(c) the military interventions and use of coercive diplomacy by the West.[7] In the first decade of the twenty-first century, the West threatened, militarily intervened in and sanctioned nine countries: Afghanistan, Iraq, Lebanon, Palestine, Libya, Pakistan, Yemen, Syria and Iran. In this paper the West refers mainly to the US; its closest allies (Israel and the UK) and the members of the respective coalitions of the willing.

---

[7]This century the term "the West" is used more frequently, especially with respect to foreign policy in the MENA. The term refers to the US, the most assertive NATO members (the UK, France) and Israel.

14.1.2.8 A Two-Dimensional Approach to Relations in the MENA

The foreign policy of the West in the MENA is determined by two factors: the possession of oil and the compliance or deference to Western interests. Compliance implies an assured flow of oil at reasonable prices, purchase of sophisticated weapons, engaging in anti-terrorism and not supporting so-called radical Palestinian parties. Consequently, four types of countries can be categorized (Fig. 14.1).

- *Oil-rich and compliant countries*: Saudi Arabia, the UAE, Qatar, Kuwait and Bahrain are protected and receive diplomatic support; there is no interference in internal affairs; there is no pressure for democratization and the repression of liberation movements is not publicly pilloried. Most of these countries house American military bases. They also are buying great amounts of weapons and military services from the West. The protests of (discriminated against) Shiites in Bahrain were brutally repressed. For the past three decades, the Saudi monarchy has viewed Iran as its nemesis. Saudis have relied on the US, Arab nationalism and Sunni identity to weaken or slow Iran's power (Nasr 2011). They are especially concerned about the realization of the "Shiite crescent" stretching from Lebanon to Manama that King Abdullah of Jordan once warned against.[8] Qatar, a small country with 300,000 Qataris and one million foreign workers with Doha as its capital city, has become a key player in the Middle East. It played a significant role in driving the Arab League vote on Libya and Syria. The Arab League is now dominated by the richest countries in the Arab world (Courrier International 2011; International Herald Tribune 2011). It is easier for the West to manipulate.
- *Oil-deprived and compliant countries*: Egypt, Tunisia and Jordan are rewarded for their compliance (Egypt annually received 1.3 billion dollars) and the interference in internal affairs was minimal. Egypt and Tunisia are the only countries with relatively peaceful regime changes [revolutions]. From these experiences, the West could learn that regime changes can be achieved in less violent, less destructive ways and more humane and cost-effective ways. In both countries elections have been held. In both countries the revolutionary spirit lives, but resonates less. In Egypt, there is a widespread gloom that the country is stagnating, its economy heading towards a cliff, while the caretaker government refuses or fails to act (Afify 2011). Some 40% of the population live on less than $2 a day and among the country's unemployed as many as 90% are young and two-thirds have never worked (Dinmore 2011). "It's a surreal situation. People … were so proud after the revolution. Now the revolution is starting to get a bad name because the situation is getting worse" (Khalaf 2011). The economic support from the West is inadequate. Jordan's pro-democracy protest provides a test of the ability of the kingdom to reform a divided and resource-starved

---

[8]The West should not encourage a sectarian showdown, but support a redistribution of power among Sunnis, Shi'ites and Christians. The Shiites are not getting a fair share in the current power structure.

	Oil-rich	**Oil-poor**
Compliant	Saudi Arabia, Bahrain, United Arab Emirates, Qatar receive diplomatic and military support/ no interference in internal affairs/ no public critique of the repression of liberation movements.	Egypt, Tunisia and Jordan are rewarded (Egypt annually received 1.3 billion). Until the Arab Spring there was not much internal interference or criticism of the repression of opposition or democratic movements. Egypt and Tunisia are the only countries with relatively peaceful revolutions.
Non-compliant	Iraq, Iran, Libya are the objects of increasingly coercive diplomacy, including economic sanctions, boycotts, military threats and/or interventions. The regimes are relegated to the axis of evil and treated as rogue states.	Syria, Lebanon (especially Hezbollah)*, Palestine (especially Gaza) and Afghanistan are the object of military intervention, coercive diplomacy, and/or occupation.

*The interventions of 2006 and before.

Fig. 14.1 Drivers of Western foreign policy toward states in the MENA. *Source* The author

country. Jordan is dependent on the generosity of governments in the US, Europe and the Gulf. Money will continue to flow as long as the country fulfils its goal of joining the Gulf Cooperation Council, a Saudi-led group of oil-rich nations (Buck 2011). Everything in the resource-starved countries revolves around youth unemployment, the price of food and the bad economic climate. The pruning of the public sector will hit many parts of society hard.

- *Oil-rich and non-compliant countries*: Iraq, Iran and Libya have been the object of coercive diplomacy, including the implementation of sanctions and the use or threat of armed violence. The regimes were relegated to the axis of evil and depicted as rogue states. Their leaders were labelled enemies of civilization, which gave the West an excuse to intervene in the region. Gaddafi, the head of a third-rate military country, was an ideal scapegoat.

  - *Iraq*: For over thirty years, Iraq has suffered from political repression, wars and/or sanctions. Far more innocent men, women and children have died in Iraq over the past decades as result of UN sanctions and the unwillingness of the Iraqi regime to cooperate with the UN than in the course of the war that led to these sanctions (Brown 2010: 223). The Iraq sanctions were collective and included a near total financial and trade embargo. They began on 6 August 1990, four days after Iraq's invasion of Kuwait, and stayed largely in force until May 2003 (after Saddam Hussein was forced from power), with certain portions, including reparations to Kuwait, persisting later. Richard Garfield, a Columbia University nursing professor, cited figures of 345,000–530,000 for the entire 1990–2002 period for sanctions-related deaths ("Iraq sanctions" n.d.). Conservative estimates for 2003–2011 put the figures in

Iraq at approximately 135,000 dead, 140,000 wounded, 1,700,000 internally displaced and 1,800,000 refugees.

- *Libya*: The intervention in Libya turned international attention away from the failing operations in Afghanistan, Iraq, Gaza and Pakistan, and it was presented as a humanitarian mission. In fact, the military coalition of the willing escalated an internal war in order to install a more compliant government with a democratic demeanour and privileged access to oil contracts. The war ended with the wounded and brutally sodomized body of Gaddafi. The number of people wounded and killed in the destruction is not known. The transition is expected to be difficult because of the proliferation of weapons in the country as well as the political contradictions between the rebels within the transitional government and the conflicting interests of foreign powers still busy in Libya. Weak and failed states are easier to penetrate; outside powers have democratic proxies and interfere in national affairs at will. The Economist (2011c) worries about the impact of other players like the Gulf States.
- *Iran*: In 1979, Iran saw the first peaceful regime-change in the Middle East. Since then the country has been the recipient of an increasing number of sanctions (economic, military, diplomatic); it has been stigmatized as a rogue state belonging to the axis of evil; it is surrounded by US military bases and forces and has frequently been threatened with military strikes. There were demands for stronger 'crippling' sanctions (embargo against oil exports; seize Iranian assets in the US; blacklist Iran's central bank; make contact with Iranian officials illegal etc.) and for the bombing of its nuclear installations. The country has not forgotten the CIA coup that ousted Prime Minister Mohammed Mossadegh in 1953 or America's green light to Saddam Hussein's war against Iran in 1980, which caused a million casualties. Iran is an authoritarian capitalist country with a human rights record that does not differ from America's authoritarian friends. Is has not fought offensive wars for the last three hundred years. It is the last oil-rich country that is not planning to comply with the demands of the West in the region. It is blamed for its support of Hamas in Palestine and Hezbollah in Libya, for having the intention to destroy Israel, and for building nuclear weapons. The first claim is true. The second is a distortion of what was meant by the destruction of Zionism. Ahmadi-Nejad clarified in an interview that this did not imply the physical destruction of the state of Israel; he compared it with the end of apartheid in South Africa and communism in Eastern Europe. Concerning the nuclear issue, an IAEA report warned that Iran appeared to be on a structured path to building a nuclear weapon – a setback for the West, because it would make it more difficult to constrain the only remaining non-compliant country. Analysts who speak of a "ticking bomb" are not very serious. The main motivation of countries to acquire nuclear weapons is deterrence; to prevent hostile action by another state. The military policies of the West in the Middle East are creating a serious insecurity dilemma and heighten the feelings of insecurity in Iran. Efforts for acquisition can be

expected from isolated and insecure countries like Israel or Pakistan. The depiction of Iran as a devilish country is exaggerated. It is a pretext for pursuing a policy that facilitates the arms trade; the deployment of a mix of increasingly sophisticated sea- and land-based missile interceptors around Europe and the friends in the Middle East; further militarization and control of the region; and containing and constraining Iran. This policy will not lead to sustainable security, but promises a more dangerous world (Butt 2011).

- *Oil-deprived and non-compliant countries*: Syria, Lebanon and Palestine have frequently been the object of military aggression and coercive diplomacy.

    - *Palestine*: Palestinians are the longest victimized people in the region. They have been the object of irredentism, segregation, military repression and all kinds of sanctions. Israel, an ally of the West, could, at least until now, do what it wanted. The country possesses the mightiest, most sophisticated and experienced army in the region, it guards the biggest open-air prison in the world and it pursues a policy of irredentism and segregation in the West Bank and Gaza. Each year America transfers three billion dollars of aid. Since 1980, it has received approximately 100 billion dollars. The "peace negotiations" were an extension of the coercive diplomacy of Israel. Recently the Palestinians were seeking affirmation of statehood at the UN within the 1967 borders (22% of the historical Palestine). The vast majority of the UN members are ready to swing behind the claim to statehood. The pressure of the US and Israel blocked recognition by the UN Security Council. In the UNESCO however, Palestine became a full member. 104 countries voted in favour (representing 77.6% of the world's population) and 14 voted against (7.3%) (Mabubani 2011). Consequently, the US decided to stop funding the UNESCO (80 million a year or 22% of its operating budget). The recognition of Palestine as a state is an important precondition for genuine peace negotiations; it is considered non-negotiable.

    - *Syria*: The conflict is escalating and has turned into a violent civil war. The international community is divided, but initially it seemed to abstain from military intervention. The West has asked Assad to step down and imposed punitive sanctions on Syria's oil sector. Russia insists that both the government and the armed rebels cease fire. Navi Pillay, the High Commissioner for Human Rights, was expected to call for Mr. Assad to be referred to the International Court (Field 2011). Some analysts voiced concerns that a crackdown on the Syrian oil sector could result in some kind of collective punishment as experienced in Iraq, while countries such as Sweden generally dislike sanctions. The UK has been worried that the sanctions will weaken the regime's ability to provide for the 30% of the Syrian people who rely on it for jobs, pensions and housing (Field/Barker 2011). These punitive measures weaken both the regime and the state, and could turn it into an anocracy.

    - *Afghanistan*: The story of Afghanistan is well documented. The tenth anniversary of the war and the reconstruction efforts were disappointing. The war has passed Vietnam as the American military's longest war. The situation is at stalemate; the security has deteriorated; the economy depends on foreign aid and most people remain in deep poverty. Between 14,000 and 18,000 civilians have died because of the fighting; there are 415,000 displaced people and 2,913,000 refugees (2011) (De Morgen 2011; The Guardian Weekly 2011a, b).

Characteristics of Western policy in the MENA Western policy vis-à-vis the Middle East and North Africa could be depicted as follows:

- The aim is to control the region to secure access to oil at reasonable prices and assure markets for military and other products.
- The control is achieved by (a) allying with Israel and compliant Arab countries, (b) increasing military presence, (c) turning non-compliant states into weak ones and containing and constraining Iran.
- To achieve support for the interventions after 9/11, the US and its main allies (Israel and the UK) (a) exaggerate threats and manipulate fears; (b) insist that there is no time left for further deliberation and that immediate responses are necessary; (c) stress that there are no alternatives to the use of coercive diplomacy and armed intervention to deal with difficult Arabs (Enzenberger 2011)[9]; (d) polarize the world by dividing countries into two groups: those who are with us and those who are against us: (e) sell foreign involvement through appeals to human rights and values; (f) highlight and embellish international support for the interventions; and (g) remind their own citizens of the exceptionalism of their country. Conservative American and Israeli politicians especially suffer from this political syndrome (Johns 2011).

14.1.2.9 Assessing Success and Failure?

The success or failure could be assessed in two ways:

(1) by distilling and clustering the success and failure arguments from the discourse in the West and intuitively estimating their weight;
(2) by assessing the success/failure of the interventions more systematically on the basis of a set of objective criteria.

14.1.2.10 Success and Failure Discourses

A preliminary synthesis of opinions/arguments favouring or criticizing the policy of the West in the MENA looks as follows (see Fig. 14.2). This is a difficult debate

---

[9]TINA = there is no alternative.

	Successes/strengths	Failures/weaknesses
Security	Homeland security has improved International terrorism has diminished / A1 Qaeda is weakened/ Bin Laden killed. Suicide terrorism has curtailed.	The root causes (direct and indirect occupation) have not been addressed The security is unsustainable.
Economy	Access to oil is secured. There are new reconstruction and investment projects. The situation is good for high-tech arms trade (drones).	The wars in Iraq and Afghanistan are very expensive (more than $3 trillion).
Political	The Israeli government can continue its irredentist policy and delay genuine peace negotiations. We enabled the development of democracy.	The US and Israel have became garrison states. The irredentist and segregationist policy of Israel is counterproductive and has a very negative impact on the image of the West. The impact of Israel on the policy of the West in the MENA is disproportionately high and dangerous. Big powers should never allow smaller allies to shape their security policy. The situation is comparable with the role of weaker allies (e.g. Serbia) before the First World War.
Military	Military presence and control is assured. The MENA was an ideal testing/ training ground for new weaponry and doctrines.	We have experienced the limits of military power. We have not seen the results. Destroyed and weakened states do not guarantee stability. Our policies have produced many negative impacts (insecurity dilemmas, arms escalation and humiliation).
Cost-effectiveness	We succeeded in developing low-risk military interventions with light boot prints: nobody was killed in Libya. Air warfare (missiles and drones) reduced casualties considerably.	The wars caused an enormous amount of casualties and destruction in the target countries. The risk-free interventions could lower the threshold for initiating wars. The opportunity costs are enormous. These interventions and especially the Israel-Palestine conflict usurp too much diplomatic time that is urgently needed to deal with other challenges and opportunities in the world.
	Successes/strengths	*Failures/weaknesses*
Ethical	"Surgical strikes" reduce collateral damage. We fought for democracy/human rights: the emancipation of women.	The suffering of the people is very high (casualties/internally displaced/ refugees/destruction of houses and infrastructure/reduction of life expectancy.
Soft power	We have helped to spread freedom; people are thankful.	We have lost a lot of soft power in the region. The West has become a negative international role model in international relations and international law.

Fig. 14.2 A preliminary summary of arguments depicting the MENA policies as a success or failure. *Source* The author

full of taboos and conformity pressure. A silent majority of people, however, disapproves of the MENA policy.

14.1.3 Systematic Assessment on the Basis of Objective Criteria

A more systematic assessment of the reasonableness of the MENA policy can be made on the basis of seven objective criteria.

1. *Were the aims realized and the national interests enhanced?* At first sight most military operations may appear successful: A1 Qaeda is weakened, Bin Laden murdered, suicide terrorism reduced, the Gaddafi regime toppled, etc. A second look, however, leads to less positive conclusions. The occupation and destructive interventions have fuelled negative attitudes (hate) and the potential violence has not been eradicated. The difficult exit policy mirrors these problems. The latest Congressional Budget Office (CBO) projection calculates the possible war costs until 2021. Evaluating the achievement of the aims is difficult because some aims are made explicit; others are not made explicit and operate in the shadow of war. Most military interventions and coercive diplomacy seem to cater more to special interests than to the national interest of the intervening countries. They benefit the military industrial complex (MIC), think tanks and consultants, private security services, reconstruction business and foreign lobbies. It is difficult to end wars when influential groups are conflict profiteers.
2. *Were the interventions cost-effective?* There are human costs, economic costs, time costs and opportunity costs. Our own human costs have been drastically reduced by (a) better arms and intelligence, (b) the use of missiles and drones and (c) the delegation of security work and fighting to private security forces and proxies (Gettleman et al. 2011). Claims of cost-effectiveness have been seriously disputed on three grounds: (1) the price of operations was higher than anticipated, (2) the duration of the wars was longer than expected, (3) the high opportunity costs and (4) the availability of alternatives. The cost of the wars in Afghanistan and Iraq were higher than expected. In the latest Congressional Budget Office projection, the war costs for Iraq and Afghanistan from 2012 to 2021 could total another $496 billion. The total war costs would thereby amount to $1.8 trillion, while Joseph Stiglitz estimates the true costs to be $3 trillion (Stiglitz/Bilmes 2008). This is twenty times the yearly budget of the European Union. Most military interventions lasted longer than anticipated. The opportunity costs were very high. A small part of the $3 trillion could facilitate the difficult transitions in Egypt and Tunisia. Some of that money could have been used to develop proactive violence prevention and sustainable peacebuilding. There were feasible and more cost-effective alternatives. The diplomatic demarches of Turkey and the African Union were sidetracked. Diplomacy has been given a secondary role in the Middle East and has been reduced to coercive

diplomacy. In contrast to the financial support given to defence and intelligence, the budget for diplomacy in the West is small. Intelligence and military operations are no substitute for the critical work of diplomats on the ground. Good diplomacy is essential for dealing with conflicts in a constructive way. America closed its embassy in Bagdad and severed diplomatic relations with Iraq in 1991. "The 30-year freeze in diplomatic relations with Iran produced a US government that knows precious little about a country that is integral to stabilizing American interest in non-proliferation, terrorism, Afghanistan, Iraq, energy security and the Israeli-Palestinian conflict" (Marashi 2011).

3. *Was the use of violence proportional and appreciated by the citizens of the target country?* Despite the technical and verbal efforts to reduce and hide the collateral violence, the wars have been highly destructive. They significantly reduced the human security of the citizens in the target countries. In contrast to the precise data on the casualties of the allied forces, the statistics of the dead, wounded, internally displaced, refugees and destruction of the target countries are unreliable. Putting together the conservative numbers of war dead in Afghanistan and Iraq, in uniform and out, the total is estimated at 236,000. The number can be multiplied by two or three to estimate the seriously wounded. Millions of people have been displaced indefinitely and they are living in inadequate conditions (Watson Institute n.d.). In the Lebanese war of 2006, large parts of the Lebanese civilian infrastructure were destroyed, including 400 miles (640 km) of roads, 73 bridges and 31 other targets, such as Beirut's Rafic Hariri International Airport, ports, water and sewage treatment plants, electrical facilities, 25 fuel stations, 900 commercial structures, up to 350 schools and two hospitals, and 15,000 homes. Some 130,000 more homes were damaged ("2006 Lebanon war" n.d.). In all of the conflicts (with the exception of Lebanon and the Occupied Palestinian Territories), it is very difficult to find reliable data on death and destruction. In *Cultural Cleansing in Iraq*, editor Raymond Baker describes the wilful inaction of the occupying forces, which led to the ravaging of one of the world's oldest recorded cultures. The authors documented the destruction of unprotected museums and libraries, the targeted assassination of over 400 academics, and the forced flight of thousands of doctors, lawyers, artists and other intellectuals (Baker 2010). The suffering in target countries is not only caused by military interventions, but also by the imposition of sanctions (collective and/or targeted economic and financial as well as non-economic sanctions) (Farrall 2007). Comprehensive and collective economic and financial sanctions imposed by major players over a long period can have debilitating impacts. Five countries have been the object of very strong sanction regimes: Palestine (since 1967 and earlier), Iraq, Iran (since 1979), Libya and Syria. Opinion polls show a low percentage of Arab respondents who express their either 'very' or 'somewhat' favourable feelings toward America and Israel, which have not improved. The relatively positive attitude towards the Obama Middle East policy in 2009 dropped significantly a year later (Telhami 2010). In April 2011, the *International Herald Tribune* wrote that the use of drones in Pakistan and the killing of many innocents made people furious. This

was reflected in an opinion poll about threat perceptions in Afghanistan: 7% identified the Taliban, 6.4% A1 Qaeda, and 68% the US-led war on terrorism (International Herald Tribune 2011).[10]

4. *Have the chances of sustainable peace and security increased?* The verdict is negative, (a) because the root causes of the conflicts have not been addressed, (b) because of the unforeseen negative impacts and (c) because of the double-effect doctrine. The "war on terror" dealt with the symptoms of terrorism. It protected the homeland and identified and destroyed terrorists and terrorist networks, but denied the root causes (the Israeli-Palestinian conflict, the external support of authoritarian regimes, and the installation of military bases on Arab soil). The deteriorating economic situation, food price inflation and the high number of urbanized unemployed youth cause a great deal of the instability in the Middle East (Mackintosh 2011). The second reason for not achieving sustainable peace is the (anticipated or unanticipated) negative impacts on the conflict and peace dynamics of military intervention, collateral damage, coercive diplomacy, occupation, segregation, the support of friendly authoritarian regimes and the exclusion of countries and communities from peace negotiations. The lack of concern about the negative side-effects tends to be enhanced by the doctrine of double effects. The doctrine (or principle) of double effect is often invoked to make activities that cause serious harm, such as collateral damage and destruction, more permissible because they are expected to bring good results. "This reasoning is summarized with the claim that sometimes it is permissible to bring about a merely foreseen harmful event that it would be impermissible if done intentionally" (Brown 2010: 232–234; McIntyre 2004).

5. *Did the interventions have a high level of external legitimacy?* Overall, the international legitimacy of the military interventions and coercive diplomacy has been low. Approval by the Security Council is not a reliable indicator of international approval. It is more a reflection of power than of global public opinion. Ten members supported the UNSC resolution 1973 (demanding an immediate ceasefire in Libya, including an end to the current attacks against civilians and imposing a no-fly zone and sanctions on the Gaddafi regime and its supporters). Five important countries (Brazil, Russian, India, China and Germany) abstained. Even among the nine states of the 22-member Arab League that voted in support of a no-fly zone (the rest were absent or voted against), there was visible discontent with NATO for stretching the UN resolution (Steele 2011). When the aim of the Libya campaign shifted from protection of the people to toppling Gaddafi international support dwindled. Russia and China vetoed a resolution condemning Syria. Western policy in the MENA is also blamed for the use of double standards. Examples can be found in the Israeli-Palestinian conflict, the nuclear question, the imposition of democratization, the use of torture, the International Criminal Court and the war on terrorism. The US and the EU decided to consider the UN recognition of

---

[10]*International Herald Tribune* (1 April 2011).

Palestine only at the end of successful peace negotiations. This policy goes against one of the basic principles of peace negotiation (the recognition of all the primary parties). Instead, the US and the EU should assure the installation of the necessary preconditions for 'genuine' peace negotiations (Reychler 2011b). The depiction of Iran as potentially the greatest nuclear threat in the Middle East is another example of double standards. What about Pakistan and Israel or the nuclear powers arms spending-spree? A report of the British American Security Council warns that the US will spend $700 billion on the nuclear weapons industry over the next decade, while Russia will spend at least 70 billion on delivery systems alone. Other countries, including China, India, Israel, France and Pakistan, are expected to devote formidable sums to tactical and strategic missile systems. Several countries are assigning roles to nuclear weapons that go well beyond deterrence (Norton-Taylor 2011). The democratization campaign is also tainted by double standards. The attitude of the West towards democracy is positive, but it has some particularities: (a) only friendly democracies are welcome; democratically elected groups, such as the FIS in Algeria, Hezbollah in Lebanon and Hamas in Palestine, are ostracized and sanctioned; (b) it supports compliant authoritarian regimes; (c) it tolerates the segregation of the Palestinians and delays the recognition of an independent Palestinian State; and (d) it firmly opposes the democratization of the international system and of international governmental organizations, such as the UN Security Council, the IMF and the World Bank.

6. *Were the ethical standards (of the civilized world) respected?* Several ethical principles were not respected. Questions have been raised about right intention as well as the use of proportional violence, torture, the protection of innocent civilians, the consideration of alternative options, and internal and external legitimacy. No comprehensive or comparative reports have been published on the ethics of the wars. However, three findings stand out: the *disproportionate killing and destruction of anti-terrorism,* the *high percentage of civilians killed and maimed* and the application of a principle of *"10 to 100 eyes for an eye".* Most interventions are portrayed as anti-terrorism operations. Anti-terrorism is much more destructive and violent than political terrorism. It distinguishes itself by: (a) a higher number of innocent civilian casualties, (b) the predominance of collective sanctions, (c) disproportionate destruction, (d) the application of the "ten eyes for an eye" principle, (e) the preference for operations with low-risk for soldiers, such as bombardments with fighter planes, drones or missiles and the subcontracting of security talks to private security firms, and (f) the relatively high security of the citizens of the intervening countries. During these wars people from the West could go on vacation and enjoy its security fully. The culmination and duration of military intervention and/or coercive diplomacy has in some areas drastically reduced human security and inhibited economic and political development. The Iraqis, the Afghanis and the Palestinian people have suffered the most damage. In Iraq, Afghanistan, Gaza (2008) and Lebanon (2003) 92%, 61%, 79% and 61% respectively of the people killed were civilians. The ratio between the people of the intervening state killed and those of the

target countries are: 1:28 Allied Forces-Iraq, 1:7 Allied Forces-Afghanistan, 1:109 Israel-Gaza, 1:11 Israel-Lebanon.[11] Ethical questions could also be raised with respect to selective humanitarianism (the different weight given to fellow citizens and the citizens of the target country) and the doctrine on the pre-emptive and preventive uses of force.

7. *Did the interveners learn from previous interventions?* Yes, there was learning, but the learning curve is flatter than steep. The military interventions and coercive diplomacy seem to obey Abraham Maslow's law of the hammer: "It is tempting, if the only tool you have is a hammer, to treat everything as if it were a nail" (Maslow 1966: 15). There is the conviction that more bombing and military intervention can end the problem and deliver our peace. Albert Einstein is reported to have said that insanity consists of doing the same thing over and over again and expecting different results. Not much has been learned about violence prevention and sustainable security or peacebuilding. Most of the attention goes to crisis management and the development of low-risk interventions (using drones, delegating land operations to armed rebels and subcontracting private security firms).

14.1.4 The End of Pax Occidentalis

The policy of the West in the MENA during the twenty-first century has been a history of failures. The coercive diplomacy and military interventions did not produce sustainable security and peace. Such a verdict should normally provoke changes towards policies that are more adapted to the reality of the MENA. The exit policies in Iraq, Afghanistan and Libya could be seen as indicators of such change. However, in each of these countries the West keeps a military presence; Israel has no exit policy. The militarization of the region can be expected to last as long as powerful groups with strong lobbies benefit from it. These conflict profiteers are the military industrial complex, oil businesses, reconstruction firms, private security firms, favourable think tanks and media and foreign lobbies. An unexpected side-effect of the wars is the liberation movements and the political revolutions. The liberation movements aspire not only to remove internal repression, but also the external occupation and interference in the region. In addition, the relatively non-violent revolutions in Tunisia and Egypt show that endogenous non-violent struggles can be effective and much more cost-effective.

For the last couple of decades, the Pax Occidentalis (the Western foreign and security paradigm) has been challenged and the limits of its logic are being exposed not only by the unsuccessful military interventions in the Middle East and North Africa, but also by other seismic events like the panic in the financial system since

---

[11]Figures obtained during Luc Reychler's ongoing research at the University of Leuven into the statistics of the Ethics of War.

the end of 2008, the problems related to climate change, the nuclear accident of Fukushima on 11 March 2011, Wikileaks and the attack on Murdoch's media empire (Reychler 2011b). "Operation unified protector" in Libya was a welcome distraction, turning international attention away from failures. Although initially declared a success, such a conclusion is turning out to be premature. There are a lot of missing data and the killing of Gaddafi cannot be equated with a happy end.

14.1.5 Towards a More Adaptive Diplomacy

Within the decline of Western influence in the MENA, however, lie seeds of hope for healthier relations in the future (The Economist 2011b). To improve these relations, more adaptive diplomatic thinking will be needed, with a higher level of sophisticated realism, transparency, accountability, proactive and long-term planning, and reconciliation of competing interests at different system levels (human, sub-national, national, regional, global and planetary).

- *Sophisticated realism* aims at sustainable security and peacebuilding. It categorizes countries in terms of interest and power relations, not in terms of medieval "good and evil" categories. It knows the limits of coercive diplomacy and military intervention. It does not reward regimes that repress, colonize and segregate. Above all, it does not allow national or foreign lobbies with particular interests to determine its security strategy.
- *Transparency and accountability* is vital for the development of successful foreign and security policies. As mentioned earlier, the foreign policy process is the least democratically controlled activity in most countries of the West. Internal and external interest groups whose particular interests do not always correspond with the national interest influence it. The least transparent activities are the covert actions of intelligence organizations and the secret forces. Illustrative of this are the American denial of the shooting of an Iranian passenger aeroplane above the Persian Gulf in 1988 and the covert actions to topple unfriendly regimes (Fisk 2011). Transparency also implies the non-use of non-transparent language (Reychler 2011b: 9).
- *Proactive and long-term planning*: The factor of time (the t-factor) is one of the most underestimated factors in conflict transformation.[12] Time is not always healing: proactive violence prevention can prevent escalation; many protracted conflicts are histories of missed opportunities; planning peace involves decisions about priorities, sequencing and synchronicity; there are last chances (Gardner 2009). According to Thomas Friedman, Israel feels relatively secure after beating Arafat, Hezbollah and Hamas in three short wars. The brutality of these interventions has given Israel some kind of time-out. The international

---

[12]This is explored further in Luc Reychler's ongoing research into the role of time on the dynamics of conflict and peace.

community will no longer accept another war with a large number of civilian victims. Therefore, the Israeli government must urgently adapt its policy and support the establishment of an independent and viable Palestine. The international community, including the Arab World, has offered the solution on a silver plate. In economics, time is money. In conflicts, it can be the difference between life and death, peace and war. This protracted conflict is a lose-lose affair; it undermines the interest and legitimacy of the West in the region.

- *Reconciling competing interests or reasons at different system level*: In the end the building of sustainable security and peace requires genuine negotiations with all the key stakeholders in the region, including actors who have been stigmatized as terrorists, radicals or evildoers, such as Iran, the Taliban, Hezbollah and Hamas. The creation of sustainable security and peace requires an intensifying application of knowledge and rationality. The "escalation of reason", as Pinker (2012) calls it, can force inimical actors to recognize the futility of violence and reframe it as a problem to be solved rather than a contest to be won (Cookson 2011). Tomorrow, a reasonable foreign country will have to reconcile the interest of the nation with the competing human, regional, global and planetary interests.

More adaptive and successful diplomacy is not only the responsibility of politicians and the diplomatic corps, but also of academics and researchers. The latter can make a difference by (a) researching what is needed for sustainable peacebuilding, (b) anticipating the positive and negative impacts, including the costs and benefits of different policy options, and (c) helping to raise the accountability of decision-makers and -shapers for their policies.

14.2 Conclusion

The policy of America and her closest allies after 9/11 turned the world into a wasteland of buried reason. That is how Tadjbakhsh (2011) describes what was done to the Middle East. I could not say it better. The policy of the West in the Middle East regressed from *raison* to *déraison d'état*. Reason was replaced by military muscle and coercive diplomacy. The war on terror inhibited progress. It gave authoritarian states and Israel an excuse to keep people from achieving inalienable rights like liberty and the pursuit of happiness. So much time, money and human life has been wasted! While fighting wars, most Westerners could enjoy their safety and go on vacation. The West failed to see the connections between terrorism and its policy in the Middle East. It allowed the irredentist and repressive policy of Israel, supported regimes with poor human rights records, and enlarged the Western military boot-print in the region. Instead of resolving these problems, the war was framed as a clash of civilizations: Islam versus the Judeo-Christian world. Anti-terrorism became more destructive and deadly than terrorism. It claimed "ten eyes for an eye"; violated human rights, and even curtailed intellectual freedom. You were with us or with the terrorists. The commemorations of 9/11 were moments of nationalism, respectful silence and expressions of rage. Missing from these events were reflections on the overreactions to 9/11 that harmed so many people. It is high time to unlearn the simplifications of the polarizing world-view and do more forward thinking (Nader 2011). This implies the need for several changes:

1. Sophisticated realism that invests in building sustainable security and peace.
2. Transparency and accountability in the shaping and making of foreign policy.
3. Calling out those in the media and think tanks who become the mouthpiece of politicians responsible for the *déraison d'état*.
4. Maintaining the health and vigour of our own society; not becoming a garrison state (Gaddis 2011).[13]
5. Dealing more effectively with 'enemies' and 'friends' who violate international law and undermine and inhibit the building of sustainable security and peace in the MENA.

It is high time we stop pursuing national interests in an unreasonable way. It wastes people, money and opportunities to build sustainable security and peace. Lau-Tzu warned that arms are the tools of those who oppose wise rule; they should only be used as a last resort. "Every victory is a funeral; when you win a war, you celebrate by mourning" (Dyer 2008: 61–63).

References

Afify, H. (2011). Egyptians' Hopes Turn Feeble. *International Herald Tribune* (3 November): 8.
Bacevich, A. J. (2008). *The Limits of Power: The End to American Exceptionalism.* New York: Metropolitan Books.
Baker, R. (Ed.). (2010). *Cultural Cleansing in Iraq: Why Musea are Looted, Libraries Burned and Academics Murdered.* London: Pluto Press.
Bauwens, W., & Reychler, L. (1994). *The Art of Conflict Prevention.* London: Brassey's.
Brown, C. (2010). *Practical Judgement in International Political Theory.* London: Routledge.
Buck, T. (2011). Rifts in the Valley. *Financial Times* (16 August): 7.
Butt, Y. (2011). The Delusion of Missile Defense. *International Herald Tribune* (21 November): 6.
Cookson, C. (2011). Peace Process: despite appearances, the human race is losing its appetite for violence. *Financial Times* (8 October): 18.
Courrier International. (2011). Et le vainqueur pour 2022 est…le Qatar. *Courrier International* (19 October): 13–17.
De Morgen. (2011). Afghanistan 10 jaar na de invasie: bedrogen door westerse beloften. *De Morgen* (7 October): 15.
Deng, F., Kimaro, S., Lyons, T., Rothchild, D., & Zartman, W. (1996). *Sovereignty as Responsibility: Conflict Management in Africa.* Washington, DC: The Brookings Institution.
Diamond, J. (2005). *Collapse: How Societies Choose to Fail or Succeed.* New York: Viking.

---

[13]*The Economist* (12 November 2011), p. 78.

Dinmore, G. (2011). Fear of a Clash of Civilizations Appears to be Overdone. *Financial Times* (9 September): 2.

Dyer, W. (2008). *Living the Wisdom of the Tao*. Carlsbad, CA: Hay House.

Edgecliffe-Johnson, A. (2011). Interview with Michael Moore. *Financial Times* (1 September): 3.

Elliott, L. (2011). Jobs Crisis Threatens Global Wave of Unrest, Warns ILO. *The Guardian Weekly* (4 November): 16.

Enzenberger, H. M. (2011). Giftige Wolk boven Europa. *De Morgen* (17 September): 18.

Farrall, J. M. (2007). *United Nations Sanctions and the Rule of Law*. Cambridge: Cambridge Press.

Ferguson, N. (2011). America's Oh Sh*t Moment. *Newsweek* (11 November): 41.

Field, A. (2011). West Unites in Calls for Assad to Step Down as US imposes Sanctions on Syria's Oil Sector. *Financial Times* (19 August): 1.

Field, A., & Barker, A. (2011). US Sanctions Need European Allies' Backing to Hurt Syria. *Financial Times* (19 November): 4.

Fisk, R. (2011). Hoe Rupert Murdoch mij de deur uitwerkte. *De Morgen* (14 July): 24.

Gaddis, J. L. (2011). *George F. Kennan: An American Life*. New York: Penguin Press.

Gardner, D. (2009). *Last Chance: The Middle East in the Balance*. London: I. B. Tauris.

Gettleman, J., et al. (2011): US Proxy War in Somalia. *International Herald Tribune* (12 August): 2.

Henig, R. M. (2011). Why is this Guy Smiling? *Newsweek*.

Human Security Centre. (2005). *Human Security Report 2005*. Retrieved from https://css.ethz.ch/en/services/digital-library/publications/publication.html/13773.

International Herald Tribune. (2011). Bigger sphere of influence for tiny Qatar. *International Herald Tribune* (15 November): 1–5.

Jakobsen, P. V. (2007). *Coercive Diplomacy*. In A. Collins (Ed.), *Contemporary Security Studies* (pp. 225–247). Oxford: Oxford University Press.

Janis, I. (1972). *Victims of Groupthink*. Boston, MA: Houghton Mifflin.

Jervis, R. (1976). *Perception and Misperception*. Princeton, NJ: Princeton University.

Johns, J. H. (2011). Before We Bomb Iran. *International Herald Tribune* (16 November): 8.

Kennedy, G. (1999). *The Planetary Interest: A New Concept for the Global Age*. London: UCL Press.

Kerr, P. (2007). Human Security. In A. Collins (Ed.), *Contemporary Security Studies* (pp. 91–108). Oxford: Oxford University Press.

Khalaf, R. (2011). Egypt's Unfinished Revolution. *Financial Times* (29 October): 19.

Kristof, N. (2011). Balance on the Middle East. *International Herald Tribune* (5 August): 7.

Krugman, P. (2011a). Bombs, Bridges and Jobs. *International Herald Tribune* (1 November): 7.

Krugman, P. (2011b). Oligarchy, American Style. *International Herald Tribune* (5 November): 7.

Luce, E. (2011). Why America is Embracing Protest. *Financial Times* (31 October): 9.

Mabubani, K. (2011). A Letter to Netanyahu: Time is No Longer on Israel's Side. *Financial Times* (11 November): 9.

Machiavelli, N. (1532). *Il Principe*. Rome: Antonio Blado d'Asola.

Mackintosh, James. (2011). *Financial Times* (12 November): 16.

Mancassola, M. (2011). Wikileaks. *Courrier International* (27 July): 42.

Marashi, R. (2011). America's Real Iran Problem. *International Herald Tribune* (11 November): 6.

Maslow, A. H. (1966). *The Psychology of Science: A Reconnaissance*. New York: Harper and Row.

McIntyre, A. (2004). Doctrine of Double Effect. In: *The Stanford Encyclopedia of Philosophy*. Retrieved from https://plato.stanford.edu/entries/double-effect/.

Mishra, P. (2011). A Dead End Looms for Our Young. *The Guardian Weekly* (2 September): 21.

Nader, R. (2011). Painful 9/11 Lessons. *USA Today* (1 September): 10A.

Nasr, V. (2011). When the Arab Spring Turns Ugly. *International Herald Tribune* (27 August): 6.

Norton-Taylor, Richard. (2011). An Attack on Iran Would Be Disastrous. *The Guardian Weekly* (4 November): 11.

Nye, J. (2008). *The Powers to Lead*. Oxford: Oxford University Press.

Occupy Wall Street (OWS). (2011). in en.wikipedia.org/wiki/Occupy_Wall_Street (8 November).

Pignal, S. (2011). Growth Pessimism Triggers Warning of Recession. *Financial Times* (11 November): 4.

Pinker, S. (2012). *The Better Angels of Our Nature: Why Violence Has Declined*. New York: Penguin Books.

Preston, P. (2011). Contradictions of Freedom. *The Guardian Weekly* (14 October): 39.

Reid, T. R. (2004). *The United States of Europe*. New York: Penguin Books.

Reychler, L. (2010). Intellectual Solidarity, Peace and Psychological Walls. *Paper presented at the International Conference on Peacebuilding, Reconciliation and Globalization in an Interdependent World*, Berlin.

Reychler, L. (2011a). Failing Foreign policy – Foreign Policy Failures: Assessing the Impact of the West on Peace and Development in the MENA. Leuven: KU Leuven.

Reychler, L. (2011b). Time for Peace: Europe's Responsibility to Build Sustainable Peace. Paper presented at the *EUPRA conference*, Tampere.

Reychler, L. (2012). Raison ou Déraison d'État: Coercive Diplomacy in the Middle East and North Africa. In A. Fabian (Ed.), *A Peaceful World is Possible* (pp. 217–241). Sopron, Austria: University of West Hungary Press.

Rieff, D. (2011). R2P, RIP. *International Herald Tribune* (8 November): 6.

Senneth, R., & Sassen, S. (2011). Britain's Broken Windows. *International Herald Tribune* (12 August): 6.

Steele, J. (2011). Rebels must show restraint. *The Guardian Weekly* (26 August): 19.

Stiglitz, J., & Bilmes, L. (2008). *The Three Trillion Dollar War: The True Costs of the Iraq Conflict*. New York: Norton and Company.

Stockholm International Peace Research Institute. (2010). *SIPRI Yearbook 2010: Armaments, Disarmament and International Security*. Oxford, England: Oxford University Press.

Stoessinger, J. G. (2005). *Why Nations Go To War* (9th edn.). Boston, MA: Wadsworth Thompson Learning.

Tadjbakhsh, S. (2011). *After 9/11: A Wasteland of Buried Reason* (10 September). Retrieved from https://www.opendemocracy.net/en/after-911-wasteland-of-buried-reason/.

Telhami, S. (2010). *The View from the Middle East: The 2010 Arab Public Opinion Poll* (5 August). Retrieved from https://www.brookings.edu/events/the-view-from-the-middle-east-the-2010-arab-public-opinion-poll/.

Thakur, R. (2011). *The Responsibility to Protect: Norms, Laws and the Use of Force in International Politics*. London: Routledge.

The Economist. (2011a). Rage against the machine. *The Economist* (22 October): 13.

The Economist. (2011b). Crescent Moon, Waning West. *The Economist* (29 October): 14.

The Economist. (2011c). Libya and its Friends: All Too Friendly. *The Economist* (12 November): 50.

The Guardian Weekly. (2011a). The Limits of Power. *The Guardian Weekly* (14 October): 22.

The Guardian Weekly. (2011b). US had a simplistic view of Afghanistan. *The Guardian Weekly* (14 October): 12.

The Guardian Weekly. (2011c). Minister's Resignation Exposes Tory Links to American Right. *The Guardian Weekly* (21 October): 14.

Tuchman, B. (1984). *The March of Folly. From Troy to Vietnam*. London: Sphere Books.

Watson Institute. (n.d.). *Costs of War*. Retrieved from http://costsofwar.org.

Wikipedia. (n.d.). 2006 Lebanon war. Retrieved from http://en.wikipedia.org/wiki/2006_Lebanon_War.

Wikipedia. (n.d.). Iraq sanctions. Retrieved from https://en.wikipedia.org/wiki/Sanctions_against_Iraq#Estimates_of_deaths_due_to_sanctions.

Chapter 15
Time, Temporament, and Sustainable Peace: The Essential Role of Time in Conflict and Peace

Abstract Time is the most precious resource we have (This article is a long version of the keynote speech the author gave at the *International Peace Research Association* (IPRA) in Istanbul in August 2014, on the occasion of IPRA's 50th anniversary. This text was first published as: "Time, Temporament, and Sustainable Peace: The Essential Role of Time in Conflict and Peace", in: *Asian Journal of Peacebuilding*, 3,1 (2015): 19–41. The permission to republish this text here was granted by the original copyright-holder.). It is irreversible and non-renewable. It makes the difference, more than ever, between the best and worst scenarios of climate change, energy competition, economic development, poverty, and security. Despite this, an incredible amount of time is wasted, especially the time of others and of nature. These latter resources are needed to prevent violence, build sustainable security, and ensure the well-being of all. Therefore, is it high time to radically change the way we deal with time and to develop a more adaptive 'Temporament'? This article defines time, surveys temporal deficiencies, and presents the parameters of a more responsible way of dealing with time in conflict transformation.

Keywords Sustainable peacebuilding · Temporament · Time · Democratic fascism · Failed foreign policy

15.1 Introduction

Something is fundamentally wrong with the way we deal with conflicts. Instead of contributing to sustainable peacebuilding, the foreign policy of the West has led to more insecurity abroad and at home. This has been especially true since 9/11 in the Middle East and North Africa. The obsession with military supremacy has transformed diplomacy into coercive diplomacy. Characteristic of the latter is the absence of proactive conflict prevention, a high level of temporal inadequacies, and the enormous waste of time, especially the time of others and of nature. Without a more adaptive 'temporament' (the manner of thinking, feeling and behaving towards time) we will not be able to deal successfully with global challenges or achieve sustainable development and peace.

The first part of this article defines sustainable peace in terms of outcomes and necessary preconditions, given that sustainable peace is an indispensable precondition for the survival of humanity. The second part looks at the impact of Western foreign policy on the global peacebuilding process. In the Middle East and North Africa, especially since 9/11, the impact has been disastrous. The third part highlights time as an essential factor in conflict transformation and peacebuilding. It also offers a multidimensional definition of time and stresses the importance of taking a big-picture view of time. Most people focus on one or two facets of time but overlook others. The fourth part of the article overviews several types of temporal inadequacies. The prevailing temporament is clearly not fit for dealing with the global challenges or achieving sustainable development and peace. The last part of the article deals with how to upgrade our temporal behaviour. It defines the parameters of a more adaptive temporament. This discussion of time is based on the research done for my book, *Time for Peace* (2015), which includes a questionnaire to assess one's own or another actor's temporament. Most of the examples relate to the Middle East and North Africa, and to the Gaza war, which killed some 2100 Palestinians and 73 Israelis, and destroyed 17,000 buildings in the Gaza area.

15.2   The Imperative of Sustainable Peace

In our globalized world, building sustainable peace has become imperative. If we do not improve the human climate or handle conflicts more constructively, we will not be able to deal with the interlocking global crises successfully. Sustainable peace is an old dream that has become an indispensable precondition for the survival of humanity. The opportunity costs of the unresolved violent conflicts and military interventions are too high. Too many opportunities to deal more effectively with the global crises are lost. Military interventions in the name of democracy, regime change, human rights, or in pursuit of Netanyahu's "sustainable quiet,"[1] do not further sustainable security and peace.

Since terms like 'peace' and 'sustainability' have many different meanings, it is imperative to provide an operational definition. In this article, "sustainable peace" refers to a situation with a very low level of direct and indirect violence. It distinguishes itself from other types of peace by the absence not only of physical violence, but also of structural, psychological, cultural, environmental, and temporal violence. It is the most cost-effective means of violence prevention.[2]

---

[1] A term used and coined by Israeli Prime Minister Netanyahu in an interview during the Gaza war of 2014. https://www.haaretz.com/netanyahu-gaza-must-be-demilitarized-1.5257042 (accessed 19 February 2015).

[2] A complete cost assessment considers humanitarian, economic, political, material, social, cultural, psychological, ecological, and spiritual costs (Reychler/Paffenholz 2001: 4).

Sustainable peace requires the installation of five peacebuilding blocks plus one, which is called the peacebuilding pentagon (Reychler 2006).

The first building block focuses on the establishment of *an effective communication, consultation, and negotiation system at different levels* between the conflicting parties or party members. In contrast to the negotiation styles used in most international organizations, the European Union's negotiation style, for example, is predominantly integrative. Ample time and creativity is invested in generating mutually beneficial agreements. Without win-win agreements, the Union could disintegrate and become a 'disunion'.

The second building block consists of *peacebuilding structures.* In order to achieve a sustainable peace, (conflict) countries have to install political, economic, and security structures and institutions that sustain peace. The political reform process aims at the establishment of political structures with a high level of legitimacy. The legitimacy status is influenced by: (1) the effectiveness of a regime to deliver vital basic needs, such as security, health services and jobs; and (2) the democratic nature of the decision-making process. Initially, an authoritarian regime with high-quality leaders and technocrats can obtain a high legitimacy score, but in the end, consolidated democracies are the best support for sustainable peacebuilding. It is crucial to note that the transition from one state (e.g., non-democratic structures) to another (e.g., a consolidated democratic environment) is not without difficulties; the devil is in the transition (Reychler 1999). The economic reform process envisions the establishment of an economic environment that stimulates sustainable development, eliminates gross vertical and horizontal inequalities, and develops positive expectations about the future. The security structure safeguards and/or increases the population's objective and subjective security by effectively dealing with both internal and external threats. This implies a cooperative security system producing a high level of human security, collective defence and security, and proactive conflict prevention efforts (Cohen/Mihalka 2001: 69).

The third necessary building block for establishing a sustainable peace process is an *integrative climate,* which is the software of peacebuilding (Reychler/Langer 2003: 53–73). This peacebuilding block highlights the importance of a favourable social-psychological environment. Although climate is less tangible and observable than the other building blocks, it can be assessed by the consequences. The presence of an integrative or disintegrative climate can be assessed by the prevailing attitudes, behaviour, and institutions. Characteristic of an integrative climate are expectations of an attractive future as a consequence of cooperation, the development of a 'we-ness' feeling or multiple loyalties, reconciliation, trust, social capital, and the dismantlement of senti-mental walls.

The fourth building block consists of *systems supporting the development and installation of the other peacebuilding blocks:* (1) an effective and legitimate legal system, (2) an empowering educational system, (3) peace media, (4) a well-functioning health system, and (5) humanitarian aid in the immediate post-war phase.

The fifth building block is *a supportive regional and international environment.* The stability of a peace process is often dependent on the behaviour and interests of neighbouring countries or regional powers. They can have a positive influence on the peace process by providing political legitimacy or support, by assisting with the demobilization and demilitarization process, or by facilitating and stimulating regional trade and economic integration. However, these same actors can also stifle progress towards stability, for example, by supporting certain groups that do not subscribe to the peace process. Likewise, the larger international community plays a crucial role in most post-conflict countries. The international community, by means of UN agencies or other international (non-) governmental organizations, can provide crucial resources and funding or even take direct responsibility for a wide variety of tasks, such as the (physical) rebuilding process, political transformation, humanitarian aid, and development cooperation.

The installation of all these building blocks requires *a critical mass of peace-building leadership* (Reychler/Stellamans 2005) in different domains (politics, diplomacy, defence, economics, education, media, religion, health, etc.) and at different levels: the elite, mid-level, and grass-roots (Lederach 1997). Peacebuilding leadership envisions a shared, clear, and mutually attractive peaceful future for all who want to cooperate; these leaders do everything to identify and gain a full understanding of the challenge confronting them; they frame the conflict in a reflexive way; their change behaviour is adaptive, integrative, and flexible; they are well acquainted with non-violent methods; they use a mix of intentional and consequential ethics and objectives; and they are courageous men and women with high levels of integrity (Reychler/Stellamans 2003: 1–49).

All of these peacebuilding blocks are essential and interlocking. The lagging of one or more can seriously impede the pace of the peacebuilding process.

15.3 Failing Foreign Policy

What is the impact of Western foreign policy on peacebuilding? Instead of finger-pointing at the so-called "undemocratic and less civilized world", let us look at the three fingers pointing back at ourselves. After 9/11 the diplomatic landscape changed drastically. The art of diplomacy became a different creature:

- It transformed into a type of coercive diplomacy that makes use of diplomatic isolation, sanctions, threats, armed interventions, and psychological warfare. In the twenty-first century, democratic countries are fighting most international wars in the world, in particular in the Middle East and North Africa. Interventions in the name of security, anti-terrorism, anti-weapons of mass destruction (WMD), regime change, human rights, and democratization have brought instability, human suffering, and material destruction, leaving weak and failing states. All of this could be judged as "failed foreign policy". Powerful interest groups, however, consider these interventions successes; they prefer weak and unstable states over stable regimes that can resist Western interference. The prime goal of the West in the Middle East is to achieve absolute security by means of offensive and defensive military dominance and regime change. The democratic West is the best-armed group of countries in the world. The main difference between authoritarian and democratic regimes is that the former commit violence against their own citizens, whereas the latter do so against other people.

Domestic interference, or government-to-people diplomacy, became a major tool for the new diplomacy. The support and manipulation of "square democracy" in Iran, Egypt, Libya, Syria, and Ukraine are prime examples of this. The scenarios are similar: protests are staged on a square in the limelight of the international news media; the opposition is supported and portrayed as the voice of the people; and the regime currently holding office is encouraged to step down. When the new regime does not meet Western expectations, it is ousted by a *coup d'etat*, repression, isolation, or by a new square democracy protest. Think of the elimination of the elected Hamas government in Palestine and of the Morsi-led government in Egypt. Regarding the former, when a Palestinian unity government was created, Israel claimed that it would never accept a peace agreement. Hamas needed to be destroyed first and then elections could he held.

Foreign policy decision-making has become less democratic. Winston Churchill's observation that "democracy is the worst form of government, except for all others that have been tried" applies to domestic politics, not foreign policy. Worrisome, especially in regard to the West's foreign policy towards the Middle East and North Africa (MENA), is the appearance of democratic fascism. This may sound like an exaggeration, but when judged by the main parameters of fascism, the foreign policy of the West and its allies in the MENA shows signs of creeping fascism. It is chilling to see how the foreign policy of the West gradually changed into an Ionescan *Rhinoceros* type of behaviour (Ionescu 1960).

First, foreign policy in the Middle East is extremely militaristic. The aim is military superiority at a price amounting to hundreds of thousands of casualties and billions of dollars. During the last Gaza war, Israel's biggest-selling newspaper called for returning Gaza to the Stone Age (Regev 2014). Second, foreign policy has become extremely patriotic. Citizens are rallied around the Haj to defeat terrorists and evil rogue regimes. There are first-, second- and third-class victims. Enemies deserve to become victims; they are collateral damage. "As Palestinian children are killed, that may seem like a lot to stomach, but it is no less necessary", (Jones 2014: 18). Third, the 'enemy' is dehumanized and stigmatized as cruel, evil, vicious, backward or a terrorist. One seems to forget that anti-terrorism and state terrorism cause disproportionately more violence and destruction (in some cases ten, a hundred or ten thousand times more) than most terrorist acts. Fourth, this policy seeks full control of the region and is expansionistic. Much of the territory of Palestine since 1967 has become exclusive zones or colonized areas, and when the Israeli settlers left Gaza in 2005 it was turned into a ghetto of 1,800,000 people. Fifth, propaganda and spin doctors define realities and responsibilities. For example, "Our domestic interference is justified; others are responsible for the rack and ruin. We are fighting in a civilized way whereas 'they' are barbaric". Gazan people would receive telephone calls to inform them that their houses would be bombed five minutes later. If they did not leave their homes in time, the inhabitants would be considered fully responsible for their own deaths. ISIS fighters' slicing of Westerners'[3] throats has rightly been depicted as full, barbaric horror. But what about the tens of thousands of innocent civilians, adults and children who are pierced, torn to pieces, burned, poisoned, or suffocated by intelligent weapons? Part of the propaganda is the staging of scapegoats, such as Iran, Hamas or, more recently, President Putin. There is a systematic denial of the policy of domination and of the responsibility for the negative consequences of interventions. Dissidents arc labelled unpatriotic, traitors, leftists, peaceniks, and are ostracized and threatened with death or job loss.

- Finally, the new diplomacy is shaped by the media revolution and by fear. There is not only the culture of instant news, but also the phenomenon that, except in extreme circumstances, it is scarier to follow an event on TV than actually being there covering it (Brooks 2014: 13). David Axelrod points to the political and news media culture that has gone well beyond healthy scepticism and scrutiny. "There's an impetus to create fear and then market and exploit it. And that's true on the part of the media, and that's true on the part of the politicians" (cited in Backer 2014: 5). A Swedish conflict analyst, Jan Oberg, commenting on the submarine hysteria in the coastal waters of Sweden in October 2014, observes that during the last twenty to thirty years the quality of media work and commentary in the fields of security, defence and peace has steadily declined. There is less professional knowledge, research, and independent analysis; there is less resistance to marketing and psychological operations (PSYOPS), which is done by the military to boost its legitimacy in the eyes of the paying public; there is much more uniformity; the disease exists of journalists interviewing other journalists as if they were experts; and there is a fierce struggle for sheer survival in the digitalized media world with fewer owners in the business (Oberg 2014). The escalation of fear after 9/11, and more recently due to the ISIS and Ebola crises, has been amplified by the feeling that boundaries have become more porous and that government policies are less effective in preventing threats and keeping problems and death at a distance. According to Higgins (2014: 1), this hysteria and overreaction "springs from a paradox at the heart of the West". The more we master the world through science, technology (and economic and military dominance), the more frightened we are of those things we cannot control, understand, and anticipate. "We live in very secure societies and like to think we know what will happen tomorrow. There is no place ... for the unknown" (Higgins 2014: 4). All of this has transformed the art of diplomacy into rapid-reaction diplomacy (Seib 2012). On the whole, the impact of foreign policy on peacebuilding in the MENA has been disastrous.

---

[3]The promise of military intervention encouraged the armed rebellion and the refusal to negotiate. The NATO intervention also increased the number of casualties significantly.

15.4   The Essential Role of Time

15.4.1   The Relevance of Time

Time is an essential factor in conflict transformation and peacebuilding. First, it provides a more sensitive and comprehensive measure of violence. Attention is given not only to fast violence (killing), but also to slow, chronic violence that results from structural, psychological, ecological, cultural and other means of violence. Analysts frequently overlook temporal violence. Second, time is an important aspect of the conflict environment. Think of the impact of a crisis environment or of historical trauma. The appetite to join or not join the Arab Spring revolutions was strongly influenced by memory. In Algeria, the civil war against the Islamists in the 1990s – at the cost of 200,000 lives – inhibited revolutionary escapades. Third, time is an aspect of descriptive, explanatory, and predictive analysis. There is, for example, the *sticky past*, events or developments in the past that have an impact on today's behaviour and are difficult to erase or neutralize, such as demographic and ecological trends, or traumatic experiences. There is also the *malleable past*, relating to lessons learned, the manipulation of history, and dealing with a violent past. Fourth, time is a major component of the planning and implementation of peace. Decisions have to be taken with respect to sequencing, prioritization, entry and exit, and the nature of change. Fifth, time is an important aspect of evaluation. It provides criteria for distinguishing bad and good temporal behaviour. Good temporal behaviour implies a high level of temporal democracy and temporal empathy, deference to nature's time, and adequate efforts to build sustainable peace. Finally, time is a tool of power and influence.

15.4.2   The Dimensions of Time

While I was writing *Time for Peace,* some people asked me if the book was about philosophy or history. 1 told them that it dealt with existential time. We are all heading to the future and to death, or the end of our time. To save our own lifetimes, or that of loved ones, most people would consider killing if no other options are seen. I also reminded them that history is very important, but it is only one facet of temporal reality. Time is a multi-dimensional phenomenon. A thorough temporal analysis investigates the big picture, the cross-impacts between the different dimensions, and the emotions of time. A narrow temporal analysis can lead to serious distortions and misunderstandings. The temporal dimensions are divided into five principal and four transformative dimensions. The principal dimensions categorize the fundamental components of time in conflict and peace. The transformative dimensions draw attention to variables that can alter temporal perception and behaviour.

15.4.2.1 Principal Dimensions

Existential Time

Existential time is about life and death. The duration and quality of a lifetime can be reduced by fast (armed) violence or slow (structural) violence. People who have been subjected to armed and long-term structural violence may decide to risk or even offer their lives to achieve freedom. Someone confided to me that he had played with the idea of considering himself dead. This made it easier to live with difficult circumstances and to risk his life for freedom, as he was dead anyway. In the current Gaza conflict, the issue is not only how to bring about a ceasefire, but also a cease-occupation. The price of turning a blind eye to freedom aspirations is often a surprise (Table 15.1).

Time Orientation: Past, Present and Future

After a bloody conflict, a peace process must deal with *the past, the present, and the future*. The three time orientations are interdependent. It is very difficult to deal with the past successfully when the parties involved in a conflict cannot imagine a better common future.

Time Modes

This dimension relates to change, succession, continuity, turning points, duration, and sustainability. Each of these modes can be further broken down. Change can vary in terms of its speed or pace, its momentum, magnitude, pattern, and its visibility or invisibility. The Gaza conflict entails a temporal confrontation between Israelis, who prefer the status quo or slow, step-by-step change, and Palestinians, who demand radical and immediate change, and an independent state here and now.

Table 15.1 Principal and transformative dimensions of time. *Source* The author

Principal dimensions		Transformative dimensions	
1.	Existential time	6.	Manipulation of time
2.	Time orientation	7.	Temporal equality or inequality
3.	Time modes	8.	Temporal empathy
4.	Anticipation	9.	Temporal efficacy
5.	Temporal management options			

Anticipation

Anticipating or not anticipating crises and the negative impacts of interventions: The recent regime change in Libya is an example of tremendous temporal misconduct or violence. 9/11 may have been the most visible turning point in US foreign policy, but the armed regime change in Libya altered the international landscape tremendously.

Temporal Management

Temporal management deals with the timing of interventions, proactive or reactive conflict prevention, the prioritization of efforts, sequencing, synergy, and coherence. Today's diplomatic landscape is dominated by a reactive approach to conflicts. As long as an anti-terrorist policy does not deal with the root causes, it is likely to remain expensive, counter-productive, and incapable of delivering lasting results.

15.4.2.2   Transformative Dimensions

Manipulation

Manipulation and framing of the perception of time: The Gaza war of 2014 has been the subject of a great deal of temporal manipulation. The bombardments of Gaza by Israel were justified as reactions to Palestinian missiles. Crucial elements in the causal chain, such as the transformation of Gaza into a mega prison camp and the imprisonment of a great number of political leaders, are left out of the picture. The same is true with regard to Iranian-American relations. America tends to delete from the narrative: the CIA-assisted coup of an elected, secular government in 1953; giving the green light to Saddam Hussein's war against Iran in 1980; and shooting down an Iranian civilian airliner in 1988.

Temporal Equality

Temporal equality and inequality: In a genuine democracy the time of each citizen is equally valuable. In the Occupied Territories, the time of the Palestinians is controlled and wasted by checkpoints, slow and unpredictable administrative procedures, political imprisonment, and so forth.

Temporal Empathy

Temporal empathy is the capacity and will to discern how others think and feel about time. Each culture has its own orientation to the past, present, and future; a preference for sequential or synchronic organization of its activities; particular uses of short- or long-term time horizons; and a distinctive sense of temporal control, and of the value of human, natural, and transcendental time.

Temporal Efficacy

Temporal efficacy is the opposite of determinism, fatalism, powerlessness, defeatism, despair. It indicates a reasonable confidence in one's ability to understand the significance of time and to deal with time in ways that further one's interests.

To fully grasp conflict and peace behaviour, attention should be given to the big picture of time. First, this implies assessing the impact and cross-impact of every temporal dimension. It also means paying attention to temporal emotions. Time is not only an abstract, conceptual experience. Emotion is central to the experience of time. The past can be very painful. People can become victims of future shock, which can lead to maladaptive behaviour, such as denial, obsessive nostalgia for previously successful, adaptive routines, or the use of super simplifiers (Toffler 1970). Significant changes can trigger strong emotions, such as the feeling of unpredictability or surprise, anger directed against those who resist or push too hard, denial and disillusionment. Change can also raise high expectations, hope, euphoria, enthusiasm, anticipation, happiness, love, and fulfilment (Cloke/ Goldsmith 2002). While researching temporal emotions, I was surprised not only by the importance, but also the great variety, of temporal emotions. Twenty-four could be identified. Each emotion has its opposite (hope/despair, trust/distrust, optimism/pessimism), but also subcategories: smart and blind trust, generalized and particularized trust, moralistic trust, strategic trust, deterrence-based trust, and identification-based trust. The strongest feelings accompany death or closeness to death; the best and the worst emotions take over when wars break out, such that war itself becomes the cause of more war. Finally, the big picture entails paying attention to the influence of both secular and religious time. Despite the process of secularization, religion remains an important factor in human behaviour. For the majority of people in the world, religion is a major part of private and collective identity. Religious institutions promote moral principles and defend their interests. They have also become part of the globalization process in which people compare and evaluate the limits and possibilities of different belief systems. This can lead to friction and conflict, but also to the growth of eclecticism or shared (poly-) religious truth (Galtung/MacQueen 2008) (Table 15.2).[4]

---

[4]See also Wikipedia. "Double bind", at: http://en.wikipedia.org/wiki/Double_bind (accessed 5 January 2015).

Table 15.2 Temporal inadequacies. *Source* The author

Temporal insensitivity
1. Weak appreciation of the role of time in conflict and peace processes
2. Neglect or denial of temporal violence
3. A wide gap between the value of my time and your time
4. Low level of temporal discernment and empathy
Temporal malpractice
5. An unsatisfactory and unbalanced orientation to the past, present and the future
6. Strong propensity for reactive conflict prevention
7. Incoherent temporal management
8. Low investment in sustainable development and peace
Unethical temporal behaviour
9. Low accountability for temporal misconduct and violence
10. Undemocratic time
Low temporal efficacy and reflexivity
11. Temporal inefficacy
12. Low reflexivity, invention and learning

15.5 Temporal Inadequacies

Conflict transformation and peacebuilding efforts continue to be seriously hampered by multiple temporal inadequacies. In *Time for Peace,* twelve inadequacies in the ways people deal with time are identified. The inadequacies are split into four groups dealing with problems of temporal sensitivity, temporal praxis, temporal ethics, and temporal efficacy.

15.5.1 Low Temporal Sensitivity

1. Weak appreciation of the role of time in conflicts. The military intervention in Libya (March–October 2011) illustrates several temporal deficiencies that significantly raised the human and material costs. This intervention, although not recognized as such, has been a major turning point in international relations. Despite the fact that the intervention was lauded as a success and role model for humanitarian intervention, it can also be judged a case of temporal folly or temporal misconduct. Kuperman (2013) calls it a negative model of humanitarian intervention; it suffered from serious temporal deficiencies.

   The first of these was the manipulation of the perception of the past, present, and future. The media diabolized Gaddafi's violence as a lunatic massacre, reminded its audience of the Lockerbie bombing, warned them of a new Srebrenica or Rwanda, depicted the protests as non-violent, and exaggerated the initial death toll by a factor of ten. In fact, large-scale violence was initiated by one section of the protestors, and the government never responded with indiscriminate force. Second, a strong sense of urgency and existential crisis was created in which no time could be wasted by delays or diplomatic niceties to prevent or stop a possible human disaster. Third, several opportunities to deal with the conflict in more constructive and effective ways were missed. Offers by Venezuela, the African Union (AU) and Turkey for mediation towards a ceasefire were refused. The primary objective of the rebels and NATO was to overthrow the regime. Fourth, regime change was prioritized as the preferred outcome, even if this escalated and extended the civil war and thereby raised the threat to the Libyans (Kuperman 2013: 115).

Fifth, there was a failure to anticipate and prevent negative external side-effects or harm. The war spilled over into Mali and also had a negative effect on the non-Western part of the international community. Russia, China and the AU felt betrayed by a transformation of the Responsibility to Protect (R2P) principle of the intervention into regime change. The most negative external impact was felt in Syria. The success of the rebels in Libya raised the expectations of the Syrian armed opposition. With external support the opposition expected a quick, decisive victory and so negotiations with Assad were out of the question. On the other side, the Syrian regime was prepared to do anything to avoid what they had seen in Iraq and Libya (the hanging of Hussein, the lynching of Gaddafi, and the degradation of the countries into broken and weak states). Major international players, especially Russia, China and the African Union, refused to support military intervention in Syria, even to protect civilians.

2. *Neglect or denial of temporal violence*: Temporal violence refers to a quantitative and qualitative depreciation of one's life expectancy as a consequence of protracted conflict, long-term sanctions, structural violence, ecological deterioration, or the killing and wasting of others' time. Everybody has the right to be fully alive right now. Corrupt regimes and failed countries waste a disproportionate amount of the time of their citizens. Colonization goes hand in hand with considerable temporal violence. In the remaining Palestinian Territories (approximately 22% of the mandatory Palestine of 1947), Palestinians undergo long-term structural violence (also called apartheid, colonization, sanctions), and thousands are political prisoners (Brown 1998). The time frames of Gaza's inhabitants are enclosed, their economic and educational opportunities curtailed, and a great deal of their precious time wasted at checkpoints and in the absence of corridors between isolated parts of the territory. Another example of temporal violence is poverty. The term "extreme poverty", used by the rich and powerful, veils the impact of gross inequality and the long-term violence experienced by the less poor.

3. *A wide gap between the value of my time and their time*: Earlier we distinguished between first-, second- and third-class victims. Second- and third-class victims are those on the other side. Our victims are prioritized as first-class.

Threats to 'us' are strategic or existential threats, which justify the disproportionate use of firepower, torture and rapid dominance. Innocent citizens on the other side are labelled collateral damage, as the regrettable outcome of 'unintentional' but predictable incidents. After shooting down an Iranian airbus in 1988 (in the same year as the Pan Am Lockerbie bombing), the US government regretted the loss of human lives and paid reparations, but never apologized or acknowledged wrongdoing. And in the climate debate, future generations have no voice.

4. Low empathy for temporal differences: There are many temporal cultures. Academic disciplines conceive of time in a variety of ways. The professionals involved in conflict transformation and peacebuilding have different views about priority-setting and sequencing. People think and act differently in peacetime and wartime. The temporal culture of the rich and strong contrasts significantly with the time experienced by the poor and weak. Religious time differs from secular time. Despite these differences, there is a tendency to interpret the temporal experiences of others through one's own familiar temporal lenses and/ or to impose our temporal culture on others. On the whole, the will and ability to view and feel how the other party conceives and experiences time and temporal violence is low.

15.5.2 Temporal Malpractice

5. An unbalanced orientation to the past, present and future: Temporal orientations can be considered inadequate when the needs and challenges of the past, present and future are handled insufficiently and in unbalanced ways. For example, the past can be repressed or put between brackets in order to build a new future, or inadequate efforts may be made to raise hope for a better and common future. In many conflicts, the violent past is not dealt with satisfactorily. On 27 February 2012, the Spanish Supreme Court exonerated Judge Baltasar Garzón of abusing his authority (prevarication) in an investigation into disappearances during the Spanish Civil War. Previously, Garzón had successfully issued arrest warrants against the Chilean dictator Pinochet and the Argentine military command for their responsibility in genocide, torture and state terrorism. The trouble came when he applied his theory that neither statutes of limitation nor amnesty laws could preclude investigations into crimes against humanity in his own country. No one has ever been held accountable for crimes during that civil war, and up to 150,000 dead remain unidentified in unmarked graves (Roht-Arriaza 2012).

Several problems relate to the handling of the present. George Loewenstein says that the ceaseless influx of information has conditioned our decision-making machinery to what is latest, not what is more important or more interesting (the 'recency' effect). "We pay a lot of attention to the most recent information, discounting what came earlier" (Begley 2011). Decisions are driven by what is urgent rather than what is important. Finally, there are future orientations that inhibit adaptive change. Think of the short-term thinking of the majority of politicians or the use of worst-case scenarios. Judt (2012) criticized the state of Israel for its use of fear, especially "the fear that Israel could be wiped off the face of the earth", in order to justify the continuation of an unavailing policy. The fear, he argued, is not a genuine one, but a politically calculated, rhetorical fear. Many governing elites are not seizing the moment to create responsible financial systems, sustainable economies, and a more equal world, or to deal with environmental deterioration and the shrinking access to vital human necessities, such as food and water (George 2010).

6. *Propensity for reactive conflict prevention*: Proactive violence prevention is at a low point. Following 9/11, traditional diplomacy was high on force and low on diplomacy. The very word diplomacy became unfashionable on Capitol Hill and in some European capitals, such as Tony Blair's London: the drums of confrontation, toughness and inflexibility prevailed (Cohen 2013). Concepts like preventive diplomacy and conflict prevention were removed and replaced by layers of economic sanctions, diplomatic pressure and isolation, military threats, pre-emptive and preventive wars. The Middle East turned into a living museum of defunct diplomacy. In reactive policy most, if not all, the attention went on the symptoms and not the root causes. The war on terror aimed at killing or incapacitating the terrorists and the organizations or networks behind them. Frequently, negative side-effects were not anticipated, were denied or considered necessary evils. Think of the American support of Muslim fighters in the Afghan-Soviet war, or the emergence and growing strength of Hezbollah during the eighteen years of Israeli occupation of southern Lebanon. Measures are taken to eliminate risks to 'our' people by the installation of stringent border controls, the building of protective walls, fighting from the air, launching missiles from distant bases, and engaging private security corporations and mercenaries. In the short term, reactive conflict prevention may seem successful, but in the long term it can be counter-productive and overly expensive.

7. *Incoherent temporal management*: Achieving a coherent peacebuilding process is difficult. Coherence depends on the confluence of decisions taken with respect to the six components of the peacebuilding architecture: (1) inclusion or exclusion of the major stakeholders; (2) definition of the end state – the peace they want to install and the theoretical assumptions about how to get there; (3) assessment of the conflict at the baseline and the peacebuilding deficiencies; (4) analysis of the context, including power relations, the willingness to build peace, and the peacebuilding resources; (5) the nature of the peacebuilding process, including temporal issues, such as when to intervene and when to stop, priority-setting and sequencing of operations, and the creation of synergies; and (6) monitoring and evaluation of the results.

8. *A shortage of investment in sustainable development and peace*: Despite the fact that sustainable development and peace resonate with strongly held convictions about the present and the future, their realization has proven to be highly elusive (Adger/Jordan 2009).

15.5.3 Unethical Temporal Behaviour

9. *Low accountability for temporal misconduct*: Decision-makers and shapers are accountable if they are expected to explain their decisions and believe they can be rewarded or punished as a result. When they are held accountable for their decisions, they are likely to be more careful and will be more likely to procure and evaluate recommendations or policies in a more holistic manner (Mintz and De Rouen 2011: 30–33). For Mintz and De Rouen, a holistic search means reviewing all the information on alternative courses of action, the dimensions that influence the decision, and the implications of each alternative. The level of unaccountability and immunity for temporal misconduct in foreign policy is high. Temporal misconduct can involve: the neglect of early warning signals of genocidal behaviour or civil war, the absence of adequate preventive or damage-limiting measures, exaggerating threats and thereby the manipulation of fear, negligence towards the negative impacts of economic and political interference, the defence of policies with inappropriate historical analogies, killing and wasting the other's time, giving more time to military and coercive intervention than to diplomatic efforts to stop violence, and so on.

10. *The undemocratic control of time*: Controlling time has always been a key to power. Giordano Nanni illustrated the linkage of power and time in his book, *The Colonization of Time* (2012). Nanni examines British rituals and concepts of time imposed on other cultures as fundamental components of colonization during the nineteenth century. Today, time remains an important source of political, economic, and military power. Jeremy Rifkin observes that some people's time is more valuable or expendable than others': millions starve while a minority lives in splendour. The rich and powerful tend to shape the (preferred) future world order (Kapur 2014). Jaron Lanier, a philosopher and computer specialist, argues that the corporations with the newest and fastest computers, using data gathered for free from the public, are able to calculate ways to avoid risk, thus making society riskier for everybody else. "Instead of leaving a greater number of us in excellent financial health, the effect of digital technologies – and the companies behind them – is to concentrate wealth, and challenge [the] livelihoods of an ever increasing number of people" (Lanier 2013). Temporal autonomy, or discretionary time, which is unequally distributed, is a salient measure of freedom and democracy (Whillans 2011).

15.5.4 Low Temporal Efficacy and Reflexivity

11. *Temporal inefficacy*: Several developments have increased feelings of discomfort and temporal inefficacy in different parts of the world. There is growing pressure to be efficient and meet the fast pace of life. There are complaints about the world changing at lightning speed (Benkler 2011), the increasing

scarcity of time or "time famine", fast information and communication facilities, and short response times. The world is plagued by chronic crises. More than a billion people must try to survive on less than $1 a day. Urban youngsters with poor economic prospects are impatient, and slow political change has led to revolutionary protests in Sarajevo, Kiev and Cairo. Indicators of temporal inefficacy are the emotions of fear in the West and humiliation in the Muslim world, as described by Moisi (2009).

12. *Low on reflexivity, invention, and learning*: Our brains need more space and time to have new strategic insights and ideas.[5] The lack of understanding of rapid political changes seems to be sublimated by moralizing about international affairs. Secular missionaries pursue national interests in the name of the responsibility to protect, human rights, democratization, freedom, liberalization. However, interpreting Lao Tzu, Ralph Alan Dale reminds us that the course of events does not simply follow our wishes and prayers:

> The harder we try to force events to conform to our moralization, the less likely our success. On the other hand, the more we yield to the rhythms of life, the greater our fruition. How often Lao Tzu bids us to put aside our ideological predilections so that we may be free to ebb and flow with the new opportunities of every pregnant moment (Lao 2002: 172).

Temporal inadequacies can result from poor temporal intelligence, the arrogance of power and greed, pseudo-democratic decision-making, a low level of accountability for temporal misconduct and violence, the denial of temporal problems, the sanctioning of dissident voices, and the blaming of others for negative impacts.

Although people pay ample attention to time, they tend to focus on some aspects, for example the past and the near future, and overlook the other aspects of time. Temporal inadequacies have also been attributed to stupidity. Tuchman (1984) calls it wooden-headedness. It consists of assessing a situation in terms of preconceived and fixed notions while ignoring or rejecting any contrary signs. She epitomizes this type of self-deception with a historian's statement about Philip II of Spain: "No experience of the failure of his policy could shake his belief in its essential intelligence" (Tuchman 1984: 7). The pursuit of power, prestige and greed can intensify and prolong violence. Nixon and Kissinger's New Year's wish of 1973, to end the Vietnam War with a "peace with honour", protracted the terrible war for more than two years. During the 1980s, America supported brutal pro-Western dictatorships and opposed truly popular governments and opposition movements in Latin America. Exclusive democracies characterized by segregation, apartheid and repression commit temporal violence to control second-class citizens. When elections are held in the Occupied Territories, elected collaboration is favoured and elected resistance repressed. Liberal democracies that do not provide a

---

[5]Shock therapy involved a sudden privatization of Russia's 225,000 or so state owned businesses, a sudden release of price and currency controls, withdrawal of state subsidies, and sudden trade liberalization. See Smith (n.d.) and Wikipedia, "Shock therapy", https://en.wikipedia.org/wiki/Shock_therapy_(economics), accessed on 20 March 2020.

minimum of social security can create gross inequalities and poverty. Another factor that enhances temporal inadequacy is the lack of political accountability for temporal misconduct and violence.

Problematic also, especially in the Middle East, is the denial of double-bind policies by the West and Israel. Both strongly avoid recognizing and confronting their conflicting aims, such as pursuing military dominance and at the same time expecting sustainable security and peace, or Israel's colonization of the occupied Palestinian territories and its expectation of "sustainable quiet". These policies are confounded by inherent unresolvable dilemmas. To deal with emotionally distressing dilemmas, resolute efforts are made to deny responsibility, silence critics, and blame and punish others for the negative impacts. Governments who are not willing to confront an unresolvable dilemma will neither resolve it nor opt out of the situation (McNally 2012; Gisha 2009).[6] The people who suffer most from the negative impacts of double-bind policies[7] are those in the area where the interventions take place. A double bind is also present in our relationship with nature. On the one hand, we do try to preserve the natural environment; on the other, we wish to continue economic growth although our standard of living disrupts nature and our relationship with it (Wedge 2011). Human beings act in destructive ways towards other human beings and fragile ecological systems because we do not want to see the impact of our behaviour upon others, the environment and, in the end, on our own lives. To overcome these policies we must throw light on the contradictions, place the problem in a larger temporal context, and protest against these double-bind policies.

15.6 Towards a More Adaptive Temporament

Temporal behaviour can be upgraded by: (1) installing a more effective accounting or monitoring and evaluation system of temporal behaviour, (2) codifying gross temporal misconduct and temporal violence in international criminal law (the responsible decision-makers and shapers should be made accountable), (3) making people more conscious of the limits of the prevailing temporament, and (4) developing a more adaptive temporament. The prevailing temporament today – the manner of thinking, feeling and behaving towards time – is clearly not fit for dealing with the challenges of the world or achieving sustainable development and peace. An adaptive temporament can be defined by twelve parameters.

---

[6]The legitimacy of a government depends on effectiveness and democracy. Political legitimacy = democracy × effectiveness.

[7]For Rothman (1997: 33–52), this is a slowed-down and self-conscious analysis of the interactive nature of reactions that allows actors to be proactive agents in a conflict instead of reactive victims, and which furthers analytic empathy.

15.6.1 Temporal Sensitivity

1. *A high level of appreciation of time*: Appreciation of time refers to the value people attach to the role of time in the pursuit of one's interests and needs, especially with respect to economic well-being and security. Time is not only money, but also well-being and security. People with a high appreciation tend to pay ample attention to the role of time and try to use time in ways that further their life expectancy (Table 15.3).

2. *Discernment of temporal violence*: An adaptive temporament is attentive to temporal violence. This involves an awareness that the quality and quantity of life can be reduced by both fast and slow killing, long-term poverty, wasting and imprisoning the time of adversaries and dissidents, unequal opportunities, allowing conflicts to become protracted, not allocating the necessary time and means for conflict prevention, denying or not anticipating negative side-effects of our actions, too-little-too-late responses, criminal negligence, failing to address root causes, and so forth. People with an adaptive temperament pay serious attention to the role of time in conflict analysis and conflict transformation.

3. *An inclusive approach to time*: Inclusion implies that not only my/our time is considered and valued in decision-making, but also that of other stakeholders, including past and future generations. Inclusion also implies recognition of the biological and physical clocks of nature. Religious people demand respect for sacred and transcendental time. Those with an exclusive time approach focus

Table 15.3 Parameters of an adaptive temperament. *Source* The author

Temporal sensitivity
1. High awareness and appreciation of time
2. Discernment of temporal violence
3. An inclusive approach to time
4. Recognition of, and empathy with, different temporal cultures, interests, and needs
Good temporal practice
5. Constructive and balanced orientation to the past, present, and future
6. Strong propensity for proactive violence prevention
7. Synergetic temporal planning and implementation
8. Enabling sustainable development and peace
Ethical temporal behaviour
9. An ethical approach to time
10. Democratization of time
Temporal efficacy and reflexivity
11. A sense of temporal efficacy
12. Reflexivity and adaptive leadership

solely on their own lifetimes and life expectations; others' lifetimes are secondary. People with an inclusive temporal approach respect the time of others and of nature.

4. *Recognition of, and empathy with, different temporal cultures*: There are no universally accepted time frames. In fact, there are many temporal cultures. The way people deal with time is influenced by their culture, but also by the interests at stake, power, professional outlook, age, generation, gender, and religion. Temporal empathy (cognitive and emotional) refers to the will and ability to view and feel how the other conceives and experiences time and temporal violence. High temporal empathy correlates positively with a high level of understanding and compassion. This enables conflict transition and peacebuilding. Low temporal empathy leads to inconsiderate, indifferent, and often disproportionate or vindictive temporal behaviour.

15.6.2 Temporal Practice

5. *A balanced orientation to the past, present and future*: People with an adaptive temporament tackle the needs of the past, present and future in a sufficient and balanced way. An unbalanced approach deals with these needs insufficiently and in an unbalanced way. For example, the past can be repressed or put between brackets in order to build up something new in the present, or inadequate efforts can be made to raise hope for a better and common future. Dealing with the past without an attractive common future is a tantalizing experience. Bracketing the past for some time is possible, but denying or forgetting it can spoil the future.

6. *Propensity for proactive conflict prevention*: Serious efforts are made to anticipate future threats and opportunities by means of scenarios and other forecasting methods. Analysts are acquainted with such concepts as theory-based methods for anticipating civil wars, genocidal behaviour, ripeness of conflict, and also with research on tipping points, black swans, decisive moments, and so on. Efforts are also made to anticipate the positive and negative impacts of interventions. They are acquainted with the newest impact assessment methodologies. Proactive conflict prevention is a high priority; special attention goes to the identification and elimination of the root causes and to the building of sustainable peace. When prevention fails, effective crisis management, and damage limitation skills and facilities are available.

7. *Synergetic planning and implementation*: An adaptive temporament furthers synergy and coherence in conflict transformation and peacebuilding. This involves decisions with respect to: the nature of the involvement and coordination; clarity and consensus about the end state or peace sought as well as the road map to get there; conflict analysis and assessment of the peacebuilding deficiencies at the baseline; entry and exit timing; the prioritization and sequencing of military, diplomatic, political, economic, educational, and other

interventions; the pacing or the speed of external intervention; the preference for slow, gradual or fast and radical change (such as military shock and awe interventions, or Jeffrey Sachs' economic shock therapy for Russia after the Cold War); and the anticipation of possible negative side-effects and measures to stop or reduce such negative impacts.

8. *Enabling sustainable development and peace*: Sustainable development and peacebuilding are considered vital for humanity and our planet. Sustainability requires a lot of cooperation, or agreement to work together for mutual benefit (Mainelli/Harris 2011: 43), and more emphasis on social enterprise (Smith 2012). Robert Axelrod, in *The Evolution of Cooperation* (2006), writes that cooperation depends on "the shadow of the future" or the expectation that interactions in the future might be affected by the quality of current ones (Mainelli/Harris 2011: 43; Axelrod 2006; Benkler 2011).

15.6.3 Temporal Ethics

9. *An ethical approach to time*: Ethical time deals with the normative assumptions underlying temporal thinking and behaviour. It judges the negative consequences of intervention, but also of non-intervention, of delay and criminal negligence, and demands more accountability for temporal misconduct and temporal violence. It favours and advocates sufficient investment in sustainable development and peacebuilding.

10. *Democratization of time*: Time is dealt with as a political issue. Everybody's time is considered equally valuable and no one person's time is regarded as more expendable than another's (Rifkin 1987). People cannot be disempowered by manipulating time, by destroying artefacts and documents from the past, or by controlling their future. A genuine democracy, at both national and international levels, respects the time of all citizens.

15.6.4 Temporal Efficacy and Reflexivity

11. *A sense of temporal efficacy*: Temporal efficacy is the opposite of determinism, fatalism, temporal disorientation and stress. There is reasonable confidence in understanding the role of time and dealing with time in ways that further one's interests and the pursuit of conflict transformation and sustainable peace. Temporal efficacy demands a great deal of practical experience, learning from history, imagining alternative futures, self-esteem, and courage. In tough conflicts it requires us to embrace death – not necessarily physical death, but death of the ego. When we are too invested in our egos, we cannot collaborate, change, adapt or mediate in peace negotiations (Warner/Schmincke 2009).

Temporal efficacy should not be confused with temporal hubris (the feeling that one is the future), which leads to security and foreign policy follies.

12. *Reflexivity and adaptive leadership*: One of the most frequently heard complaints concerns the lack of time and/or too much time pressure. In essence, the problem is the allocation of enough time for important issues and decisions, and for developing good judgement and legitimate political governance at different systemic levels. An adaptive temperament frees time for: (1) broader and deeper understandings of conflict dynamics and seeing the big picture; (2) reflexive framing of conflicts; (3) developing more accurate measures of temporal misconduct and their costs and benefits; (4) imagining and planning mutually attractive and sustainable common world futures; (5) anticipating the positive and negative consequences of interventions or policies; (6) resolving difficult inter-temporal problems faced by people who are both far-sighted planners and myopic doers by aligning incentive systems and/or imposing rules (Mainelli/Harris 2011: 72); and (7) diplomatic work and the improvement of conflict management systems.

15.7   Conclusion

Reading this chapter has likely taken approximately forty minutes, or 2400 s, of your life. That time is now gone. I hope it was worth it. However, as one Sufi saying goes, even "when the heart weeps for what it has lost, the soul laughs for what it has found" (Atwater 1999). This chapter reminds us that in our fast-changing globalized world, the way we deal with time will more than ever determine the success or failure of dealing with global crises and the building of sustainable peace and development. Despite the urgency of handling time more effectively, too much time is wasted. The prevailing way of dealing with conflict is glutted with temporal inadequacies. Illustrative of this is the foreign policy of the West in the Middle East and North Africa after 9/11. The impact has been disastrous. A thorough and comprehensive analysis and evaluation of temporal behaviour in foreign policy is needed. This implies considering the big picture: the temporal dimensions and emotions, as well as both secular and religious time. Assessing the temporal behaviour of the stakeholders in conflict and peace should be an essential part of monitoring and evaluation. It could also help to identify the temporal inadequacies more systematically and steepen the learning curve of peacebuilders. It could advance the accounting for the costs of temporal misconduct and the accountability of policy-makers. Above all it should make us more aware that today's prevailing political temperament stands in the way of sustainable peace and security. Thus, take the time to get to know your temperament and start making it more adaptive.

References

Adger, W. N., & Jordan, A. (Eds.). (2009). *Governing Sustainability*. Cambridge: Cambridge University Press.

Atwater, P. M. H. (1999). *Future Memory*. Charlottesville, VA: Hampton Roads Publishing Company.

Axelrod, R. (2006). *The Evolution of Cooperation*. New York: Basic Books.

Backer, P. (2014). Confidence Ebbs as Americans See Missteps Mount. *International New York Times* (23 October).

Begley, S. (2011). I Can't Think. *Newsweek* (7 March).

Benkler, Y. (2011). *The Penguin and the Leviathan: The Triumph of Cooperation over Self-Interest*. New York: Crown Business.

Brooks, D. (2014). The Quality of Fear. *International New York Times* (22 October).

Brown, A. (1998). Doing Time: The Extended Present of the Long-Term Prisoner. *Time & Society, 7*(1): 93–103.

Cloke, K, & Goldsmith, L. (2002). *The End of Management and the Rise of Organizational Democracy*. San Francisco: Jossey-Bass.

Cohen, R., & Mihalka. M. (2001). *Cooperative Security: New Horizons for International Order*. Garmisch-Partenkirchen: George C. Marshall European Center for Security Studies.

Cohen, R. (2013). Diplomacy is Dead. *International Herald Tribune* (22 January).

Emanuel, G. (2013). Time Poverty. *State of Opportunity* (3 September). Retrieved from https://stateofopportunity.michiganradio.org/post/time-poverty (accessed 15 February 2015).

Galtung, J., & MacQueen, G. (2008). *Globalizing God: Religion, Spirituality and Peace*. Grenzach-Whylen: Transcend University Press.

George, S. (2010). *Whose Crisis, Whose Future?* Cambridge: Polity.

GISHA – Legal Center for Freedom and Movement (2009). *Restrictions and Removal: Israel's Double Bind Policy for Palestinian Holders of Gaza IDs in the West Bank. Fact Sheet*. Tel Aviv: GISHA. Retrieved from https://gisha.org/ (accessed 5 January 2015).

Higgens, A. (2014). Fear Itself Rather than Risk Underlies Ebola Alarm. *International New York Times* (18 October).

Ionescu, F. (1960). *Rhinoceros*. New York: Grove Press.

Jones, O. (2014). Occupying can often Corrupt an Occupier. *The Guardian Weekly* (25 July): 18.

Judt, T. (2012). *Thinking the Twentieth Century*. London: Random House.

Kapur, S. (2014). Scholar Behind Viral 'Oligarchy' Study Tells You What It Means. *TPM* (22 April).

Kuperman, A. (2013). A Model Humanitarian Intervention? Reassessing NATOs Libya Campaign. *International Security, 38*(1): 105–136.

Lanier, J. (2013). *Who Owns the Future?* New York: Allen Lane.

Lao T. (2002). *Tao Te Ching: A New Translation and Commentary,* Translated by Ralph Alan Dale. New York: Karnes & Noble Books.

Lederach, J. P. (1997). *Building Peace: Sustainable Reconciliation in Divided Societies*. Washington, DC: United States Institute of Peace Press.

Mainelli. M., & Harris, I. (2011). *The Price of Fish: A New Approach to Wicked Economics and Better Decisions*. London: Nicholas Brea Publishers.

McNally, K. (2012). Global Security and the Intervention Double Bind. *Global Policy* (23 October).

Mintz, A., & DeRouen Jr., K. (2011). *Understanding Foreign Policy Decision-Making*. Cambridge: Cambridge University Press.

Moisi, D. (2009). *De Geopolitiek van Emotie* [The Geopolitics of Emotion]. Amsterdam: Nieuw Amsterdam.

Nanni. G. (2012). *The Colonization of Time: Ritual, Routine and Resistance in the British Empire*. Manchester: Manchester University Press.

Oberg, J. (2014). Sweden's Submergency. *TFF Associates & Themes Blog* (24 October). Lund, Sweden: Transnational Foundation for Peace & Future Research.

Regev, A. (2014). Return Gaza to the Stone Age. *Israel Hayom* (9 July).

Reychler, L. (1999). *Democratic Peace-building and Conflict Prevention: The Devil is in the Transition*. Leuven: Leuven University Press.

Reychler, L., & Paffenholz, T. (2001). *Peace Building: A Field Guide*. Boulder: Lynne Rienner.

Reychler, L., & Langer, A. (2003). The Software of Peace Building. *Canadian Journal of Peace Studies, 35*(2): 53–73.

Reychler, L., & Stellamans, A. (2003). *Peace Building Leaders and Spoilers* (Cahiers of the Centre for Peace Research and Strategic Studies No. 66). Leuven: Centre for Peace Research and Strategic Studies.

Reychler, L., & Stellamans, A. (2005). *Researching Peace Building Leadership* (Cahiers of the Centre for Peace Research and Strategic Studies No. 71). Leuven: Centre for Peace Research and Strategic Studies.

Reychler, L. (2006). Challenges of Peace Research. *International Journal of Peace Studies, 11*(1): 1–16.

Reychler, L. (2015). *Time for Peace: The Essential Role of Time in Conflict and Peace Processes*. Brisbane: University of Queensland Press.

Rifkin, J. (1987). *Time Wars: The Primary Conflict in Human History*. New York: Henry Holt and Company.

Roht-Arriaza, N. (2012). The Spanish Civil War, Amnesty, and the Trials of Judge Garzon. *Insights: American Society of Internal Law, 16*(24): 1–5.

Rothman, J. (1997). *Resolving Identity-based Conflict in Nations, Organizations and Communities*. San Francisco: Jossey-Bass.

Schmidt, E., & Cohen, J. (2014). *The New Digital Age: Transforming Nations, Businesses, and Our Lives*. New York: Vintage Books.

Seib, P. (2012). *Real-Time Diplomacy: Politics and Power in the Social Media Era*. New York: Palgrave MacMillan.

Smith, E. (2012). Won Over by Social Enterprise. *Financial Times* (25 March). Accessed 5 January 2015).

Smitha, F. E. (n.d.). Transition to a Market Economy, End of the Cold War and Soviet Union (9 of 9). *Macro History and World Timeline*. Retrieved from http://www.fsmitha.com/h2/ch33-9.htm (accessed 21 January 2015).

Tolfler, A. (1970). *Future Shock*. New York: Bantam Books.

Tuchman, B. (1984). *The March of Folly: From Troy to Vietnam*. London: Abacus-Shere Books.

Warner, C., & Schmincke, D. (2009). *High Altitude Leadership: What the Worlds Most Forbidding Peaks Teach Us*. San Francisco: Jossey-Kass.

Wedge, M. (2011). The Double Binds of Everyday Life: Everything We Do to Grow Our Economy Disrupts the Environment. *Psychology Today* (13 October).

Whillans, J. (2011). Book Review: Time Poverty: The Unequal Distribution of Temporal Autonomy. *Time & Society, 20*(1): 137–140.

On the University of Leuven/KU Leuven

The *Katholieke Universiteit Leuven* (*Catholic University of Leuven*) is a research university in the Dutch-speaking town of Leuven in Flanders, Belgium. It conducts teaching, research, and services in the sciences, engineering, humanities, medicine, law, and social sciences. KU Leuven is the largest university in Belgium and the Low Countries. In 2017–18, more than 58,000 students were enrolled. Its primary language of instruction is Dutch, although several programmes are taught in English, particularly graduate degrees.

KU Leuven consistently ranks among the top 100 universities in the world. As of 2016–2017, it ranks 40th globally according to *Times Higher Education*, 79th according to *QS World University Rankings*, and 93rd according to the *Academic Ranking of World Universities*. According to Thomson Reuters, in 2016, 2017 and 2018, KU Leuven researchers have filed more patents than researchers at any other university in Europe; its patents are also the most cited by external academics. As such, KU Leuven was ranked first in the publication's annual list of Europe's most innovative universities for those three years. A number of its programmes rank within the top 100 in the world, according to QS World University Rankings by Subject. It is the highest-ranked university from the Low Countries.

The old University of Leuven was founded at the centre of the historic town of Leuven in 1425, making it Belgium's first university. After being closed in 1797 during the Napoleonic period, the Catholic University of Leuven was re-founded in 1834, and is frequently, but controversially, identified as a continuation of the older institution. In 1968, the Catholic University of Leuven split into the Dutch-language Katholieke Universiteit Leuven and the French-language Université catholique de Louvain (UCLouvain), which moved to Louvain-la-Neuve in Wallonia. Historically, the Catholic University of Leuven has been a major contributor to the development of Catholic theology. It is considered the oldest existant Catholic university. Although Catholic in heritage, it operates independently from the Church. KU Leuven is open to students from different faiths.

In 1968, tensions between the Dutch-speaking and French-speaking communities led to the splitting of the bilingual Catholic University of Leuven into two

L. Reychler and A. Langer (eds), *Luc Reychler: A Pioneer in Sustainable Peacebuilding Architecture*, Pioneers in Arts, Humanities, Science, Engineering, Practice 24, https://doi.org/10.1007/978-3-030-40208-2

"sister" universities, with the Dutch-language university becoming a fully functioning independent institution in Leuven in 1970, and the *Université catholique de Louvain* departing to a newly built greenfield campus site in the French-speaking part of Belgium. In 1972, KU Leuven set up a separate entity, Leuven Research and Development (LRD), to support industrial and commercial applications of university research.

KU Leuven is a member of the Coimbra Group (a network of leading European universities) as well as the LERU Group (League of European Research Universities). Since November 2014, KU Leuven's Faculty of Economics and Business has been accredited by the European Quality Improvement System, which is a leading accreditation system specializing in higher education institutions of management and business administration.

Since August 2017, the university has been led by Luc Sels, who replaced former rector Rik Torfs. The Belgian archbishop, André-Joseph Léonard, is the current *Grand Chancellor* and a member of the university board.

KU Leuven is dedicated to Mary, the mother of Jesus, under her traditional attribute as "Seat of Wisdom", and organizes an annual celebration on 2 February in her honour. On that day, the university also awards its honorary doctorates. The seal used by the university shows the medieval statue Our Lady of Leuven in a *vesica piscis* shape.

Centre for Research on Peace and Development (CRPD)

CRPD is a multidisciplinary research centre at the Faculty of Social Sciences of Leuven University (KU Leuven). CRPD's multidisciplinary team conducts conceptual, empirical and applied research with the aim of improving our understanding of the causes of violent conflicts and the challenges of sustainable peacebuilding.

CRPD was created in 2011. It builds upon the pioneering success of the university's previous *Centre for Peace Research and Strategic Studies* (CPRS), which was initiated in 1982 as one of the first independent peace research centres in Europe. In recognition of the importance of peace research as an autonomous field of study as well as the conviction that understanding the causes of conflict and the challenges of sustainable peacebuilding requires a multidisciplinary approach, the Faculty of Social Sciences, under the leadership of the former Dean, Professor Katlijn Malfliet, recently decided to institute CRPD as an independent research centre at Faculty level.

We consider peace and development to be two sides of the same coin, with peace being a precondition for societal development and progress, including economic, political and human dimensions, which in turn are crucial for bringing about durable peace. The Centre's overall objective goes beyond the analysis of how to break the conflict trap and includes an investigation into how the peace-development nexus can be promoted.

CRPD brings together researchers from the Faculty of Social Sciences with different disciplinary backgrounds, including political scientists, public management experts, anthropologists, sociologists and communication experts. Although there has been a decline in the number of violent intra-state conflicts around the world since the mid-1990s, in a large number of countries (especially developing ones), violent conflicts, communal tensions and political repression continue to cause immense human suffering and undermine development efforts. Improving our understanding of how to resolve violent conflicts and foster more peaceful relations

L. Reychler and A. Langer (eds.), *Luc Reychler: A Pioneer in Sustainable Peacebuilding Architecture*, Pioneers in Arts, Humanities, Science, Engineering, Practice 24, https://doi.org/10.1007/978-3-030-40208-2

within and between communities, societies and countries therefore remains a crucial research issue.

Research on conflict and sustainable peacebuilding is a broad field of study, which is reflected in the Centre's three main research clusters:

- Governance
- Conflict causes and dynamics
- Challenges of sustainable peacebuilding.

CRPD is also privileged to house the UNESCO Chair in Building Sustainable Peace, which was established in 2007 at KU Leuven's Faculty of Social Sciences. The purpose of the Chair is to promote an integrated system of research, training information and documentation in the field of building sustainable peace. It also serves as a means of facilitating collaboration between high-level, internationally recognized researchers and teaching staff of KU Leuven and other institutions in Belgium and elsewhere in Europe and North America, and in other regions of the world.

About the Editors

Luc Reychler, Ph.D. (Harvard, 1976), is Emeritus Professor of International Relations at the Faculty of Social Sciences at the University of Leuven. He was previously Director of the *Centre for Peace Research and Strategic Studies* (CPRS) and *Secretary General of the International Peace Research Association* (IPRA) from 2004 to 2008. He has published widely on sustainable peacebuilding architecture, planning and evaluation of violence prevention and peace-building interventions, and multilateral negotiations. His books include *Patterns of Diplomatic Thinking: A Cross-national Study of Structural and Social-psychological Determinants* (New York: Praeger, 1979), *Nieuwe muren: Overleven in een andere wereld (New walls: surviving in another world)* (Leuven: Leuven University Press, 1994), *Een wereld veilig voor conflict: Handboek voor vredesonderzoek (A world safe for conflict)* (Leuven-Apeldoorn: Garant), *Le défi de la paix au Burundi (The challenge of peace in Burundi)* (Paris: L'Harmattan, 1999), *Democratic Peace-building and Conflict Prevention: the Devil is in the Transition* (Leuven: Leuven University Press, 1999), *Peacebuilding: A Field Guide* (Boulder, CO: Lynne Riener, 2001), *De volgende genocide* ["The next genocide"], (Leuven: Leuven University Press, 2004), with Thania Paffenholz, *Aid for Peace: a Guide for Planning and Evaluation in Conflict Zones* (Baden-Baden: Nomos, 2007), and *Time for Peace: The Essential Role of Time in Conflict and Peace Processes* (Australia: University of Queensland Press, 2015). Since becoming Emeritus Professor in October 2010, he has dealt with ethics, good governance and conflict resolution in sports, especially Taekwondo. He has also begun to research the role of humour in deep-rooted conflicts: how humour can free people from defeatism and despair; how it can turn anger into constructive energy; how it can be used to show the new clothes of the emperors that are invisible to those who are unfit for their positions, stupid, or incompetent.

353

L. Reychler and A. Langer (eds.), *Luc Reychler: A Pioneer in Sustainable Peacebuilding Architecture*, Pioneers in Arts, Humanities, Science, Engineering, Practice 24, https://doi.org/10.1007/978-3-030-40208-2

Address: Luc Reychler, Ph.D., Emeritus Professor, Faculty of Social Sciences, University of Leuven, Parkstraat 45—Box 3602; 3000 Leuven, Belgium. *Email*: luc.reychler@soc.kuleuven.be

Professor *Arnim Langer* is Director of the Centre for Research on Peace and Development (CRPD), Chair Holder of the UNESCO Chair in Building Sustainable Peace, and Professor of International Politics at the University of Leuven (KU Leuven). He is further a Research Associate at the Oxford Department of International Development (ODID) at the University of Oxford and an Honorary Research Fellow at the School of Social Sciences at the University of Western Australia (UWA) in Perth. He was recently awarded a highly prestigious Humboldt Fellowship for Experienced Researchers, which he took up in September 2017 at the University of Heidelberg in Germany.

His research focuses on group behaviour and identity formation, the causes and consequences of violent conflict, the dynamics and persistence of horizontal inequalities, post-conflict economic reconstruction processes for sustainable peacebuilding and peace education in post-conflict countries, as well as national service programmes in Africa. He has conducted extensive field research and is overseeing large research projects on these topics in a range of countries, in particular Côte d'Ivoire, Ghana, Indonesia, Nigeria, Uganda, Kenya and the Democratic Republic of the Congo. He has published extensively on issues of ethnicity, inequality and conflict, including in a range of top journals such as *Political Analysis*, *World Development*, *Foreign Affairs* and *Social Indicators Research*.

Address: Professor Arnim Langer, Centre for Research on Peace and Development (CRPD), Parkstraat 45—box 3602, 3000 Leuven, Belgium. *Email*: arnim.langer@kuleuven.be *Website*: https://www.kuleuven.be/wieiswie/en/person/00036734.

Index

A
A critical mass of peacebuilding leadership, 247
Adaptive diplomacy, 63
Adaptive leadership, 345
Anticipation, 333
Appreciation of time, 41
Arms control efforts, 70
Arms dynamics, 70
Art of conflict prevention, 159

B
Bosnia and Herzegovina, 69
Boulding, Elise, ix
Boulding, Kenneth, ix
Burton, John, ix
Burundi, 66

C
Central America, 73
Centre for Research onPeace and Development (CRPD), xiii
Challenges of peace research, 241
Coercive diplomacy, 66
Collective defence, 71
Collective security, 71
Collective security system, 160
Commitment to future cooperation, 228
Common security thinking, 170
Conceptions of Peace, 87
Conceptualization of peace, 92
Conflict analysis, 282
Conflict avoidance, 161
Conflict dynamics, 175, 213
Conflict Impact Assessment, 63

Conflict Impact Assessment System (CIAS), 72
Conflict management, 90
Conflict prevention, 161
Conflict resolution, 129, 161
Conflict settlement, 161
Constructive management of conflicts, 176
Constructive transformation of conflicts, 72
Cooperative security, 71
Cultural violence, 180
Cumbersome decision-making, 169

D
Definition of peace, 280
Degree of peacefulness, 99
Democratic fascism, 325
Democratic peacebuilding architecture, 198
Democratic transition approaches, 204
Democratization, 73, 200, 281
Depoliticization of religion, 181
Desecularization of the world, 179
Deterrent diplomacy, 161
Deutsch, Karl, 17
Diagnosis of a conflict, 211
Dimensions of security, 70
Dimensions of time, 331
Diplomacy, 63
Diplomatic history, 64
Diplomatic thinking, 63, 85
Disarmament, 119
Disintegrative climate, 225
Domain of peace, 87

E
Early warning system, 72
Effective decision-making, 169, 172

355

Effectiveness of a pacifist strategy, 129
Elaine Exum, D., 20
Electoral assistance, 206
Empirical-analytical, 250
Epistemic violence, 260
Essential role of time, 331
European Union, 43, 71
Existential time, 41, 75
Eyskens, Marc, 27

F
Fabric of violence, 243
Failed foreign policy, 325
Field diplomacy, 189
Field diplomats, 65
First World War, 36
Foreign-policy thinking, 104
Freedom house, 283

G
Galtung, Johan, ix, 90, 180
Genocide, 68
Globalization, 43
Gulf War, 159

H
Human security, 81, 229

I
Imo summoque concors, 6
Incoherent temporal management, 338
Indicators of peace, 96
Inhibitors of peace, 99
Institutional learning, 205
Integrative climate, 224
Intellectual solidarity, 257
Interactive problem-solving diplomacy, 187
international peace, 106
International Peace Research Association
    (IPRA), xi
Interpretative research, 250
Isard, Walter, ix

K
Kelman, Herbert, ix

L
Levels of security, 70
Low-intensity violence, 179

M
Master of Conflict and Sustainable Peace
    (MaCSP), 24
Mediation, 66

Middle East, 66, 73, 175
Military Industrial Complex (MIC), 71
Modes of violence, 67
Moral-political climate, 183
Multiple loyalties, 229

N
Negative peace, 90
Negotiation, 66
New diplomacy, 168
New World Order, 175
Non-interference, 307
Non-violence, 63
Non-violent strategy, 90
North Africa, 66
North Atlantic Treaty Organization (NATO),
    44, 71

O
Objective and subjective security, 70
Operationalization of the peacebuilding blocks,
    81
Oral and visual humour, 79
Organization of Security and Cooperation in
    Europe (CSCE), 71

P
Pacifism, 184
Pacifist strategy, 67, 130
Paradoxes in peace thinking, 85
Patterns of diplomatic thinking, 64
Paxzalvers, xi
Peace action research, 250
Peacebuilders, 76
Peacebuilding, 63
Peacebuilding architecture, 279
Peacebuilding climate, 223
Peacebuilding coordination, 288
Peacebuilding deficiency assessment, 282
Peacebuilding leadership, 253
Peacebuilding mindsets, 70
Peacebuilding potential, 284
Peacebuilding software, 223
Peace enforcement, 163
Peace-enhancing structures, 220
Peace negotiations, 275
Peace quacks, xi, 253
Peace Research II, xi
Peace Research Institute (PRIO), 64
Peacespoilers, 76
Perceived security, 221
Political economics of democratic transition,
    202
Political-strategic thinking, 70

Positive peace, 90
Post-Cold-War, 71
Post-Cold War diplomacy, 160
Post-conflict phase, 210
Power-sharing, 281
Preconditions for a sustainable peace, 219
Preconditions for building sustainable peace, 259
Preferred World Order, 85
Preventive diplomacy, 274
Proactive conflict prevention, 63
Proactive policy-making, 171
Proactive violence prevention, 209
Psychological walls, 257

R
Rapoport, Anatol, ix
Reactive conflict management, 168
Reactive conflict prevention, 63
Reactive violence prevention, 209
Reason of the state, 301
Reconciliation, 65, 223, 281
Reflective peace research, 251
Relative deprivation, 43
Religion, 78, 175
Religious wars, 176
Responsibility to protect, 307
Role expectancy, 131
Role of humour in conflict transformation, 79
Role of religion in conflicts, 176
Rwanda, 66

S
Second world war, 306
Secular and religious time, 41, 76
Securitization, 71
Security Building Measures (CSBMs), 163
Security dilemma, 71
Security organization, 160
Security studies, 70
Selection of the temporal strategy, 76
Sentimental walls, 70
Simultaneity dilemma, 202
Skinner, B.F., 16
Social capital, 81, 230
Software of peacebuilding, 81
Sovereignty as responsibility, 307
Structural violence, 73, 180
Sustainable peace, 65, 75
Sustainable peacebuilding, 258
Symbolic reparations, 228

T
Temporal behaviour, 76
Temporal democracy, 41
Temporal discernment and empathy, 76
Temporal efficacy, 41, 76, 334, 345
Temporal emotions, 41, 76
Temporal empathy, 41, 334
Temporal equality, 76, 333
Temporal ethics, 344
Temporal inadequacies, 325, 335, 340
Temporal malpractice, 337
Temporal management, 333
Temporal management options, 41, 76
Temporal misconduct, 339
Temporal practice, 343
Temporal sensitivity, 335, 342
Temporal violence, 336
Temporament, 35, 75, 325
Terrorism, 42
Theory-practice gap, 251
Thomas schelling, 67
Time, 63, 75, 325
Time management, 63
Time modes, 41, 332
Track II diplomacy, 188
Track II peacemaking, 187
Traditional diplomacy, 168
Transformative dimensions, 333

U
Undemocratic control of time, 339
UNESCO Chairin Building Sustainable Peace, xiii
Unethical temporal behaviour, 339
UnitedNations (UN), 71
Unstable peace, 164

V
Value of peace, 88
Violence, 63, 210
Violence prevention, 210

W
Western European Union (WEU), 71
Widening and deepening peacebuilding, 273
World security, 71

Y
Yugoslavia, 164

Z
Zones of peace, 87